www.wadsworth.com

THE SOCIOLOGY OF
HEALTH, ILLNESS, AND HEALTH CARE

—

A Critical Approach

—

SECOND EDITION

Rose Weitz
Arizona State University

Wadsworth
Thomson Learning™

Australia • Canada • Mexico • Singapore • Spain • United Kingdom • United States

Sociology Editor: *Lin Marshall*
Assistant Editor: *Dee Dee Zobian*
Marketing Manager: *Matthew Wright*
Project Editor: *Jerilyn Emori*
Print Buyer: *Tandra Jorgensen*
Permissions Editor: *Joohee Lee*
Production Service: *Forbes Mill Press*
Photo Researcher: *Terri Wright Design*
Copy Editor: *Robin Gold*

Cover Designer: *Qin-Zhong Yu / Qya Design Studio*
Cover Images: *PhotoDisc: S. Meltzer/PhotoLink, Scott T. Baxter*
Cover Printer: *Phoenix Color Corp.*
Compositor: *Wolf Creek Press & Forbes Mill Press*
Printer: *The Maple-Vail Book Manufacturing Group*

Wadsworth/Thomson Learning
10 Davis Drive
Belmont, CA 94002-3098
USA

For more information about our products, contact us:
Thomson Learning Academic Resource Center
1-800-423-0563
http://www.wadsworth.com

International Headquarters
Thomson Learning
International Division
290 Harbor Drive, 2nd Floor
Stamford, CT 06902-7477
USA

UK/Europe/Middle East/South Africa
Thomson Learning
Berkshire House
168-173 High Holborn
London WC1V 7AA
United Kingdom

Asia
Thomson Learning
60 Albert Street, #15-01
Albert Complex
Singapore 189969

Canada
Nelson Thomson Learning
1120 Birchmount Road
Toronto, Ontario M1K 5G4
Canada

Library of Congress Cataloging-in-Publication Data

Weitz, Rose
    The sociology of health, illness, and health care : a critical approach / Rose Weitz —
2nd. ed.
    p.   cm.
Includes bibliographical references and index.
ISBN 0-534-24756-3
1. Social medicine. 2. Social medicine—United States. 3. Medical personnel and patient. 4. Social control. I. Title.

RA418.3.U6 W46 2000
306.4'61—dc21   00-042301

 This book is printed on acid-free recycled paper.

To Miriam and Lara,
and in memory of Marty

CONTENTS

Preface

During the last quarter century, the sociology of health, illness, and health care has changed dramatically. Begun primarily by sociologists who worked closely with doctors, taking for granted doctors' assumptions about health and health care and primarily asking questions that doctors deemed important, the field has shifted toward asking a very different set of questions. Some of these new questions have challenged doctors' assumptions, whereas others have focused on issues that lie outside most doctors' areas of interest or expertise, such as whether increases in income inequality can affect average life expectancy within a nation or how individuals can develop meaningful lives despite chronic illness.

I entered graduate school during this shift, drawn by the prospects of studying how health and illness are socially created and defined and how gender, ethnicity, social class, and, more broadly, power affect both the health care system and individual experiences of health and illness. As a result, over the years I have researched such topics as how medical values affect doctors' use of genetic testing, how midwives and doctors have battled for control of childbirth, and how social ideas about AIDS affect the lives of those who have this disease.

Although I had no trouble incorporating the new vision of the sociology of health, illness, and health care into my research, I consistently found myself frustrated by the lack of a textbook that would help me incorporate it into my teaching. Instead, most textbooks still seemed to reflect older ideas about the field and to take for granted medical definitions of the situation. Most basically, the books assumed that doctors define illness according to objective biological criteria and, therefore, failed to question whether political and social forces underlie the process of defining illnesses. Similarly, most textbooks ignored existing power relationships rather than investigating the sources, nature, and health consequences of those relationships. For example, the textbooks gave relatively little attention to how doctors gained control over health care or how the power of industrialized nations has affected health in developing nations. As a result, the available textbooks used sociology primarily to answer questions posed by those working in the health care field, such as what social factors lead to heart disease and why patients might ignore their doctors' orders. Consequently, these textbooks often seemed to offer a surprisingly unsociological perspective, with their coverage of some topics differing in only minor ways from that found in health education textbooks.

In addition to failing to raise critical questions about health, illness, and health care, the available textbooks seemed unlikely to encourage students to engage with the materials and to question either the presented materials or their own assumptions, such as the assumptions that the United States has the world's best health care system, that life expectancy increased during the twentieth century primarily because of medical advances, or that all Americans receive the same quality of health care regardless of their ethnicity, gender, or social class. Instead, the textbooks primarily gave students already-processed information to memorize.

My purpose in writing this textbook was to fill these gaps by presenting a critical approach to the sociology of health, illness, and health care. This did not and does not mean presenting research findings in a biased fashion or presenting only research that supported my preexisting assumptions, but it does mean bringing a set of critical skills to bear in evaluating and interpreting the available research findings and in figuring out how to pull these findings together into a coherent "story" in each chapter. In addition, I hoped to tell these stories in a manner that would engage students— whether in sociology classes, medical schools, or nursing schools—and stimulate students to learn actively and think independently. These remain the primary goals of this second edition. Both these goals led me to decide not to try to please all sides or cover all possible topics, as I believe such a strategy leads both to the intellectual homogenization that makes many textbooks seem lifeless and the grab-bag approach that makes them hard to follow.

THE CRITICAL APPROACH

The critical approach, as I have defined it, means using the "sociological imagination" to question previously taken-for-granted aspects of social life. For example, most of the available textbooks in the sociology of health, illness, and health care in essence have examined the issue of patients who do not comply with prescribed medical regimens through doctors' eyes, starting from the assumption that patients should do so. More broadly, previous textbooks have highlighted the concept of a sick role—a concept that embodies medical and social assumptions regarding "proper" illnesses and "proper" patients and that downplays all aspects of individuals' lives other than the time they spend as patients.

In contrast, I emphasize recent research that questions all such assumptions. For example, I discuss patient compliance by examining recent research about how patients view medical regimens and compliance, why doctors sometimes have promoted medical regimens and procedures (such as DES for pregnant women) that later proved dangerous, and how doctors' tendency to cut short patients' questions can foster patient noncompliance. Similarly, this textbook explains the concept of a sick role but pays more attention to the broader experience of illness—a topic that has generated far more sociological research than has the sick role in the last decade.

COVERAGE

Although I have tried in this book to present a coherent critical view, I have not sacrificed coverage of topics teachers have come to expect. Consequently, this book covers essentially all the topics—both microlevel and macrolevel— that have become standard over the years, including doctor-patient relationships, the nature of the U.S. health care system, and the social distribution of illness. In addition, I include several topics that usually receive little coverage, including bioethics, mental illness, the medical value system, the experience of illness and disability, and the social sources of illness in both the developing and industrialized nations. As a result, this text includes more materials than most teachers can cover effectively in a semester. To assist those who choose to skip some chapters, each important term is printed in bold and defined the first time it appears in the text and is printed in bold without a definition the first time it appears in each subsequent chapter, alerting students that they can find a definition in the book's glossary.

In addition, reflecting my belief that sociology neither can nor should exist in isolation but must be informed by and in turn inform other related fields, each chapter includes a historical overview. For example, the chapter on health care institutions discusses the political and social forces that led to the development of the modern hospital, the chapter on medicine as a profession discusses how and why the status of medicine grew so dramatically after 1850, and the chapter on the meaning of illness discusses how people throughout history have explained and responded to illness and ill persons. These discussions provide a context to help students understand the current status of, respectively, hospitals, doctors, and ill persons.

NEW TO THIS EDITION

More Accessible Approach

To make the book more accessible to those who are unfamiliar with a critical perspective, small changes have been made throughout the book to move it toward a more balanced approach without sacrificing its distinctive perspective and engaging voice.

"Making a Difference"

To help students see how sociological knowledge can translate into effective social action, half of the chapters now include boxes describing the work of nonprofit organizations that are using sociological insights to "make a difference" in health and health care. For the same reason, I have woven throughout the book descriptions of positive changes that have occurred in recent years in health and health care, such as the rise of more-humanistic training in medical schools.

New and Expanded Topics

- The Internet (Chapters 1, 6, and 11)
- Globalization (Chapters 2 and 4)
- Legal battles over tobacco manufacturing and sale (Chapter 2)
- Relative income differentials and health status (Chapters 2 and 4)
- Breastfeeding and health in the industrialized nations (Chapter 2)
- Health status of migrant farmworkers (Chapter 2)
- Health care workers' "cultural competence" and its impact on minority patients (Chapters 2 and 11).
- Diarrheal diseases (Chapter 4)
- AIDS in Africa (Chapter 4)
- Changing definitions of Attention Deficit Disorder (Chapter 5)
- The Human Genome Project and the rise of the "genetic paradigm" (Chapter 5)
- The Americans with Disabilities Act (Chapter 6)
- Illness behavior and self-care (Chapter 6)
- Alternative health care (Chapters 6, 11, and 12)
- New treatments for mental illness (Chapter 7)
- Managed care (Chapters 7 to 13)
- Market incentives and health care internationally (Chapter 9)
- The Mexican health care system (Chapter 9)
- The demise of the Health Care Security Act and the rise of state-level health care reforms (Chapter 9)
- Assisted-living facilities (Chapter 10)
- Specialized nursing (Chapter 12)
- Pharmaceutical care (Chapter 12)
- Funding priorities in health care (Chapter 13)
- Enhancing human traits and "cosmetic psychopharmacology" (Chapter 13)

Up-to-Date Coverage of All Topics

Throughout the textbook, I have thoroughly updated not only statistics and discussions of topical issues (like health care reform) but also all reviews of the theoretical and empirical literature. As a result, almost half of the references in this new edition are dated from 1995 or later, and the reader can assume that all statistics are the latest available.

Improved Integration of International Issues

To help students understand the importance and relevance of international health issues, discussions of international health and health care are now more clearly linked to U.S. concerns, such as how diseases in developing nations can spread to the United States and why many U.S. residents now travel to Mexico for health care.

Using Internet Resources

In addition to expanding the discussion of the Internet within the body of several chapters, I have included Internet exercises for students in each chapter and Internet addresses for all nonprofit organizations described in the textbook. Readers of this textbook now also have access to the wide variety of tools and resources available at Wadsworth Publishing's sociology Web site (sociology.wadsworth.com). This textbook has its own Web page at that site, which contains updated Web links for all Internet sites discussed in the textbook as well as links to other health Web sites that might be useful for students and instructors. Finally, when ordering this textbook, professors may request that their students receive free access via the Wadsworth Web site to InfoTrac©, an online archive offering full-text versions of hundreds of scholarly articles, many on health-related topics.

PEDAGOGICAL FEATURES

Chapter Openings

Unfortunately, many students take courses only to fill a requirement. As a result, the first problem teachers face is interesting students in the topic. For this reason, each chapter opens with a vignette taken from a sociological or literary source and chosen to spark students' interest in the topic by demonstrating that the topic has real consequences for real people—that, for example, stigma is not simply an abstract concept but something that can cost ill persons their friends, jobs, and social standing.

Ethical Debates

To teach students that ethical dilemmas pervade health care, each chapter includes a discussion of a relevant ethical debate. The debates are complex enough that students must use critical thinking skills to assess them; teachers can use these debates for classroom discussions, group exercises, or written assignments.

Chapter Conclusions

Each chapter in this textbook ends not with a summary that reiterates the materials but with conclusions that discuss the implications of the chapter and point the reader towards new questions and issues. These conclusions should stimulate critical thinking rather than rote memorization and can serve as the basis for class discussions.

Student Aids

Each chapter includes study questions and suggestions for further reading. The book also includes an extensive bibliography and a glossary that defines all important terms.

In addition, and in the hopes that this book will mark the beginning and not the end of students' interest in the field, each chapter includes a description of nonprofit, activist organizations that can provide students with both more information and opportunities for personal involvement. Internet addresses as well as phone numbers and street addresses are provided for all organizations.

Internet Features

As described in the previous section, each chapter includes Internet exercises for students, while Wadsworth Publishing's Web site offers access to a variety of sociology resources and this book's Web site provides updated links to numerous useful sites. Also as described, if professors request it when purchasing this textbook, their students can obtain free access to the InfoTrac© online archive of scholarly publications.

"Making a Difference" Inserts

Also as mentioned earlier, several of the chapters include boxed inserts describing actions that nonprofit groups have taken to improve health or health care.

Instructor's Manual

For each chapter, the Instructor's Manual contains a detailed summary, a set of multiple-choice questions, and a list of relevant narrative and documentary films. In addition, the Manual includes several questions for each chapter that require critical-thinking skills to answer and that teachers can use for essay exams, written assignments, in-class discussions, or group projects. To guarantee the quality of the Instructor's Manual, I wrote everything in it rather than relying on student assistants. If you prefer to receive the Manual as a computer file, please send your request to me via email at rose.weitz@asu.edu.

Critical Thinking

In this textbook, I have aimed not only to present a large body of data in a coherent fashion but also to create an intellectually rigorous textbook that will stimulate students to think critically. I have tried to keep this purpose in mind in writing each chapter. The chapter conclusions, ethical debates, and essay questions all encourage students to use critical thinking.

ACKNOWLEDGMENTS

In writing the first and second editions of this textbook, I have benefited enormously from the generous assistance of my colleagues. I am very fortunate to work with several exceptional sociologists here at Arizona State University who share my interest in health issues—Deborah Sullivan, Jennie Jacobs Kronenfeld, Verna Keith, and Victor Agadjanian. Each of these individuals critiqued chapters for me. I am also exceptionally fortunate to have had the assistance of several research assistants—Melinda Konicke, Christopher Lisowski, Diane Sicotte, Caroleena Von Trapp, and, especially, Karl Bryant, Lisa Comer, and Amy Weinberg.

Because, of necessity, this textbook covers a wealth of topics that range far broader than my own areas of expertise, I have had to rely heavily on the kindness of strangers in writing it. One of the most rewarding aspects of writing this book has been the pleasure of receiving information, ideas, critiques, and references from individuals I did not previously know and who, subsequently, have become friends as well as colleagues. My thanks to the many new and old friends who gave me the benefit of their expert advice: Emily Abel (University of California—Los Angeles), James Akré (World Health Organization), Ofra Anson (University of the Negev), Judy Aulette (University of North Carolina, Charlotte), Miriam Axelrod, James Bachman (Valparaiso University), Paul Basch (Stanford University), Phil Brown (Brown University), Peter Conrad (Brandeis University), Timothy Diamond (California State University—Los Angeles), Luis Durán (National Institute of Public Health of Mexico), Michael Farrall (Creighton University), Kitty Felker, Arthur Frank (University of Alberta), Frederic W. Hafferty (University of Minnesota—Duluth), Harlan Hahn (University of Southern California), Paul Higgins (University of South Carolina), Allan Horwitz (Rutgers University), Bradford Kirkman-Liff (Arizona State University), Judith Lorber (City University of New York), William Magee (University of Toronto), Judy Mayo, Peggy McDonough (York University), Cindy Miller, Jeanine Mount (University of Wisconsin), Marilynn M. Rosenthal (University of Michigan), Beth Rushing (Kent State University),

Wendy Simonds (Georgia State University), Clemencia Vargas (Centers for Disease Control and Prevention), Robert Weaver and his students, especially Cheryl Kratzer (Youngstown State University), Daniel Whitaker, David R. Williams (University of Michigan), Irving Kenneth Zola (Brandeis University), and Robert Zussman (University of Massachusetts-Amherst). This book undoubtedly would have been better if I had paid closer attention to their comments. I apologize sincerely if I have left anyone off this list.

This book also has been substantially improved by the suggestions of the reviewers: Karen Bettez (Boston College), Linda Belgrave (University of Miami), Karen Frederick (St. Anselm College), Stephen Glazier (University of Nebraska), Linda Grant (University of Georgia), Janet Hankin (Wayne State University), Alan Henderson (California State University—Long Beach), Frances Hoffman (University of Missouri), Gary Tiedman (Oregon State University), Diana Torrez (University of North Texas), and Diane Zablotsky (University of North Carolina—Charlotte). Finally, I would like to express my appreciation to Serina Beauparlant, Susan Shook, and Eve Howard, the Wadsworth editors who made the two editions of this book possible.

Social Factors and Illness

—

CHAPTER ONE
Introduction

CHAPTER TWO
The Social Sources of Illness

CHAPTER THREE
The Social Distribution of Illness in the United States

CHAPTER FOUR
Illness in the Developing Nations

—

Illness is a fact of life. Everyone experiences illness sooner or later, and everyone eventually must cope with illness among close friends and relatives.

To the ill individual, illness can seem a purely internal and personal experience. Yet illness is also a social phenomenon, with social roots and social consequences. In this first part, I demonstrate the role social factors play in fostering illness within societies and in determining which groups in a given society will experience which illnesses with which consequences.

Chapter 1 introduces the sociological perspective and illustrates how sociology can help us understand issues related to health, illness, and health care. In the subsequent chapters, I discuss the role social forces play in causing illness and in determining who gets ill. Chapter 2 provides a brief history of illness in the Western world; I describe how patterns of illness have changed over time and assess the relative roles of social factors and medical advances in those changes. I then look at the social sources of illness in the contemporary United States. In Chapter 3, I investigate how four social factors—age, sex, social class, and race or ethnicity—affect the distribution of illness in the United States and explore why some social groups bear a greater burden of illness than others. Finally, in Chapter 4, I describe the very different pattern of illnesses found in poorer countries and how social forces—from the low status of women to the rise of migrant labor—foster illness in these countries.

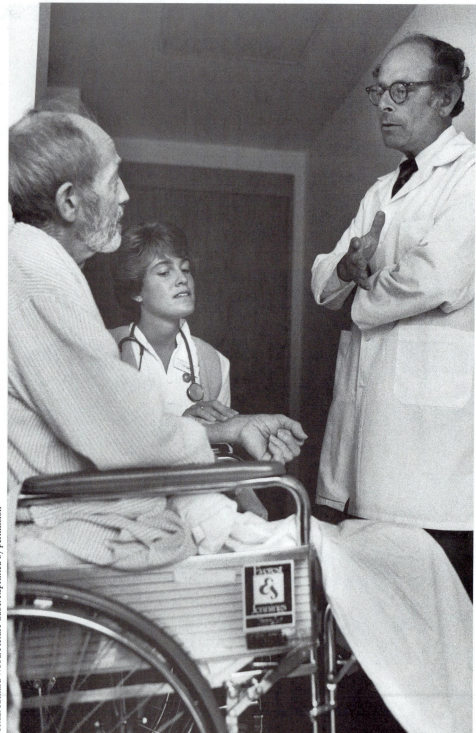

CHAPTER 1

Introduction

—

Five years ago, at the age of 46, my friend Lara learned she had breast cancer.

Once her doctor concluded from Lara's mammogram (a form of x-ray) that a lump in her breast seemed cancerous, events had followed in quick succession. The next day, a surgeon removed a piece of the suspicious lump for testing. A few days later, Lara learned that the lump indeed was cancerous. That week, she got her affairs in order and signed a "living will" specifying the circumstances in which she would want all treatment stopped and a "medical power of attorney" giving me legal authority to make medical decisions for her if she physically could not do so herself. These two documents, she hoped, would protect her from aggressive medical treatments that might prolong her suffering without improving her quality of life or chances of survival.

Two weeks after the initial tests, her surgeon removed the rest of the lump as well as the lymph nodes under her arm (where breast cancer most often spreads). The surgery went well, but the subsequent laboratory tests showed that the cancer indeed had spread to some of Lara's lymph nodes.

Yet in many ways, Lara was fortunate. Her breast cancer was detected at a relatively early stage, giving her about a 65 percent chance of surviving. Although she had no spouse or children to turn to, her friends proved uniformly supportive. She received health insurance through her employer and had no fears of losing either her job or her insurance.

Nevertheless, cancer changed Lara's life irrevocably, making it, at times, a nightmare. Having breast cancer shook Lara's faith in her body and changed her sense of her physical self. At the same time, her illness threatened her relationships with others. Despite the very positive responses she received from friends and co-workers, she nevertheless feared they would drift away as her illness continued or that she would chase *them away with her all-too-reasonable complaints, worries, and needs. Although she had far better health insurance than many Americans have, her debts for items not covered by insurance nonetheless mounted. In addition, she had to spend many hours and much of her limited energy fighting her insurance company for permission to receive relaxation training and certain anti-nausea drugs to help her cope with the side effects of chemotherapy. Finally, the chemotherapy made her so ill she often found it difficult to function. In addition, chemotherapy proved so toxic that it damaged her veins with each painful intravenous treatment. As a result, her doctors suggested inserting a semipermanent plastic tube into her chest wall and administering the chemotherapy through the tube instead. Although doing so would have reduced the pain she experienced, in the end Lara rejected the suggestion because she felt she could not cope with the feeling that—with this sign of her illness physically attached to her body at all times—she would have truly become a cancer patient, rather than merely someone for whom cancer was one part of her life.*

After a year of surgery, chemotherapy, and, subsequently, radiation, Lara's physical traumas ended, although it took another year before she regained her former energy. This summer we celebrated her fifth year free of any signs of cancer.

THE SOCIOLOGY OF HEALTH, ILLNESS, AND HEALTH CARE: AN OVERVIEW

Lara's story demonstrates the diverse ways illness affects individuals' lives. It also demonstrates the diverse range of topics that sociologists of health, illness, and health care can study. First, sociologists can study how social forces promote health and illness and why some social groups suffer more illness than others do. For example, researchers have explored whether working conditions in U.S. factories help explain why poorer Americans get certain cancers more often than wealthier Americans do. Similarly, sociologists can study how historical changes in patterns of social life can explain changes in patterns of illness. For example, to understand why rates of breast cancer have increased, some researchers have studied the impact of women's changing social roles, and others have studied the impact of political forces that promote increased meat consumption. Second, instead of studying broad patterns of illness, sociologists can study the experiences of those, like Lara, who live with illness on a day-to-day basis—exploring, for example, how illness affects individuals' sense of identity, relationships with family, or ideas about the causes of illness. Third, sociologists can focus on the impact of social factors on health care providers. For example, some sociologists have analyzed how the status and power of different occupations have shifted over time, and others have investigated how power affects interactions between health care occupations (such as doctors and nurses). Still others have examined interactions between health care workers and patients, asking, for example, how doctors can maintain control in their discussions with patients or whether doctors treat male and female patients differently. Finally, sociologists can analyze the health care system as a whole. Sociologists have examined how health care systems have developed, compared the strengths and weaknesses of different systems, and explored how systems can be improved. For example, some have studied how and why U.S. health insurance companies sometimes make it difficult for people like Lara to get needed care, explored why European countries have succeeded far better than the United States has in providing health care to all who need it, and examined whether any of the strategies used in Europe could be implemented effectively in the United States.

The topics researched by sociologists of health, illness, and health care overlap in many ways with those studied by health psychologists, medical anthropologists, public health workers, and others. What most clearly differentiates sociologists from these other researchers is the **sociological perspective.** The next section describes that perspective.

THE SOCIOLOGICAL PERSPECTIVE

Using a sociological perspective means focusing on social patterns rather than on individual behaviors. Whereas a psychologist, for example, might help a battered wife develop a greater sense of her own self-worth so she might eventually leave her abusive husband, a sociologist likely would consider therapy a useful but inefficient means of addressing the root causes of wife abuse. Most battered wives, after all, do not have the time, money, or freedom to get help from psychologists. Moreover, even when therapy helps, it takes place only after the women have experienced physical and emotional damage. The sociologist would not deny that individual personalities play a role in wife battering, but finds it more useful to focus instead on exploring whether broad social patterns can explain why wife battering is so much more common than husband battering or why battered wives so often remain with abusive husbands. Consequently, whereas the psychologist hopes to enable the individual battered wife eventually to leave her husband, the sociologist hopes to uncover the knowledge that politicians, social workers, activists, and others need to implement programs that will prevent wife abuse.

As this example demonstrates, using the sociological perspective means framing problems as *public issues,* rather than merely as *personal troubles.* According to C. Wright Mills (1959:8–9), the sociologist who first drew attention to this dichotomy:

Troubles occur within the character of the individual and within the range of his immediate relations with others; they have to do with his self and with those limited areas of social life of which he is directly and personally aware. Accordingly, the statements and the resolutions of troubles properly lie within the individual as a biographical entity and within the scope of his immediate milieu. . . . *Issues* have to do with matters that transcend these local environments of the individual and the range of his inner life. They have to do with the organization of many such milieux into the institutions of an historical society as a whole.

For example, whenever a child dies from leukemia, it is a tragedy and a personal trouble for the child's family. If, on the other hand, several children in one small neighborhood die of leukemia during the same year, it could suggest a broader social issue such as toxic contamination of the neighborhood water system. The sociological perspective, then, departs radically from the popular American philosophy that individuals create their own fates and that anyone can succeed if he or she tries hard enough.

The sociological perspective can help us identify critical research questions that might otherwise go unasked. For example, in the book *Forgive and Remember: Managing Medical Failure*, sociologist Charles Bosk (1979:62–63) described a situation he observed one day on "rounds," the time each day when recently graduated doctors (known as **residents**) and their supervisors jointly examine patients:

Dr. Arthur [the supervising doctor] was examining the incision [surgical cut] of Mrs. Anders, a young woman who had just received her second mastectomy. After reassuring her that everything was fine, everyone left her room. We walked a bit down the hall and Arthur exploded: "That wound looks like a walking piece of dogshit. We don't close wounds with continuous suture on this service. We worked for hours giving this lady the best possible operation and then you screw it up on the closure. That's not how we close wounds on this service, do you understand? These are the fine points that separate good surgeons from butchers, and that's what you are here to learn. I never want to see another wound closed like that. Never!" Arthur then was silent, he walked a few feet, and then he began speaking again: "I don't give a shit how Dr. Henry [another supervising doctor] does it on the Charlie Service or how Dr. Gray does it on Dogface; when you're on my service, you'll do it the way I want."

Dr. Arthur and the residents he supervised undoubtedly viewed this situation as a personal trouble requiring a personal solution—the residents seeking to appease Dr. Arthur, and Dr. Arthur seeking to intimidate and shame the residents into doing things the way he considered best. Similarly, depending on their viewpoint, most nonsociological observers probably would view this as a story about either careless residents or an autocratic supervisor. Sociologists, however, would first ask whether such interactions among doctors occur often. If they do, sociologists then would look for the social patterns underlying such interactions, rather than focusing on the personalities of these particular individuals. So, for example, based on his observations in this and other cases, Charles Bosk discovered that cultural expectations within the medical world regarding authority, medical errors, and the importance of personal, surgical experience had enabled Dr. Arthur and the other supervising doctors to humiliate residents publicly and to set policies based more on personal preferences than on scientific data.

Whereas Bosk investigated health issues within hospitals, David Kirp used a sociological perspective to explore health issues in the community. For the book *Learning by Heart: AIDS and Schoolchildren in America's Communities,* Kirp (1989) observed a half dozen communities around the country to determine why they responded in such different ways to the presence of schoolchildren who had **acquired immunodeficiency syndrome (AIDS).** The following events occurred relatively early in the history of the AIDS epidemic, in Swansea, Massachusetts:

> For the members of the Swansea, Massachusetts, school committee, habituated to brief and sparsely attended bimonthly meetings in the century-old red brick administration building, the evening of September 11, 1985, was an eye-opener. More than seven hundred people, almost all parents, filled the high school auditorium, the biggest meeting place in town.
>
> The people of Swansea are usually polite in their dealings with one another, but these parents were in no mood for good manners. They demanded to know why their superintendent and their school committee had acted differently than every other school official in the entire country. Why had they allowed a thirteen-year-old boy with AIDS—a boy named Mark, known and liked by many of the people, but now fatally tainted in their eyes—to remain in school?
>
> Why, the parents asked, had people they trusted—a school committee they had elected, most of whose members were natives of Swansea, and a superintendent who had been a fixture in their schools for nearly three decades—exposed their children to the bizarre terror of AIDS (Kirp, 1989:16–17)?

As in the case of Dr. Arthur and his residents, we could view the Swansea furor simply as an isolated event caused, depending on one's viewpoint, by either an unthinking and arrogant school board or uneducated and heartless parents. Probably the school board and the parents saw the problem in these terms and therefore focused, respectively, on calming the parents or overturning the school board's action. By looking at the variety of ways communities responded to the presence of schoolchildren with AIDS, however, Kirp was able to identify a different set of issues—politics, power, and stereotypes—and of causal factors, such as how the media can foster fears and how popular beliefs about the meaning of illness can breed bigotry against ill persons.

In sum, the sociological perspective shifts our focus from individuals to social groups and institutions. One effect of this shift is to highlight the role of power. *Power* refers to the ability to get others to do what one wants, whether willingly or unwillingly. Power is what allowed Dr. Arthur to treat his residents so rudely and allowed some school boards to override the wishes of their communities. Because sociologists study groups rather than individuals, the sociological analysis of power focuses on why some social groups have more power than others have, how groups use their power, and the consequences of differential access to power, rather than on how specific individuals get or use power. For example, within the sociology of health, illness, and health care, researchers have examined why doctors as a group proved more successful than have nurses in obtaining the power needed to control their working conditions and how recent changes in the health care system have constrained doctors' power. Similarly, sociologists have explored how *lack* of power exposes poor persons and disadvantaged minorities to conditions that promote ill health, while limiting their access to health care.

A CRITICAL APPROACH

Although the concept of power underlies the sociological perspective, some sociologists do not emphasize power in their research and writing. Instead, some sociologists essentially take for granted the way power is distributed in our society, examining the current system without questioning why it is this way or how it might be changed. For example, some sociologists have investigated whether lower-class persons are more likely than upper-class persons to suffer mental illness without first questioning whether definitions of mental illness might reflect an upper class perspective regarding socially acceptable behaviors or whether the same behaviors might more likely be defined as symptoms of mental illness when performed by lower-class persons.

Those sociologists, on the other hand, who do *not* take for granted existing power relationships and who instead focus on the sources, nature, and consequences of power relationships can be said to use a **critical approach.** Critical sociologists recognize that, regardless of how power is measured, men typically have more power than women, adults have more power than children, whites have more power than African Americans, heterosexuals have more power than gays and lesbians, persons with socially acceptable bodies have more power than persons who are disabled, and so on. Critical sociologists who study health, illness, and health care have raised such questions as how this differential access to power affects the likelihood that members of a social group will be exposed to illness-producing conditions or will have access to quality health care.

Critical sociologists also emphasize how social institutions and popular beliefs can support or reflect existing power relationships. For example, many researchers who study the U.S. health care system have looked simply for ways to improve access to care or quality of care within that system, such as offering poor people subsidized health insurance or providing financial incentives to doctors who practice in low-income neighborhoods. Those who use a critical approach have asked instead whether we could provide better care to more people if we changed the basic structure of the system, such as by removing the profit motive from health care to reduce the costs of care for everyone.

Similarly, critical sociologists have drawn attention to how doctors' power and authority enable them to frame our ideas about health, illness, and health care. Most basically, these sociologists have questioned the very terms *health, illness,* and *disability* and have explored whether such terms reflect social values more than they reflect objectively measurable physical characteristics.

In any sociological field, therefore, those who adopt a critical approach will ask quite different research questions from those who do not. Within the sociology of health, illness, and health care, this translates in large part into whether sociologists limit their research to questions about social life that doctors consider useful—a strategy referred to as **sociology *in* medicine**—or design their research to answer questions of interest to sociologists in general—a strategy referred to as the **sociology *of* medicine** (Straus, 1957). Research using the latter strategy often challenges both medical views of the world and existing power relationships within health care.

To understand the difference between sociology in medicine and sociology of medicine, consider the sociological literature on patients who do not follow their doctors' advice. Because doctors typically define such patients as problems, over the years many sociologists, accepting medical ideas regarding what questions need asking, have sought to determine how to "bring patients to their senses" and increase their compliance with medical advice. In contrast, sociologists *of* medicine have looked at the issue of compliance through patients' eyes. As a result, they have learned that patients sometimes ignore medical advice not out of stubbornness or foolishness but because their doctors have not explained clearly either how to follow the prescribed regimens or why they should do so. In other circumstances, patients have ignored medical advice because they have concluded rationally that the emotional or financial costs of doing so outweigh the potential medical benefits. Similarly, whereas those practicing sociology *in* medicine have studied various aspects of the experience of *patienthood,* those practicing sociology *of* medicine instead have studied the broader experience of *illness,* which includes but is not limited to the experience of patienthood. The growing emphasis on sociology of medicine and on the critical approach has led to a proliferation of research on the many ways illness affects everyday life and on how ill individuals, their families, and their friends respond to illness.

CHAPTER ORGANIZATION

This textbook demonstrates the breadth of topics included in the sociology of health, illness, and health care. The text covers both micro-level issues (those occurring at the level of interactions among individuals and small groups) and macro-level issues (those occurring at the level of the society as a whole). In Part One, I discuss the role social factors play in fostering illness and in determining which social groups experience which illnesses. Chapter 2 describes the major causes of preventable deaths in the United States and how they have changed over time, including both long-standing problems such as cancer and emerging problems such as AIDS and drug-resistant tuberculosis. The chapter demonstrates how social as well as biological factors affect health and illness. Building on this basis, Chapter 3 describes how age, gender, social class, and race or ethnicity affect which Americans get ill with which illnesses. Finally, Chapter 4 explores the nature and sources of illness in the poorer countries of Asia, Africa, and Latin America.

Part Two analyzes the meaning and experience of illness and disability in the United States. In Chapter 5, I explore what we mean when we label something an illness, as well as how social groups explain both why illness occurs and why illness strikes certain individuals rather than others. This chapter also looks at the social consequences of defining behaviors and conditions as illnesses and of the socially accepted responses to illnesses and ill people. With this as a basis, in Chapter 6, I first explore the meaning of disability and then offer a sociological overview of the experience of living with illness or disability, including the experience of seeking care from either medical doctors or alternative health care providers. Chapter 7 provides a parallel assessment of mental illness, describing what we mean when we label something a mental illness, analyzing the relationship between social factors and mental illness, providing a sociological account of the diagnosis and treatment of mental illness, and exploring the experience of living with mental illness.

In Part Three, I move the analysis to a more macro-level perspective. Chapter 8 describes the basic outlines of the U.S. health care system and some of the current problems with that system. I begin Chapter 9 by suggesting some basic measures for evaluating health care systems and then use these measures to evaluate the systems found in Great Britain, Germany, Canada, the People's Republic of China, and Mexico. I conclude this chapter by asking what useful lessons the United States can take from these other countries and by assessing the prospects for health care reform within the United States. Finally, Chapter 10 examines several health care settings, including hospitals, hospices, nursing homes, and family homes.

Part Four shifts our focus from the health care system to health care providers. In Chapter 11, I analyze how doctors achieved both prestige and professional autonomy and the factors now threatening their position. The chapter also describes the process of becoming a doctor, the values embedded in medical culture, and the impact of those values on doctor-patient relationships. Chapter 12 describes the history and social position of various other mainstream and alternative health care occupations, including pharmacists, lay midwives, osteopaths, and Christian Science practitioners. Finally, in Chapter 13, I present a history and overview of bioethics, the study of ethical issues involved in the provision of health care. The chapter discusses how bioethics can inform sociological debate and how sociology can inform bioethical debate. (Reflecting the importance of bioethics to understanding health, illness, and health care, each of the preceding chapters also includes an ethical debate on a topic related to that chapter.)

All essential terms used in the book are defined in the glossary at the end of the book. Each term is defined and set in boldface type the first time it appears in the book. In case professors assign the chapters out of sequence, each term also appears in boldface type (without a definition) the first time it appears in any subsequent chapter.

Each chapter ends with lists of suggested readings, review questions, and pertinent nonprofit organizations. The suggested readings were selected not only because of the materials they cover but also because they are exceptionally well-written and interesting. The study questions were designed to provide an overview of the chapter. Readers who can answer these questions should feel confident that they understand the material. Finally, the nonprofit organizations were listed both as sources of additional information and as potential means through which readers can become personally involved in working on the issues raised in the chapters. Updated Web addresses for these organizations can be found at this textbook's Web site, located at sociology.wadsworth.com. To give readers an idea of how one can make a difference, several of the chapters include boxed discussions of activities organizations have undertaken in recent years to prevent illness or improve the lives of those who experience illness or disability.

A NOTE ON SOURCES

Printed Sources

This book is based primarily on data from three types of printed sources: medical journals, sociological journals and books, and government and United Nations statistics. Before readers can evaluate this book and the conclusions drawn in it, however, they need to know how to evaluate these sources.

The most influential medical journals in the United States are the *Journal of the American Medical Association* and the *New England Journal of Medicine.* The comparable British journals are the *British Medical Journal* and *Lancet.* These journals are most influential for several reasons: Each has been in existence for several decades, proving its worth through its longevity. Each has a large readership, indicating that doctors take them seriously enough to pay for subscriptions. Each accepts for publication only a small percentage of submitted manuscripts, suggesting that only the best articles are published. Finally, each uses peer review, sending every submitted manuscript to two or more reviewers for evaluation before the editors decide whether to publish it.

Much of the medical research presented in this textbook comes from these sources or from the *American Journal of Public Health,* published by the major professional association in public health. Because it is a specialty journal rather than a general medical journal and therefore has a smaller audience, the *American Journal of Public Health* is not as prestigious as the top medical journals. However, the standards for publication in this journal are as high as for the top medical journals, and anything published in it would be taken seriously by all health care professionals.

The most influential sociological journals in the United States are the *American Sociological Review,* the *American Journal of Sociology, Social Forces,* and, to a somewhat lesser extent, *Social Problems.* The most important journal in the sociology of health, illness, and health care is the *Journal of Health and Social Behavior.* Many of the sociological articles cited in this book come from these sources. These journals are widely respected among sociologists for the same reasons that the *New England Journal of Medicine* is widely respected among doctors.

Although all these journals—especially the medical journals—sometimes print articles based on only a few cases, most of the articles cited in this book draw on large samples. As a result, the conclusions presented in these articles are more likely to reflect trends among the population as a whole rather than individual idiosyncrasies. For the same reason, the most methodologically sophisticated articles use **random samples.** In a random sample, participants are selected in such a way that each member of a population has an equal chance of being selected (such as drawing names out of a hat, or interviewing every fifth person listed on a class roster). When a sample is randomly selected, we can be fairly certain that the selected individuals will represent statistically the population as a whole. In addition, these articles typically use statistical techniques to **control** for the impact of extraneous factors on the observed relationships. For example, researchers studying the relationship between smoking cigarettes and lung cancer can use statistical techniques to control for the impact of smoky work environments by making sure that they compare persons who work in smoky environments only to others who work in similar environments.

Finally, this book draws heavily on statistics collected by the U.S. government and by the World Health Organization (WHO), a branch of the United Nations. Because these statistics are collected by nonpartisan bureaucrats whose employment typically continues regardless of shifts in the political climate, rather than by groups with a particular political agenda, they are generally regarded as the most objective data available.

This brief discussion of sources suggests several questions readers should keep in mind while reading this book. First, ask if the data come from a reputable source. Second, ask whether the data were peer-reviewed or in some other way checked for quality or potential bias. Third, ask about the size and nature of the study's sample and whether the study controlled statistically for possible confounding factors. Fourth, ask what questions the researchers asked in collecting their data and what questions they *should* have asked. For example, countries that define infants who die during the first week after birth as stillborns will appear to have fewer infant deaths than will countries that define these as infant deaths. Finally, ask if the data presented are sufficient to justify the conclusions. If not, ask what additional data are needed to reach a firmer conclusion and how one might obtain that information from reputable sources.

Internet Sources

In addition to printed sources, some of the information used in this book was obtained through the Internet. The Internet refers to an electronic communications network, first started in the early 1980s, that connects computers around the world. It can be an excellent source for current statistics and an efficient way of learning about many topics. However, the vast wealth of materials available via the Internet and the ease with which anyone can post anything on it make it crucial that these sources be evaluated critically.

The same basic principles and questions used to evaluate printed materials can be used to evaluate materials garnered through the Internet. Most important, users must determine whether the information was provided by a reputable source; most of the information used in this book and obtained through the Internet came from either U.S. government sources or the World Health Organization. The source of an Internet page is often apparent in its address. Internet addresses for government agencies usually end with *.gov* and addresses for educational institutions usually end with *.edu*. Nonprofit organizations, like the Sierra Club or the Muscular Dystrophy Association, usually have addresses that end with *.org*. Commercial sites, on the other hand, usually end with *.com;* this applies both to sites run directly by businesses, such as General Electric (www.ge.com), and to sites run by individuals who purchase Internet access from businesses, such as addresses ending with *aol.com*. For example, in evaluating information about different treatments for cancer, one should probably give more credence to information obtained from www.healthfinder.gov, a site run by the United States government, or from www.mayohealth.org, run by the nonprofit

---

Box 1.1   **Useful Internet Sources**

---

www.healthfinder.gov: Run by the United States Department of Health and Human Services, this site offers a wide range of health information, as well as an extensive set of links to other government and nongovernmental health-related sites.

www.nim.nih.gov: This site provides access to both published and unpublished materials available at the National Library of Medicine, the largest medical library in the world.

www.mayohealth.org: Run by the Mayo Clinic, this site offers both consumer health information and the opportunity to e-mail questions to physicians.

www.who.int: Run by the World Health Organization, this Web site provides a vast array of information about health, illness, and health care around the world.

hippo.findlaw.com: This is an invaluable compendium of information on health law, policy, and regulations. Although run by a for-profit organization, "hippo" is highly regarded in the health field.

Mayo Clinic, than to information obtained from a site that ends with *.com* and that might reflect either one individual's views or the views of a business that earns its profits by selling a particular treatment.

Unfortunately, Internet sources come and go rapidly, and addresses change constantly, making it difficult to provide a reliable list of useful Internet sites. Box 1.1 suggests some currently useful Internet sources for health issues that seem likely to remain stable for the near future. Updated addresses and additional sources can be found at this textbook's Web site, located at sociology.wadsworth.com.

SUGGESTED READINGS

Mills, C. Wright. 1959. *The Sociological Imagination.* New York: Grove Press. The classic statement of the sociological perspective.

Schwalbe, Michael. 1997. *The Sociologically Examined Life: Pieces of the Conversation.* Palo Alto, CA: Mayfield Publishing. Another excellent introduction to the sociological perspective.

REVIEW QUESTIONS

What is the sociological perspective?

How do the questions sociologists ask differ from the questions asked by psychologists or by health care workers?

What does this textbook mean by a critical approach?

What are some ways a reader can tell if a journal article or Internet Web site is a reliable data source?

INTERNET EXERCISES

Check the Web site for this textbook at sociology.wadsworth.com for updated Internet addresses.

Try different Internet search strategies to find information about writing a living will. First, try one of the major search engines, such as Excite, Yahoo, or AltaVista, which can probably be accessed by clicking on the word "search" or a "search" icon on the main menu of your Internet browser. What kinds of information (quantity, quality, type) do you find? Then try using Metacrawler (www.metacrawler.com) which searches and combines results from other search engines.

If you searched for "living will," you probably found a great deal of irrelevant information. (Your search was probably more productive if you searched for "living wills." Can you figure out why?) To make your search more effective, you'll need to learn how to perform "advanced" or "power" searches. Instructions for doing so, or tips for searching, probably appear somewhere on the Web page for your browser. For example, in some browsers, to find Web pages on living wills (rather than on every document about living that includes the word "wills"), you must search for "living+wills," whereas in other browsers you would need to search for "living wills." Do your search again, using the proper syntax to specify your request. How does this affect the information you find?

Now try the same search, using Medline, the major online archive for medical and other health-related journals. You might be able to access Medline through your college library or its Web site. Otherwise, you will need to first search for and then connect to the Grateful Med Web site, the library of a major university, or the National Library of Medicine (a branch of the National Institute of Health). Check your screen, and see if it offers instructions for narrowing your search, power searches, or advanced searches. How does the information you get from Medline differ from the information you found using a Web browser?

Finally, try looking for articles on living wills in InfoTrac©, a large online archive of scholarly articles available through Wadsworth Publishing at www.infotrac-college.com/wadsworth. (You have free access to InfoTrac this semester if your professor ordered it when ordering this textbook.) If you don't find anything after searching for "living wills," try searching for "right to die" or for "advanced directives" (a general term referring to legal documents specifying what types of medical care an individual would want in a given situation).

CHAPTER 2

The Social Sources of Illness

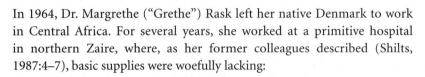

In 1964, Dr. Margrethe ("Grethe") Rask left her native Denmark to work in Central Africa. For several years, she worked at a primitive hospital in northern Zaire, where, as her former colleagues described (Shilts, 1987:4–7), basic supplies were woefully lacking:

> *You just used needles again and again until they wore out; once gloves had worn through, you risked dipping your hands in your patient's blood because that was what needed to be done. The lack of rudimentary supplies meant that a surgeon's work had risks that doctors in the developed world could not imagine. . . .*

In the early 1970s, Rask began working at a major hospital in the capital city of Kinshasa. By Christmas 1976:

> *She was thin, losing weight from a mysterious diarrhea. She had been suffering from the vague yet persistent malaise for two years now, since her time in the impoverished northern villages. In 1975, the problem had receded briefly after drug treatments, but for the past year, nothing had seemed to help. The surgeon's weight dropped further, draining and weakening her with each passing day.*
>
> *Even more alarming was the disarray in the forty-six-year-old woman's lymphatic system, the glands that play the central role in the body's never-ending fight to make itself immune from disease. All of Grethe's lymph glands were swollen and had been for nearly two years. Normally, a lymph node might swell here or there to fight this or that infection, revealing a small lump on the neck, under an arm, or perhaps in the groin. There didn't seem to be any reason for her glands to swell; there was no precise infection anywhere, much less anything that would cause such a universal enlargement of the lymph nodes all over her body. . . .*
>
> *In early 1977, it appeared that she might be getting better; at least the swelling in her lymph nodes had gone down, even as she became*

more fatigued. But she had continued working, finally taking a brief vacation in South Africa in early July.

Suddenly, she could not breathe. Terrified, Grethe flew to Copenhagen sustained on the flight by bottled oxygen. [Throughout 1977,] the top medical specialists of Denmark had tested and studied the surgeon. None, however, could fathom why the woman should, for no apparent reason, be dying. There was also the curious array of health problems that suddenly appeared. Her mouth became covered with yeast infections. Staph infections spread in her blood. Serum tests showed that something had gone awry in her immune system; her body lacked T-cells, the quarterbacks in the body's defensive line against disease. But biopsies showed she was not suffering from a lymph cancer that might explain not only the T-cell deficiency but her body's apparent inability to stave off infection. The doctors could only gravely tell her that she was suffering from progressive lung disease of unknown cause. And, yes, in answer to her blunt questions, she would die. . . .

On December 12, 1977, Margrethe P. Rask died. She was forty-seven years old.

A scant few years later, the cause of Grethe Rask's death—AIDS—would make headlines around the world. The news of a new, fatal infectious disease stunned both the medical community and the general public. Yet throughout history, new diseases have appeared and old diseases have disappeared. In this chapter, I provide a brief history of how patterns of disease have shifted over time, from the great epidemics of the past, to the late nineteenth-century decline of infectious diseases, to their modern reemergence. I then describe the current evidence regarding the main sources of premature death in the United States: tobacco, alcohol, diet and activity levels, firearms, sexual behavior, motor vehicles, microbes, illicit drugs, and toxic agents.

Before we can understand patterns of disease, however, we need to define some basic concepts scholars use in discussing this topic. The next section introduces these concepts.

AN INTRODUCTION TO EPIDEMIOLOGY

The first concept we need to define is disease. To researchers working in health care, **disease** refers to a biological problem within an organism. **Illness,** on the other hand, refers to the social experience and consequences of having a disease. So, for example, an individual who becomes infected with the polio virus has the disease we call polio. When we refer, however, to subsequent changes in that individual's sense of self and social relationships, we should properly refer to these changes as consequences of the *illness* known as polio, not the *disease*. (I will discuss the meaning of illness in more detail in Chapter 5.)

The study of the distribution of disease within a population is known as **epidemiology.** This chapter and the next focus more specifically on **social epidemiology,** or the distribution of disease within a population according to social factors (such as social class or use of tobacco) rather than biological factors (such as blood pressure or genetics). For example, whereas biologists might investigate whether heart disease is more common among those with high versus low cholesterol levels, social epidemiologists might investigate whether it is more common among smokers versus nonsmokers.

What do we mean when we say that a certain disease is "more common" among one group than another? One way is to look at the number of persons in each group that has the disease. Relying on raw numbers, however, can distort our picture of a population's health. For example, by the end of 1997, the **World Health Organization** (WHO), the United Nations agency charged with tracking and responding to health concerns, estimated that 580,000 persons were infected with the virus that causes AIDS in Brazil but only 6,300 persons were infected in the Bahamas. On the surface, these numbers suggest that Brazil has a far greater AIDS problem than the Bahamas. However, Brazil's population is much larger than that of the Bahamas. To take this difference into account, epidemiologists would look at the *rate* rather than number of AIDS cases in these two countries. **Rate** refers to the proportion of a specified population that experiences a given circumstance. The rate of any event (whether disease, disability, birth, or death) is calculated by the following formula:

$$\frac{\text{Number of events in a given period}}{\text{Specified population during that period}} \times 10^n$$

Using this formula, we find that the rate of persons known to be infected with the virus that causes AIDS (calculated as the number of infected persons in a country divided by the country's population) is 2,188 per 100,000 in the Bahamas but only 355 per 100,000 in Brazil. This tells us that AIDS affects a greater proportion of the population in the Bahamas than in Brazil and demonstrates the advantage of using rates rather than raw numbers.

Two particularly useful types of rates are incidence and prevalence rates. **Incidence** refers to the number of *new* occurrences of an event (disease, births, deaths, and so on) within a specified population during a specified time period. **Prevalence** refers to the *total* number of cases within a specified population at a specified time—both those newly diagnosed and those diagnosed in previous years but still living with the condition under study. So, for example, to calculate the *incidence* rate of lung cancer in the United States this year, we would use the formula:

$$\frac{\text{Number of } new\ cases \text{ of lung cancer diagnosed this year in U.S.}}{\text{Population of U.S. this year.}} \times 100,000$$

To calculate the *prevalence* rate of lung cancer, we would use the formula:

$$\frac{\text{Number of persons living with lung cancer in U.S. this year.}}{\text{Population of U.S. this year.}} \times 100,000$$

In general, incidence better measures the spread of **acute illnesses,** such as chicken pox and influenza, that strike suddenly and disappear quickly. Incidence also better measures rapidly spreading diseases such as AIDS. For example, to see how AIDS spread during the first decade after it was identified, we would compare its incidence in 1981 to its incidence in 1991. Prevalence, on the other hand, better measures the frequency of **chronic illnesses,** which last for many years, such as muscular dystrophy, asthma, and diabetes.

Two final terms often used in epidemiology are *morbidity* and *mortality.* **Morbidity** refers to symptoms, illnesses, and impairments; **mortality** refers to deaths. To assess the overall health of a population, epidemiologists typically calculate the rate of serious morbidity in a population (that is, the proportion suffering from serious illness), the rates of infant mortality and maternal mortality (that is, the proportion of infants and childbearing women who die during or soon after childbirth), and **life expectancy** (the average number of years individuals born in a certain year can expect to live).

The next section uses epidemiological concepts and data to describe how patterns of disease have changed over time.

A BRIEF HISTORY OF DISEASE

The European Background

All human societies, no matter how simple or small, have had diseases. However, from the perspective of Western society (and, in the long run, U.S. history), the modern history of disease begins during the Middle Ages (approximately 800 A.D. to 1300 A.D.), as commerce, trade, and urban centers began to swell (Kiple, 1993). A revolution in agricultural technology in Europe during these years not only had reduced the risk of famine but also created occasional agricultural surpluses. The resulting improvements in nutrition encouraged population growth, while the existence of surpluses spurred growth in trading. Both these factors encouraged the development of cities.

Ironically, these shifts, while reducing disability and deaths from food shortages, sparked a devastating series of epidemics. The term **epidemic** refers to any significant increase in the numbers affected by a disease *or* to the first appearance of a new disease. In the fledgling European cities, people lived in close and filthy quarters, along with rats, fleas, and lice—

perfect conditions for transmitting infectious diseases such as bubonic plague and smallpox. In addition, because city dwellers usually disposed of their sewage and refuse by tossing them out their windows, typhoid, cholera, and other water-borne diseases that live in human waste flourished. Simultaneously, the growth of long-distance trade helped epidemics spread to Europe from the Middle East, where cities had long existed and many diseases were **endemic** (that is, had established themselves within the population so they maintained a fairly stable prevalence). In addition, religious pilgrimages and crusades to Jerusalem helped spread diseases to Europe.

The resulting epidemics ravaged Europe. Waves of disease, including bubonic plague, leprosy, and smallpox, swept the continent. The worst of these was bubonic plague, popularly known as the "Black Death." Between 1347 and 1351, plague killed at least 25 million people—between 25 percent and 50 percent of Europe's population and as much as two-thirds of the population in some areas (Gottfried, 1983).

Although plague continued to strike in Europe at least once each generation until 1720, the great **pandemics** (or worldwide epidemics) began diminishing during the fifteenth and sixteenth centuries. Yet average life expectancy increased only slightly (Kiple, 1993). Mortality remained especially high in rural areas, with one-third of all individuals dying before their fifth birthday and two-thirds before reaching adulthood. In addition, although agricultural outputs remained higher than before the Middle Ages, reliance on single food crops coupled with periodic agricultural crises meant that malnutrition continued to threaten both the rural and urban poor.

By the early 1700s, however, life expectancy began to increase, at least in northern and western Europe. This change cannot be attributed to any developments in health care, for folk healers had nothing new to offer, while medical doctors and surgeons (as will be described in more detail in Chapter 11) harmed at least as often as they helped. For example, former president George Washington died after his doctors, following contemporary medical procedures, "treated" his sore throat by cutting into a vein and draining two quarts of his blood over the course of a day (Kaufman, 1971:3).

If advances in medicine did not cause the eighteenth-century decline in mortality, what did? Historians commonly trace this decline to a combination of social factors (Kiple, 1993). First, changes in warfare moved battles and soldiers away from cities, protecting citizens from both violence and the diseases that often followed in soldiers' wake. Second, the development of new crops and new lands improved the nutritional status of the population and increased its ability to resist disease. Third, women began to have children less often and at later ages, increasing both women's and children's chances of survival. Fourth, women less often engaged in long hours of strenuous field work, increasing their chances of surviving the physical stresses of childbearing. Infants, too, more often survived because mothers could more easily keep their children with them and breastfeed. (This, however, would change soon for those women who became factory workers.)

Disease in the New World

As these changes were occurring in Europe, colonization by Europeans was decimating the native peoples of the New World (Kiple, 1993). The colonizers brought with them about fourteen new diseases—including influenza, measles, smallpox, scarlet fever, yellow fever, cholera, and typhoid—that had evolved in the Old World and for which the Native Americans had no natural immunities. These diseases raced across the Americas from village to village, ravaging the Native American population and, in some cases, wiping out entire tribes. Each wave of epidemics left the population weaker, more poorly nourished, less able to support itself, and more vulnerable to the next disease. These diseases probably played a greater role than did Europeans' war technology in conquering the native tribes (Crosby, 1986).

Although the colonization of the Americas meant the death knell for many native peoples, it improved average health among the colonizers. Those who survived the hardships of the first years of settlement could draw on the New World's vast lands and agricultural resources to protect them against the malnutrition and overcrowding common in Europe. As a result, life expectancy was higher among the colonists than among their compatriots who stayed in Europe.

The Epidemiological Transition

As industrialization and urbanization increased, mortality rates rose, especially among the urban poor. The main killer was tuberculosis, followed by influenza, pneumonia, typhus, and other infectious diseases. By the late nineteenth century, however, deaths from infectious diseases began to decline rapidly in the United States as in other industrialized countries. Although infectious diseases remained common, especially among the poor, they no longer accounted for the majority of deaths. Partly as a result, infant and childhood mortality declined steeply. Between 1900 and 1930, life expectancy rose from 47 years to 60 years for whites and rose from 33 years to 48 years for African Americans (U.S. Bureau of the Census, 1975).

As infant mortality declined, families no longer felt obligated to have many children to ensure that one or two would survive long enough to get work and bring income into the household. At the same time, the national economy continued to shift from agriculture to industry, reducing couples' need to have children to work on the family farm. Similarly, employers increasingly offered pensions and other social benefits, so couples' need to have children to care for them in their old age also declined. Taken together, these trends produced a sharp decline in family size. Consequently, families could devote more resources to each child, further increasing children's chances of survival.

As infectious diseases declined in importance, chronic and degenerative diseases, which only can affect those who live long enough for symptoms to develop, gained importance. Cancer, heart disease, and stroke became major causes of mortality, while arthritis and diabetes emerged as major sources of morbidity. Increasingly, too, conditions like heart disease, stroke, and hypertension shifted from being primarily diseases of the affluent to being disproportionately diseases of the poor.

The shift from a society characterized by infectious and parasitic diseases and low life expectancy to one characterized by degenerative and chronic diseases and high life expectancy is referred to as the **epidemiological transition** (Omran, 1971). This transition seems to occur around the world once a nation's mean per capita income reaches a threshold level (in 1999 dollars) of about $6,400 (Wilkinson, 1996). As we will see in more detail in the next chapter, some countries have fully made the epidemiological transition while others have not.

Contrary to conventional wisdom, medical interventions such as vaccinations, new drugs, and new surgical techniques played little role in the epidemiological transition, which began more than 200 years ago in western societies (Leavitt and Numbers, 1985; McKeown, 1979; McKinlay and McKinlay, 1977). In a series of dramatic graphs showing how mortality from several important diseases declined over time, McKinlay and McKinlay (1977) have demonstrated that these declines by and large *preceded* the introduction of effective medical interventions (Figure 2.1). For example, the death rate for tuberculosis declined steadily from greater than 3.5 per 1,000 in 1860 to .34 per 1,000 in 1946. Yet, streptomycin, the first effective treatment for tuberculosis, was not introduced until 1947. Only polio and smallpox declined substantially after the introduction of medical interventions. Of these two, only the decline in polio can be confidently attributed to medical intervention, as we cannot separate the possible impact of inoculation on the rate of smallpox from the impact of the myriad other changes that occurred since inoculation was first widely adopted about 200 years ago.

Summing up their data, McKinlay and McKinlay (1977) estimate that medical measures account for at most 3.5 percent of the total decline in mortality since 1900. More recent studies, using fairly generous assessments of the potential impact of modern medical care on life expectancy, have concluded that medical care can explain no more than one-sixth of the increase in life expectancy during the twentieth century (Bunker et al., 1994). These findings, of course, do not negate the importance of medical care for saving individual lives and certainly do not negate the enormous improvements in

Figure 2.1   *The fall in the standardized death rate (per 1,000 population) for nine common infectious diseases in relation to specific medical measures, for the United States, 1900–1973.*

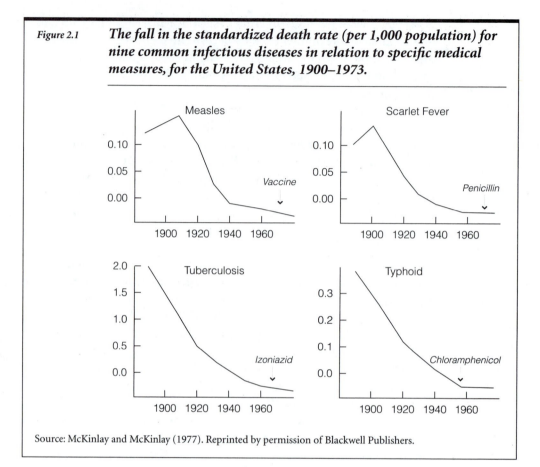

Source: McKinlay and McKinlay (1977). Reprinted by permission of Blackwell Publishers.

quality of life that medical techniques such as cataract surgery or pain relief have given the public. Nevertheless, these findings do underscore the limited effect of medical care on broad changes in mortality rates.

If medical interventions do not explain the decline in infectious diseases, what does? Research suggests that the decline in disease associated with the epidemiological transition in western societies resulted primarily from changes in the social environment (McKinlay and McKinlay, 1977). As nutrition and living conditions improved, so did individuals' ability to resist infection and to survive if they became infected. At the same time, although somewhat less importantly, public health improvements such as the development of clean water supplies and sanitary sewage systems increasingly protected individuals from exposure to disease-causing microbes.

Given the enormous improvements in life expectancy during the twentieth century, it was natural for scientists to assume that life expectancy would continue to rise steadily along with incomes. However, a highly influential book published by Richard Wilkinson (1996), based on a diverse wealth of data from studies conducted around the world, strongly suggests that increases in average income above about $6,400 (in 1999) bring only modest increases in life expectancy. Instead, further increases in life expectancy appear to occur not when *absolute incomes* increase but only when the *relative income differential* within a country narrows. In other words, if the gap in income between rich and poor narrows, as it has in Costa Rica, for example, average life expectancy increases (especially among poorer citizens). Conversely, if the income gap widens, as happened following the collapse of the former Soviet Union, average life expectancy declines. As a result, life expectancy is greatest within countries that have experienced the epidemiological transition *and have the smallest income gap between rich and poor,* like Sweden and Japan, rather than in countries like the United States, which despite its great wealth has the widest income gap among the industrialized nations (Bradsher, 1995).

After weighing all the available evidence, Wilkinson argues that the key to the better health found in societies with small income gaps is their social cohesion. These societies, he argues,

> have a strong community life. Instead of social life stopping outside the front door, public space remains a social space. The individualism and the values of the market are restrained by a social morality. People are more likely to be involved in social and voluntary activities outside the home. . . . There are fewer signs of anti-social aggressiveness, and society appears more caring. In short, the social fabric is in better condition (Wilkinson, 1996:4).

This thesis explains not only why these societies have higher life expectancies but also why, as a rule, they are especially unlikely to experience high rates of mortality from conditions linked to social stress, such as accidents, violence, and alcohol-related diseases. Importantly, Wilkinson notes that when societies reduce income inequality through increasing education, housing, and employment opportunities (as Japan and Korea did following World War II), all members of the society benefit because lower-class persons become both more economically productive and less likely to engage in criminal or violent behaviors.

The New Rise in Infectious Disease

By the second half of the twentieth century, Americans—both health care workers and the public—had come to believe infectious diseases were under control (even though they continued to rage in poorer regions of the world). In part as a result, when, on June 5, 1981, the federal government's *Morbidity and Mortality Weekly Report* published a brief article describing a curious syndrome of immune-deficiency disorders in five gay men, few paid much attention. Within a few years, however, people around the world would learn to their horror that a deadly new infectious disease, now known as AIDS, had taken root. Since then, scientists have identified other infectious diseases previously unknown in western societies, including hemorrhagic fevers and new, deadly strains of cholera and streptococcus. In addition, previously harmless microbes have become deadly; the water-borne parasite *cryptosporidium,* for example, which until recently seemed unable to harm humans, in 1993 sickened 400,000 Milwaukee residents (Altman, 1994).

To many individuals, including many doctors, who had come of age after the development of antibiotics and vaccinations, the appearance of new and more virulent infectious diseases seemed almost incomprehensible. Yet those working in the poorer nations of the world knew that infectious diseases continued to rage (Farmer, 1999). Others, also, had long regarded the reemergence of infectious disease as inevitable. Indeed, even before 1981, signs indicated that the world still needed to maintain its vigilance against infectious disease, as more and more cases of drug-resistant gonorrhea, pneumonia, meningitis, and other diseases appeared.

The renewed dangers posed by infectious disease partly reflect basic principles of natural selection. Just as natural selection favors animals whose camouflaging coloration hides them from predators so they can survive long enough to reproduce, natural selection favors those germs that can resist drug treatments. As doctors prescribed antibiotics more widely, often under pressure from patients who feel "cheated" if they do not receive a prescription at each visit (Vuckovic and Nichter, 1997), the drugs killed all susceptible variants of disease-causing germs while allowing variants resistant to the drugs to flourish.

Other forces also promoted the rise in infectious diseases (Garrett, 1994). In the same way that population growth and the rise of cities had fostered the spread of infectious diseases in Europe centuries ago, these same factors now are causing new epidemics to develop in the rapidly growing cities of Africa, Asia, and Latin America. Meanwhile, older cultural traditions often erode among those who move to these cities, making health-endangering activities like tobacco smoking and sexual experimentation more likely. At the same time, the destruction of ecosystems in these regions, as forests and farmlands are replaced by industrial sites and cities, changes the balance between human, animals, and microbes, encouraging some microbes that previously had infected only animals to begin infecting humans.

All these factors have been heightened by **globalization,** the process through which ideas, resources, and people increasingly operate in a world-wide rather than local framework. The erosion of cultural traditions in Asia, Africa, and Latin America in part reflects the increasing cultural supremacy of western ideas. Similarly, environmental changes partly stem from industries based in the west. Finally, the globalization of business investment and tourism has increased the number of westerners traveling to these regions and, in turn, has encouraged the globalization of disease

(Garrett, 1994). For example, West Nile encephalitis, a potentially fatal disease, first appeared in the United States in 1999.

The Reemergence of Tuberculosis

From a worldwide perspective, probably the most important consequence of the development of drug-resistant germs is the reemergence of tuberculosis, which kills more people yearly than any other infectious disease (Donnelly and Montgomery, 1999). During the nineteenth century, tuberculosis, known as the "white plague," killed millions. Tuberculosis is caused by the bacillus *Mycobacterium tuberculosis,* which attacks and destroys lung tissue. Individuals become infected by inhaling bacilli spread when already-infected persons sneeze or cough. Because the bacilli can survive for some time in air or dust and because one can become infected merely by breathing, tuberculosis is highly infectious. In fact, WHO estimates that one-third of the world's population is infected (although only about 10 percent of infected persons, most often those already weakened by age, other diseases, or poor living conditions, ever develop active symptoms).

The incidence rate of tuberculosis in the United States declined steadily from the late nineteenth century to the 1980s but then rose steadily until 1992, especially among immigrants and minorities. According to the U.S. **Centers for Disease Control and Prevention** (CDC), the federal agency responsible for tracking the spread of diseases in the United States, the current tuberculosis epidemic reflects the increases in (1) AIDS; (2) homelessness, poverty, and substance abuse; (3) persons lacking health care; and (4) drug-resistant strains of the disease (*Morbidity and Mortality Weekly Report,* 1993). The increase in AIDS, homelessness, and poverty beginning in the 1980s and the continued high rates of substance abuse has left more Americans with weakened immune systems, making them more likely to become infected with tuberculosis and to develop active symptoms if infected. During the same period, the numbers of persons without health insurance or access to health care have increased. As a result of these factors, those who develop active tuberculosis often do not receive consistent medical care and stop treatment once their symptoms abate rather than continuing until the bacilli are all killed.

By definition, the bacilli strains that survive are those most resistant to the drugs. Thus, a vicious cycle develops in which difficulties in treatment lead to the evolution of more resistant variants of the bacilli, which in turn makes treatment more difficult. Currently, treatment costs about $250,000 per person, takes six to eight months, and often fails (Donnelly and Montgomery, 1999). As of 1999, only 1.3 percent of U.S. cases were drug-resistant, but these numbers are growing, and drug-resistant cases have been identified in almost all fifty states.

On the other hand, even among drug users and the poor, public health workers have found that they can stem the tide of tuberculosis infection.

Studies conducted both in the United States and elsewhere suggest that even in these populations, multiple drug-resistant tuberculosis can be cured and transmission prevented if patients are given financial assistance to attend clinics and extra food so their immune systems strengthen, if clinics are moved to times and places that meet patients' needs and are staffed by people who treat patients with respect, and if staff directly observe patients taking their medications and so can identify reasons patients might not do so (Farmer, 1999). Strategies such as these have led to a steady reduction since 1992 in the tuberculosis rate in the United States.

The Emergence of AIDS

If tuberculosis illustrates how an old killer can reappear, AIDS demonstrates the continuing potential for new killer diseases to arise. Beginning in 1979, a few doctors in New York, San Francisco, and Los Angeles had noticed a small outbreak in young gay men of a deadly form of Kaposi's sarcoma, a rare cancer that normally produces only mild symptoms and that primarily affects (for unknown reasons) elderly, heterosexual men who are Italian or Jewish. Subsequently, doctors discovered five gay men who had *pneumocystis carinii* pneumonia (PCP), a rare pneumonia generally found only among persons who cannot fight infections effectively because their immune systems have been damaged by disease or chemotherapy. Diseases such as PCP are called "opportunistic infections" because they stem from microorganisms that usually live in the body harmlessly but that take advantage of the opportunity created by weakened immune systems to multiply and cause disease.

The CDC published the first report on the PCP cases in mid 1981, followed shortly by a report on Kaposi's sarcoma. At this point, no one knew what had caused these strange diseases. Obviously, however, something had destroyed the immune systems of these men, leaving them susceptible to fatal infections by virtually any microorganism in their environment.

The next year, the CDC officially coined the term **acquired immunodeficiency syndrome (AIDS)** to describe what we now know is the last, deadly stage of infection with **human immunodeficiency virus (HIV)**. Epidemiological research soon demonstrated that the disease was spread through sexual intercourse; through sharing unclean intravenous needles; through some still-unknown mechanism from mother to fetus; through blood transfusions or blood products; and, rarely, through breastmilk. The last three modes of transmission are now rare in countries where HIV blood tests, breastmilk substitutes, and drugs for reducing the risk of maternal-fetal transmission are affordable.

Despite the great fear the AIDS epidemic has engendered, studies have demonstrated conclusively that AIDS is not spread through insects, spitting, sneezing, hugging, nonsexual touching, or food preparation. Of the thousands of friends and family members who have lived in households with per-

| Table 2.1 | **Modes of Transmission for Adult and Adolescent AIDS Cases Diagnosed During 1998, United States.** | |

EXPOSURE CATEGORY	PERCENTAGE OF CASES
Men who have sex with men	35%
Injecting drug use	23
Men who have sex with men and inject drugs	4
Heterosexual contact	14
Blood recipients	1
Unknown*	23
Total	100%

*Typically, mode of transmission is unknown because the case is still under investigation, the individual refused to answer questions or died before being interviewed, or the mode of transmission, while suspected, could not be proved because the individual's past heterosexual partners had not been HIV-tested. Of cases listed as mode of transmission unknown at some point before December 1998, almost half (47 percent) were reclassified into one of the existing transmission categories following further investigation (Centers for Disease Control and Prevention, 1998a).

Source: Centers for Disease Control and Prevention (1998a).

sons who have AIDS—sharing toothbrushes, beds, dishes, and so on—only eight have become infected other than through sexual intercourse, needle-sharing, or breastfeeding (Peterman et al., 1988; Stine, 1998:195–198). Even regular sexual partners of persons with AIDS usually remain uninfected if they use latex condoms consistently and often remain uninfected even if they do not (Stine, 1998:219, 255).

Although initially identified in gay men, AIDS cannot be considered a "gay disease." Homosexual activity now accounts for less than half of new cases in the United States and less than 10 percent internationally (Stine, 1998:217). Moreover, although some lesbians have become infected with HIV through drug use or sex with men, no documented cases of female-to-female sexual transmission have occurred. Table 2.1 shows the modes of transmission for cases of AIDS diagnosed in 1998.

In the United States, it takes several years on average before HIV-infected persons experience any symptoms and more than a decade before AIDS develops. As a result, although AIDS remains an acute illness in places like Africa, many in the west now consider it a chronic disease. Because most HIV-infected persons do not, in fact have AIDS, this textbook uses the term *HIV disease* rather than AIDS except when reporting statistics based specifically on AIDS cases.

The rapid spread of HIV disease since 1981 reflects public attitudes as much as biological realities. A handful of behavioral changes could have virtually halted its spread: testing the blood supply for infection, using latex condoms and spermicide with sexual partners, and using clean needles when injecting drugs. Unfortunately, throughout the early years of the epidemic when intervention would have been most effective, the U.S. government (like most other governments) treated HIV disease as a distasteful moral issue rather than as a medical emergency. At critical junctures during the 1980s, federal officials lobbied Congress to restrict funding for HIV research and education (Epstein, 1996; Presidential Commission, 1988). Moreover, the limited funds the government provided early on for HIV education came with many strings attached, such as prohibiting explicit pictures in materials on sexual education, prohibiting language that might offend heterosexuals even in educational materials designed solely for gay men, and—even though substantial proportions of teenagers engage in sexual intercourse—refusing to fund education programs for children and young adults unless the programs taught only abstinence from sex and not how to have sex safely.

Similarly, both federal and local authorities have made it exceedingly difficult for individuals to protect themselves from infection by using intravenous needles safely. By retaining laws making it illegal to purchase or own needles and prosecuting those who distribute needles, laws encourage addicts to share needles and thus to share not only HIV infection but hepatitis and other diseases as well. At the same time, the government has refused funding to those who would teach drug users how to clean needles. Yet most research suggests that helping drug users to protect themselves reduces the incidence of HIV infection without increasing the rate of drug use (Gostin et al., 1997).

The Modern Disease Profile

Despite the recent reemergence of infectious diseases, however, these diseases still play a relatively small role in U.S. mortality rates. Table 2.2 shows the top ten causes of death in the United States and how these have changed since 1900.

As the table demonstrates, whereas the top two killers in 1900—influenza and pneumonia—were infectious diseases, the top killers currently—heart disease and cancer—are chronic diseases primarily associated with middle-aged and older populations. These diseases now far outpace infectious diseases as causes of death.

However, infectious diseases have not disappeared from the list of leading causes of death. Pneumonia and influenza remain significant and, although in 1997 AIDS fell from eighth to fourteenth place on the list, among persons ages 25 to 44 it remains the most common cause of death for

Table 2.2	Main Causes of Deaths, 1900 and 1997		
1900	**RATE/100,000**	**1997**	**RATE/100,000**
Influenza and pneumonia	202	Heart disease	271.6
Tuberculosis	194	Cancer	201.6
Gastritis	143	Cerebrovascular disease	59.7
Diseases of the heart	137	Chronic pulmonary disease	40.7
Cerebrovascular diseases	107	Accidents	35.7
Chronic kidney disease	81	Pneumonia and influenza	32.3
Accidents	72	Diabetes	23.4
Cancer	64	Suicide	11.4
Diseases of early infancy	63	Kidney disease	9.5
Diphtheria	40	Liver disease	9.4

Source: Greenberg, 1987:5; National Center for Health Statistics, 1999.

African Americans, the third most common cause for Hispanics, and the tenth most common cause for non-Hispanic whites (National Center for Health Statistics, 1999). Moreover, although new therapies for AIDS keep being developed, a research review by the World Bank (1997) concluded that no long-term controlled studies have yet identified any drugs that can increase life expectancy significantly over the long run; this remains true as of late 1999. In any event, the most effective drugs now available (the protease inhibitors) only can help those who can afford their price of $12,000 per year (as of 1999), tolerate their side-effects, and manage the required regimen of as many as 20 pills per day taken at strictly regulated times.

Finally, Table 2.2 shows the continued role social factors play in causing deaths: accidents (mostly motor vehicle accidents), chronic pulmonary disease (most commonly caused by tobacco use), liver disease (most commonly caused by alcohol use), suicide, and AIDS all reflect social behaviors rooted in particular social conditions. The remainder of this chapter discusses the role social forces play in causing mortality and morbidity.

THE SOCIAL SOURCES OF PREMATURE DEATHS

In a widely-cited article titled "A Case for Refocusing Upstream," sociologist John McKinlay (1994) offers the following oft-told tale as a metaphor for the modern doctor's dilemma:

Sometimes it feels like this. There I am standing by the shore of a swiftly flowing river and I hear the cry of a drowning man. So I jump into the river, put my arms around him, pull him to shore and apply artificial respiration. Just when he begins to breathe, there is another cry for help. So I jump into the river, reach him, pull him to shore, apply artificial respiration, and then just as he begins to breathe, another cry for help. So back in the river again, reaching, pulling, applying, breathing, and then another yell. Again and again, without end, goes the sequence. You know, I am so busy jumping in, pulling them to shore, applying artificial respiration, that I have *no* time to see who the hell is upstream pushing them all in (McKinlay, 1994:509–510).

This story illustrates the traditional emphasis within medicine on **tertiary prevention:** strategies designed to minimize physical deterioration and complications among those already ill. Tertiary prevention includes such tactics as providing kidney dialysis to persons whose kidneys no longer function or insulin to those who have diabetes. Doctors much less commonly focus on **secondary prevention:** strategies designed to reduce the prevalence of disease through early detection and prompt intervention. Examples of secondary prevention include screening patients for cervical cancer or glaucoma so these diseases can be detected at still-treatable stages. Those who focus on secondary prevention typically work in public health or in the **primary practice** fields (family practice, pediatrics, or internal medicine). Finally, only a small fraction of doctors, usually in public health or, less commonly, primary practice, focus "upstream" on **primary prevention:** strategies designed to keep people from becoming ill or disabled, such as discouraging drunk driving, lobbying for stricter highway safety regulations, and promoting vaccination.

Even when doctors, researchers, (or, for that matter, the general public) have focused on primary prevention, they typically have looked only far enough upstream to see how individual psychological or biological characteristics make some more susceptible than others to disease or unhealthy behaviors. For example, an increasing number of medical researchers now focus on the genetic roots of disease, such as a possible gene for alcoholism. Within the social sciences, meanwhile, many researchers focus on understanding how individuals choose whether to adopt behaviors believed to prevent illness, such as exercising regularly or refraining from smoking.

The most commonly used framework for understanding individual preventive health behavior is the **health belief model,** originated by Irwin Rosenstock (1966) and extended, most importantly, by Marshall Becker (1974, 1993). According to the model, four factors determine whether individuals will adopt preventive health behaviors (Table 2.3). First, individuals must believe they are susceptible to a particular health problem. Second, individuals must believe the problem they risk is a serious one. Third, individuals must believe adopting preventive measures will reduce their risks

Table 2.3	**The Health Belief Model and Medical Compliance**	
PEOPLE MOST LIKELY TO COMPLY WHEN THEY:	EXAMPLE: COMPLIANCE LIKELY	EXAMPLE: COMPLIANCE UNLIKELY
Believe they are susceptible	Fifty-year-old man with hypertension who believes he is at risk for a heart attack.	Fifteen-year-old boy diagnosed with epilepsy who has had only minor problems and does not believe he is at risk for convulsions.
Believe risk is serious	Believes that heart attack could be fatal.	Believes that convulsions would not be physically dangerous.
Believe compliance will reduce risk	Believes he can reduce risk through taking medication regularly.	Believes he doesn't really have a problem, so doesn't see how medication could help.
Have no significant barriers to compliance	Medication is affordable and has no serious or highly unpleasant side effects.	Medication makes the boy feel drowsy, dull, and set apart from his peers.

significantly. Fourth, individuals must not perceive any significant barriers to doing so. For example, individuals are most likely to adopt a low fat diet if they believe that they would otherwise face high risks of heart disease, that heart disease would substantially decrease their life expectancy, that a low fat diet would substantially reduce their risk of heart disease, and that adopting such a diet would not be too costly, inconvenient, or unpleasant. In turn, according to the health belief model, these four factors are affected by demographic variables (such as the individual's gender and age), social psychological variables (such as personality characteristics and peer group pressures), structural factors (such as access to knowledge about the problem and contact with those who experience the problem), and external cues to action (such as media campaigns about the problem or doctors' advice).

As this model suggests, although biological factors and psychological predispositions do affect the likelihood of adopting healthier behaviors, they do not occur in a vacuum. Rather, they occur in particular economic, cultural, and political settings that can make healthy behaviors or health itself either more or less possible. For example, adolescents' decisions regarding whether to drink alcohol are affected significantly by the attitudes of their friends, family, and culture in general. Similarly, the high rates of diabetes found among contemporary Native Americans in part reflect a genetic predisposition to the disease. They also, however, reflect the effects of the reservation system, with its sedentary lifestyle, ready access to fatty and sugary foods, and limited prospects for employment which make purchasing healthier foods difficult. In both cases, to blame unhealthy behavior patterns on individual choices seems oversimplistic.

As these examples suggest, truly refocusing upstream requires us to look beyond individual behavior or characteristics to what McKinlay refers to as the **manufacturers of illness:** those groups that promote illness-causing behaviors and social conditions. These groups include alcohol distributors, auto manufacturers that fight against vehicle safety standards, and politicians who vote to subsidize tobacco production.

An article by public health doctors J. Michael McGinnis and William H. Foege (1993) provides a useful starting point for refocusing upstream. The article synthesizes the available literature on the major underlying causes of premature deaths (that is, deaths caused neither by old age nor by genetic disease) to identify those causes of death that we could most readily reduce or eliminate through social or medical interventions.

McGinnis and Foege identify nine causes that, they believe, together account for 50 percent of all premature deaths. Table 2.4 shows these causes

---

Table 2.4 **Estimates of Actual Causes of Premature Death in the United States, 1990**

CAUSE	NUMBER	PERCENTAGE OF DEATHS
Tobacco	400,000	19
Diet/activity patterns	300,000	14
Alcohol	100,000	5
Microbial agents*	90,000	4
Toxic agents	60,000	3
Firearms	35,000	2
Sexual behavior	30,000	1
Motor vehicles**	25,000	1
Illicit use of drugs	20,000	<1
Total	1,060,000	50

*Does not include deaths related to HIV, tobacco, alcohol, illicit drugs, or infections caused by other diseases.

**Does not include accidents linked to alcohol or drug use.

Source: McGinnis and Foege (1993).

(listed not by disease but by the factors that cause disease) and provides estimates of their prevalence. These estimates reflect the *lowest* estimates given in the reviewed literature, so the actual prevalences could well be considerably higher. The next sections look at these nine causes of illness, focusing not on the individual behavior patterns that McGinnis and Foege emphasize in their article but on the manufacturers of illness that lie behind these individual behaviors.

Tobacco

As Table 2.4 shows, tobacco causes more premature deaths in the United States than any other legal or illegal drug. Whether smoked, chewed, or used as snuff, tobacco can cause an enormous range of disabling and fatal diseases, including heart disease, strokes, emphysema, and numerous cancers (World Health Organization, 1998c). About half of all smokers will die because of their tobacco use, with half of these dying in middle age and losing an average of 22 years from their normal life expectancy. Tobacco use also increases morbidity and mortality among those who must live and work around smokers, known as "passive smokers" (World Health Organization, 1998c). Similarly, both active and passive smoking can cause birth defects and infant mortality.

Unfortunately, quitting smoking is difficult, as use of nicotine (the active ingredient in tobacco) usually results in **addiction.** In other words, those who chew or smoke nicotine regularly, like those who use heroin regularly, experience a predictable combination of distressing physical symptoms, known as **withdrawal,** if they cease using the drug. Moreover, nicotine is more addictive than heroin: most individuals do not experience withdrawal from heroin unless they have used it often and regularly, but most experience withdrawal from nicotine after using it only a few times (Weil and Rosen, 1998).

Given nicotine's addictiveness, it is easy to understand why individuals continue smoking once they have started. But why do individuals begin smoking in the first place, especially when many initially find tobacco vile tasting and even nauseating? To answer this question, we need to look at the role of tobacco in American culture and at how tobacco manufacturers have created that role.

Although used for centuries, tobacco only recently became a truly popular drug (Ravenholt, 1993). Its first spurt in popularity came at the beginning of the twentieth century, when manufactured cigarettes and "safety matches" first made smoking easy and portable. Its popularity grew steadily until the 1960s, as a result of manufacturers' success in getting cigarettes included in soldiers' rations; continuous advertising campaigns that associated smoking with social status and sexual pleasure; and federal subsidies for tobacco growers and cigarette manufacturers.

In 1964, however, use of tobacco plummeted, especially among the middle and upper classes, following publication of a report by the U.S. Surgeon General that declared cigarette smoking a cause of lung cancer. To replace smokers who quit or died, manufacturers have labored to gain new smokers, designing advertising campaigns to convince the public to associate tobacco with positive attributes and achievements rather than with death and disability. For the last thirty years, the tobacco industry has especially targeted youths, women, and minorities (White, 1988). As summarized in an article published in the *American Journal of Public Health,* research has found that

> [t]obacco marketing does reach youth. Young people are able to name and recognize cigarette ads and can also match cigarette brand name with cigarette slogans. More than half of current adolescent smokers and approximately one quarter of nonsmoking teens own cigarette promotional items and participate in these campaigns. Moreover, there is a growing recognition that such advertising and promotion influence teen smoking. Longitudinal studies of advertising patterns and young people's tobacco use demonstrate a positive association between advertising and teenage smoking. In addition, the vast majority of adolescent smokers prefer the most heavily advertised brands and report that they would smoke the brand whose advertisements they liked most (Schooler et al., 1996).

Manufacturers also have targeted their marketing to women by playing on women's desire for equality, excitement, personal fulfillment, and weight loss (a cultural imperative for women in contemporary American culture and a major reason women smoke). This was exemplified by Virginia Slims—the name was not accidental—and its slogan, "You've come a long way, baby." To target minorities, manufacturers advertise heavily in magazines such as *Ebony* and *Jet.* They also have gained influence and visibility in minority communities by providing financial sponsorship for charitable organizations and civic and cultural events (White, 1988). In addition, until banned in 1996, manufacturers relied heavily on billboards placed low and close to the street in African-American and Hispanic (but not white) neighborhoods; these billboards were particularly good at attracting children's attention.

During the last few years, however, successful legal attacks on tobacco manufacturers have begun to erode their ability to attract new customers. Following a series of hearings that demonstrated tobacco manufacturers had known for decades that nicotine was addictive and had intentionally used manufacturing techniques to make cigarette smoking more addictive, the federal Food and Drug Administration in 1996 adopted a series of regulations on the sale and advertisement of tobacco, mostly aimed at deterring youths from smoking (Hilts, 1996). Almost simultaneously, all fifty states (some singly, some jointly) sued the five largest tobacco manufacturers to recover the money states spend on medical care for citizens who have tobacco-related diseases. Under the terms of a joint settlement reached in late 1998, the tobacco companies will pay the states $206 billion over the next 25 years. Among the many other terms of that settlement, tobacco companies may no longer use cartoon characters in advertisements and must restrict their sponsorship of sports and entertainment events.

In response to these threats, tobacco manufacturers have donated enormous sums—$2.3 million during 1997–1998 at the federal level—to political candidates to encourage them to support the manufacturers' interests (Center for Responsive Politics, www.opensecrets.org/home/index.asp, 1999). For example, tobacco manufacturers (and merchants) have pressed for passage of laws that criminalize the possession or use of tobacco *by* minors, while working against laws that criminalize the sale of tobacco *to* minors (Mosher, 1995). In this way, they hope to define tobacco use as an individual choice and place the blame on consumers rather than on the tobacco industry.

Alcohol

Like tobacco, alcohol kills far more people than all illegal drugs combined. Heavy alcohol use can cause irreversible brain damage, hepatitis, heart disease, cirrhosis of the liver, and cancers of the digestive system, while reducing the body's ability to fight infections such as tuberculosis and pneumonia. In addition, by diminishing individuals' ability to make rational choices, alcohol use contributes to deaths from drownings, fires, violence, and accidents and increases the odds of engaging in unsafe sexual behavior. Finally, withdrawal from alcohol is more dangerous than withdrawal from any other legal or illegal drug and can cause brain damage, heart failure, or stroke. Yet despite the dangers of alcohol, by law the U.S. Office of National Drug Control Policy cannot use any of its funds—$195 million during 1999—to fight problem drinking; recent efforts to change this, supported by the American Medical Association and the American Public Health Association, have met fierce resistance from alcohol manufacturers and distributors (Wren, 1999).

To ensure that the government continues to treat alcohol as a beverage rather than a drug, alcohol manufacturers contribute heavily to political campaigns, giving $2.3 million to federal candidates during the 1997–1998 election cycle alone (Center for Responsive Politics, http://www.opensecrets.org/home/index.asp, 1999). Manufacturers also have worked to define the individual drinker rather than alcohol itself as the problem by promoting the idea that alcoholism is a disease that only affects susceptible individuals, funding research on biological roots of alcoholism, and, like tobacco manufacturers, supporting laws that criminalize underage drinking while fighting laws that would criminalize the sale of alcohol to minors (Morgan, 1988; Mosher, 1995).

At the same time, alcohol manufacturers have worked diligently to sell alcohol to the public not as a drug but as a lifestyle, spending about $2 billion per year on advertising. Much of this marketing either directly or indirectly targets youths, despite voluntary industry codes that forbid manufacturers from marketing alcohol to audiences in which a majority are under age 21. During 1997–98, only four of eight manufacturers studied by the Federal Trade Commission (1999) met even this lenient standard, and manufacturers paid to have their products appear on eight of the fifteen television shows most popular with teenagers. Advertisements typically associate alcohol with adulthood, sexual adventure, status, freedom, excitement, and pleasure. Meanwhile, alcohol also sells because it offers an effective, if self-destructive, way to dull the emotional pains of daily life and the physical pains of hunger, cold, or abuse.

Diet and Exercise

The second most common cause of premature deaths, according to McGinnis and Foege (1993), is a high-fat diet and sedentary lifestyle, which may increase the odds of developing cardiovascular disease, certain cancers (of the colon, breast, and prostate), and diabetes. Nutritionists agree that Americans would be fitter, feel healthier, and perhaps live longer if they increased their activity levels; reduced consumption of fats, sugars, salt, and meat; and increased consumption of fruits, vegetables, and whole grains. Such changes would most benefit poor Americans, who are more likely than others to eat unbalanced diets that emphasize sugars and fats because those foods provide energy and satisfy hunger most cheaply (James et al., 1997).

Unfortunately, humans have a natural craving for sweet and fatty foods, and physical exercise is no longer built into most Americans' work life or transportation patterns. In addition, the tendency toward eating a sweet, fatty diet has been reinforced by both food manufacturers and the U.S. government. Manufacturers earn far greater profits from selling highly refined products loaded with fat, sugar, and salt than from selling healthier foods. As a result, advertisements for the former appear far more often than advertisements for the latter, especially on children's television.

The U.S. government, too, has helped to promote the current fatty American diet (Nestle, 1993). The U.S. Department of Agriculture (USDA) provides subsidies and price supports for dairy and meat producers, including purchasing high-fat products for free distribution to school lunch programs for poor children. In addition, beginning in 1956, the USDA heavily promoted, in schools and elsewhere, the concept of the "Basic Four" food groups: milk and milk products; meat, fish, and poultry (a group that included eggs, dried beans, and nuts); fruits and vegetables; and grains. These four food groups reflected the USDA's federally mandated goals of promoting the nation's dairy and meat industries and advising the public about diet. (Promoting public health, on the other hand, fell under the aegis of the Department of Health.)

In the 1950s, promoting meat and dairy products and advising about diet did not seem contradictory goals, for the main nutritional problem in those years was how to improve the health of low-income groups whose diets offered insufficient nutrition. By the 1970s, however, health care researchers had stopped worrying that Americans were not eating enough and had begun worrying that Americans were eating too much, especially fat and cholesterol. From this point on, the USDA would find it increasingly difficult to protect the food industries unless it ignored nutrition research on the impact of diet on health. The results of this dilemma can be seen in the USDA's current dietary guidelines. Although the guidelines now recommend eating a low fat diet and emphasize fruits, vegetables, and grains, they still recommend two to three daily servings of dairy products and an additional two to three servings of meat or other proteins, with a serving defined as 5 to 7 ounces of cooked meat. Yet both dairy products and meat often have high fat and cholesterol levels. Moreover, the majority of nonwhite Americans lack the enzyme needed to digest dairy products and suffer intestinal distress when they eat them.

Microbial Agents

The microbes that most commonly cause premature deaths in the United States (not including deaths from viruses attributable to illegal drug use or sexual behavior) are the bacteria and viruses that cause pneumococcal pneumonia, nosocomial infections (infections developed as a result of a hospital stay), legionellosis, *staphylococcal aureus* infection, hepatitis, and group A streptococcal infection. Tuberculosis may join this list in the near future.

Infectious organisms surround us all the time. Yet only rarely do individuals become infected and even more rarely do these infections lead to deaths. In what conditions does this happen?

First, individuals will not develop fatal diseases if they are vaccinated against them. Vaccinations do not exist for all diseases, but McGinnis and Foege (1993) estimate that about 13 percent of fatal infections occurring in the United States could be prevented through vaccination. Virtually all U.S. children are vaccinated before they begin school, but 28 percent do not receive all the required vaccinations by the recommended age of two years old (Centers for Disease Control and Prevention, 1997). However, vaccination rates have increased sharply since 1993, when the CDC began a major vaccination drive.

Second, even in the absence of vaccinations, individuals exposed to germs may not become infected unless they already are physically weakened. For example, by definition, nosocomial infections occur among those who already are ill enough to need hospital care. Similarly, individuals are far more susceptible to infection if age, malnutrition, poor housing, insufficient clothing, or other stressors weaken their bodies. This explains why American tourists rarely contract tropical diseases even when traveling in countries where disease is endemic and even when they are neither vaccinated nor taking prophylactic drugs.

Third, the same factors that leave some susceptible to infection help explain why, among those who do become infected with a given disease, some will die while others experience only minor health problems; measles, for example, is a minor childhood disease in the United States but a major killer in poorer countries (as Chapter 4 will describe).

Fourth, among those who become ill, death or long-term disability may not occur if individuals have ready access to good health care. For example, doctors can cure most bacterial infections in otherwise healthy individuals, while simply providing intravenous nutrition and fluids can save the lives of many infants suffering from life-threatening diarrhea.

Toxic Agents

McGinnis and Foege (1993) trace 3 percent of premature deaths to toxic agents. These agents can be divided into occupational hazards and environmental pollutants. In "light" industries like electronics, workers are often exposed to a wide variety of potentially toxic solvents, such as trichloroethylene (TCE), while in traditional industries such as mining and construction, welders often face substantially increased risks of lung cancer caused by toxic levels of chromium and nickel. Similarly, agricultural workers, as described in the next chapter, often are regularly exposed to dangerous pesticides.

Unlike occupational hazards, environmental pollution poses the greatest dangers to children, because of their still-growing bodies and immune systems, the time they spend playing outdoors, and their tendency to play on the ground and put things in their mouths. Many forms of environmental pollution threaten children (U.S. Environmental Protection Agency, 1996). For example, about 900,000 U.S. children under age 6 have elevated levels of lead in their blood from eating old house paint, which can cause retardation, learning disabilities, hearing deficiencies, hyperactivity, and other problems. Twenty-four thousand children each year are poisoned by eating pesticides, and many more are exposed to pesticides on food. Similarly, 33 percent of U.S. children now live in areas that do not meet national air quality standards, which partly explains why 4.8 million children have asthma. Finally, 10 million children under the age of 12 live within four miles of a toxic waste dump, increasing their risks of cancer and genetic defects (U.S. Environmental Protection Agency, 1996).

In the long run, the greatest environmental health threat may be global warming. During the last quarter century, carbon dioxide and synthetic gases, especially chlorofluorocarbons (CFCs) such as freon, have mushroomed. According to the Intergovernmental Panel on Climate Change, a joint venture of the World Meteorological Organization and the United Nations Environment Programme (Houghton et al., 1996), these chemical by-products of industrial manufacturing have damaged the ozone level surrounding the planet and caused temperatures to rise around the globe. Debate continues about the consequences of global warming, but many scientists suspect that global warming and the resulting damage to the ozone level will foster genetic mutations, cancers (especially skin cancer), and smog-related health problems such as bronchitis, asthma, and emphysema.

Firearms

According to McGinnis and Foege (1993), firearms account for about 2 percent of all premature deaths in the United States: 16,000 homicides, 19,000 suicides, and 1,400 accidental deaths. In some segments of the population, however, the percentage is far higher—firearms are implicated in 11 percent of deaths among children, 17 percent of deaths among persons aged 15 to 19 years, and 41 percent of deaths among African-American males aged 15 to 19.

Death from firearms is very much a U.S. phenomenon. Among young males, the rate of firearm deaths is from 12 to 273 times higher in the United States than in other industrialized nations (Kellerman et al., 1993). No other country has nearly as many privately owned firearms and, not co-incidentally, no other country has nearly as many firearm-related homicides; studies have found that having a gun in the home significantly increases the odds of suicide, of homicide, and of unintentional shooting deaths of children (Kellerman et al., 1993).

Those who support firearm ownership typically argue that having a gun protects individuals against attacks by criminals. Yet guns are far more often used against family members than against criminals. Furthermore, *even when a home is forcibly entered or a victim attempts to resist,* owning a gun increases the chances of being killed (Kellerman et al., 1993).

Interest in gun control has grown in the United States in recent years, especially following such tragedies as the 1999 massacre of 12 students and a teacher by two students at Columbine High School in Littleton, Colorado. Those favoring gun control face heavy odds against them in making legislative changes, however: between January 1997 and June 1998, the National Rifle Association and gun manufacturers spent more than $3 million lobbying Congress, while gun control groups spent only $220,000 (Wayne, 1999). Nevertheless, firearm-related violence has decreased since 1993, at least partly because of new restrictions on the sale of guns (Wintemute, 1999). Box 2.1 describes some innovative methods currently being used to curb gun violence.

Box 2.1 **Making a Difference: Physicians for Social Responsibility**

Physicians for Social Responsibility (PSR) was founded more than 30 years ago by doctors concerned about the threat nuclear arms posed to human health and life. With the decline of the Cold War, PSR has shifted its focus to working, as health professionals, toward ending other forms of violence and encouraging nonviolent means of conflict resolution. Doctors around the country belong to PSR, which has chapters at many medical schools.

In the last few years, PSR members and chapters have begun grassroots efforts aimed at educating both health care professionals and the general public about the dangers of handgun violence. The organization does not lobby to ban gun ownership but has supported laws that would keep guns away from children, dangerous individuals, and irresponsible owners. In addition, it supports actions designed to stigmatize violent gun use in the same way that drunk driving and tobacco smoking have become increasingly stigmatized in recent years. For example, student members at the University of California-Irvine recently sponsored a

---

Sexual Behavior

McGinnis and Foege (1993) attribute 1 percent of premature deaths to sexual behavior, primarily via hepatitis B, HIV disease, and cervical cancer. (Although the precise mechanisms causing cervical cancer are unknown, it occurs most often among those who have multiple sexual partners and do not use condoms, diaphragms, or spermicides.) McGinnis and Foege also include in this category infant mortality following unplanned and unwanted pregnancies, a situation occurring most commonly among teenagers and poor women.

No "manufacturer of illness" benefits from convincing people to engage in sexual activity without protecting themselves against disease or pregnancy. However, social conditions can encourage such behavior. First, those forced by economic necessity to turn to prostitution to support themselves, whether male or female, often find that they cannot suggest safer sex to clients without either losing business or risking violence. Similarly, those whose intimate relationships are not based on mutual respect and equality sometimes find that suggesting safer sex to their romantic partners results in violence or abandonment (Wingood and DiClemente, 1997). Finally, those who have learned to have little hope for the future—a sentiment particularly common among youths in communities wracked by racism and poverty—sometimes feel they have little to lose by engaging in unsafe sexual activity (Plotnick, 1992).

Other sexually active individuals, however, do fear both sexually transmitted diseases and pregnancy. For these individuals, sexual activity does

"die-in" to raise awareness of gun violence. The die-in was staged in a busy campus location, where many students were gathered for lunch, and took place to the sound of gunfire booming from loudspeakers. Die-in organizers used the event to distribute information and materials on the dangers of gun violence. Similarly, if less vividly, students at the University of Vermont Medical School sponsored a widely publicized program in which teddy bears were given to anyone who turned in a gun. The exchange was used as a forum for raising public awareness about guns, and featured a contest in which children were asked to draw posters about the dangers of guns. Finally, the Seattle PSR chapter, together with the Washington State Medical Association, the King County Prosecutor's Office, and the Seattle Police Department, has created a program called "Options, Choices, and Consequences," in which a physician-presenter visits school classes and describes what really happens when someone is shot, highlighting the differences between that reality and what children see on television.

not need to lead to disease or pregnancy if they have knowledge about safer sexual practices and access to birth control and abortion. Safer sexual practices have become more common in the last decade, but access to birth control and abortion has declined. Cuts in public funding for contraceptive services have limited options for precisely those groups—teenagers and low-income women—most at risk for unplanned pregnancies and infant mortality (Henshaw, 1995). Similarly, the federal government will not pay for abortions for women on **Medicaid** (the government-funded health insurance program for poor persons) unless the woman's life is endangered. About 20 percent of women who would have had an abortion if Medicaid paid for it now carry to term (Henshaw, 1995). Meanwhile, cutbacks in government funding for abortions, coupled with harassment and violence against abortion providers, have reduced the number and geographic distribution of abortion providers. As of 1996, 32 percent of women ages 15 to 44 lived in counties without any abortion provider (Henshaw, 1998). Other restrictions, such as requiring waiting periods or parental consent, also have limited access to abortion, especially for poor and young women whose babies are most at risk for death or disability if they carry to term.

Motor Vehicles

McGinnis and Foege (1993) attribute an additional 1 percent of all premature deaths (and almost 40 percent of deaths among those ages 15 to 24) to motor vehicle accidents, not including deaths involving drug or alcohol use.

Motor vehicle fatalities are not a necessary by-product of modern life. Rather, those fatalities reflect in part a series of decisions regarding the design of automobiles and transportation systems.

Changes in car design can reduce dramatically the chances that an accident will cause death or serious injury. The rate of deaths from motor vehicle accidents has declined substantially since 1966, when Congress established the National Highway Traffic Safety Administration (NHTSA) to regulate motor vehicle design and oversee highway safety programs. NHTSA was founded in response to the public outcry that followed publication in 1966 of Ralph Nader's book *Unsafe at Any Speed,* which documented how automobile manufacturers for years had ignored evidence of automobile safety hazards that could have been eliminated for a few dollars per car.

Automobile manufacturers have continued to fight against inexpensive improvements that could save thousands of lives yearly, such as strengthening bumpers and side doors to resist impact, covering instrument panels and roof interiors with softer materials to protect against head injuries, and redesigning gas tanks to reduce the likelihood of explosions during crashes (Nader, 1991). Equally important, legislators and government regulators have continued to exempt vans, multipurpose vehicles, and light trucks—which by 1999 accounted for 50 percent of all noncommercial vehicle sales—from passenger car safety regulations, even though most consumers use these vehicles as family cars.

One way, then, to reduce the rate of deaths and disability caused by cars is through simple changes in car design. Another way is to get people out of cars. The most basic reason why the rate of motor vehicle accidents is so much higher in the United States than in other Western nations is that U.S. residents drive far more miles per year. Although the size of the United States partially explains this difference, Americans also drive so much because they lack other options. Through a series of local and federal decisions, public transportation in this country has declined significantly since its apex in the 1920s (Yago, 1984). Trains and railroad tracks have decayed while federal dollars have subsidized highway construction and motor vehicle production. Long-distance bus systems run for profit have eliminated money-losing connections to many smaller communities. Meanwhile, cities spend billions for parking facilities, road construction, and road maintenance but offer bus service only to limited locations, during limited hours, on a limited schedule. Consequently, whereas a French citizen can use publicly subsidized trains or buses to go to any town or city in France on any given day and probably at several different times, an American citizen often has no way to go by public transportation from one town to the next; Phoenix, Arizona, for example, is now the sixth largest city in the United States but no longer has any passenger rail service.

Illicit Drugs

The last cause of premature death listed in Table 2.4 is illicit drugs. According to McGinnis and Foege (1993), illicit drugs kill users through overdose, suicide, motor vehicle injury, HIV infection, pneumonia, hepatitis, and endocarditis and kill non-users by contributing to homicide and birth defects. In addition, illicit drug use can contribute to dangerous behaviors. Box 2.2 describes the ethics of drug testing in schools and the workplace, which has emerged in response to these concerns.

The two illicit drugs that most often cause mortality and morbidity (although they are not the most commonly used illicit drugs) are heroin and cocaine (including "crack" cocaine). Both heroin and cocaine can cause physical addiction, although cocaine is usually used in quantities too small to do so (Weil and Rosen, 1998). Cocaine provides such great pleasure for so brief a time, however, that some individuals use it as often as possible, making it appear that they are addicted. As a result, both heroin and cocaine can cause people's lives to spin out of control. Although heroin causes no direct damage to the human body, cocaine can cause severe sleep disturbances, which in turn can lead to paranoia and violence (Liska, 1997; Weil and Rosen, 1998). Cocaine also may increase the risk of heart failure or stroke, although evidence for this is limited.

Whether heroin or cocaine causes birth defects or infant mortality also remains unclear. Although infants born to drug users have higher than average rates of mortality and morbidity, those problems may be caused by the mothers' poverty, malnutrition, lower education levels, or tobacco smoking rather than by their illicit drug use (Robins and Mills, 1993). In addition, the higher rates of infant mortality and morbidity among drug users may be more apparent than real, as virtually all research has used data collected inconsistently after births, and doctors naturally are more likely to collect data on mothers' drug use when babies have problems than when babies are born healthy.

Added to the inherent dangers of illicit drugs are the dangers caused by their illegality. As mentioned earlier, when drug users cannot obtain clean needles legally, they are likely to share needles and thus to increase their risks of HIV disease, hepatitis and endocarditis. Similarly, users who buy drugs on the street cannot know how powerful the drugs are and thus risk overdose. For example, a user who typically uses heroin that is 30 percent pure can die if he or she accidentally buys heroin that is 60 percent pure and thus doubles his or her normal dosage.

Pneumonia, too, results not from the drugs themselves but from the poverty and disorganized lifestyle that can either lead to drug use or result from trying to obtain steady supplies of illegal drugs at the extraordinarily high prices charged by illegal drug dealers. Similarly, violence

---

Box 2.2:   ***Ethical Debate: Is drug testing in schools and the workplace ethically justifiable?***

Since 1986, the federal government has required all federal job applicants, as well as randomly selected federal employees who hold "safety sensitive" positions, to take urine or blood tests to detect illegal drug use. In addition, many businesses use blood or urine tests to identify job applicants or current employees who use illegal drugs (DeCew, 1994), and many schools require students to test negative for illegal drugs and sometimes for alcohol and tobacco before they can participate in extracurricular activities like sports, chess clubs, and language clubs (Steinberg, 1999).

To date, U.S. courts generally have found that use of drug tests by government agencies breaches the Fourth Amendment right to privacy, unless necessary to protect public safety or unless other evidence suggests that a particular individual used drugs. Courts generally have placed no restrictions on private employers' use of drug tests and have permitted schools to require drug tests for extracurricular activities but not for academic courses.

At first glance, the benefits of drug testing seem obvious. Students, employees, and potential employees who know they will be tested may either cease using drugs or never begin doing so, thereby reducing the overall level of drug use in society. In addition, reducing drug use may reduce rates of both accidents and violence. Moreover, from a strictly financial perspective, reducing drug use could reduce absenteeism, tardiness, and insurance costs, while improving student and worker performance.

Yet drug testing does not come without a price. Those opposed to drug testing argue that testing inherently invades privacy because it involves taking urine or blood from an individual's body. Moreover, the only way to ensure a urine sample comes from a specific individual is to watch that individual urinate—an obvious invasion of Western norms of privacy. In addition, drug testing constitutes an invasion of privacy because it can reveal much more than just illegal drug use. For example, the same tests that identify use of illegal drugs can identify use of legal drugs to control epilepsy, manic

---

among heroin users results not from the drug itself (which is a sedative) but because users typically must resort to crime to pay for their drugs at street prices. Cocaine, on the other hand, can directly stimulate violent behavior.

CONCLUSIONS

Recent years have seen an increasing tendency to blame individuals for their own health problems (a topic that will be discussed further in Chapter 5). Yet as we have seen, patterns of disease stem from social conditions as much as, if not more than, they stem from individual behaviors or biological characteristics. As Marshall Becker, a sociologist and

depression, or schizophrenia. Individuals identified in this way may experience not only social embarrassment but also discrimination and even loss of employment. Finally, drug testing invades privacy because it measures not only what one does in school or on the job but also what one does during one's free time; an individual who uses drugs only in the evenings or on weekends may test positive for drugs at school or work even though the drugs no longer can affect his or her performance.

In addition, in the workplace, those who oppose drug testing also question why, if the purpose of testing is to identify workers whose performance is impaired, we measure drug use rather than performance. After all, some individuals who use drugs nevertheless will perform adequately whereas others who do not use drugs will perform poorly. Moreover, most drug-related impairment in the workplace stems from use of alcohol, yet employers usually test only for use of illegal drugs.

Finally, opponents of drug testing argue that the potential benefits of testing are far outweighed by the potential for harm when individuals are falsely labeled as drug users. As many as 40 percent of those identified as drug users by urine tests have not actually used illegal drugs. Urine tests can confuse decongestants with amphetamines, ibuprofen (Advil) with marijuana, cough syrup with morphine, and herbal teas with cocaine. The proportion of false positives is considerably lower when blood rather than urine tests are used, but the latter are more often used because they are cheaper and quicker. Similarly, schools and employers often save money by testing only once, rather than confirming test results with a second, ideally more accurate, test. Conversely, those who use illegal drugs may go undetected if they drink large amounts of water before testing; add small amounts of salt, vinegar, or bleach to their urine sample; or time their drug use so the drugs will have left their bodies before they are tested.

In sum, developing a responsible policy regarding drug testing requires us to find a balance between public safety and protection of individual rights.

one of the researchers who has done the most to help elucidate why people engage in health-endangering activities, writes (1993:4),

> I would argue, first, that health habits are acquired within social groups (i.e., family, peers, the subculture); they are often supported by powerful elements in the general society (e.g., advertising); and they have proven to be extremely difficult to change. Second, for most people, personal behavior is not the primary determinant of health status and it will not be very effective to intervene at the individual level without concomitant attempts to alter the broader economic, political, cultural, and structural components of society that act to encourage, produce, and support poor health. The "lifestyle" approach [which focuses solely on individual decisions and behaviors] enables us to ignore the more difficult,

but at least equally important, problem of the social environment which both creates some lifestyles and inhibits the initiation and/or maintenance of others.

In sum, improving the health of the population will require us to look beyond individual behavior to broader social structural issues—to look, in C. Wright Mills's terms, for public issues rather than personal troubles. Once we do so, we can focus our energies on such problems as restraining the "manufacturers of illness" and ensuring that public health considerations rather than moral values drive health policy.

SUGGESTED READINGS

Ryan, Frank. 1992. *The Forgotten Plague: How the Battle Against Tuberculosis was Won—and Lost.* Boston: Little, Brown. Traces the history of tuberculosis from the nineteenth century, when few scientists believed drugs would ever cure any illness, to the recent reemergence of tuberculosis.

Stine, Gerald J. 1998. *Acquired Immune Deficiency Syndrome: Biological, Medical, Social, and Legal Issues.* Upper Saddle River, N.J.: Prentice-Hall. An excellent overview of AIDS in the United States.

Tesh, Sylvia. 1988. *Hidden Arguments: Political Ideology and Disease Prevention Policy.* New Brunswick, N.J.: Rutgers University Press. Analyzes how ideas regarding the causes of health and illness affect public health policies.

Weil, Andrew, and Winifred Rosen. 1998. *From Chocolate to Morphine.* Rev. ed. New York: Houghton Mifflin. An iconoclastic review of both legal and illegal psychoactive drugs, co-authored by a famous medical school professor.

GETTING INVOLVED

Handgun Control. 1225 I Street NW, Suite 1100, Washington, DC 20005. 202–898–0792. www.handguncontrol.org. The most influential national organization lobbying for stricter legal limits on handgun ownership.

Planned Parenthood Federation of America. 810 7th Avenue, New York, NY 10019. 212–541–7800. www.plannedparenthood.org. The nation's foremost organization working for reproductive freedom.

Students Against Drunk Driving. P.O. Box 800, Marlborough, MA 01752. 508–481–3568. www.saddonline.com. Organization created by and for students to educate about the dangers of drunk driving.

REVIEW QUESTIONS

What is the difference between morbidity and mortality, incidence and prevalence, and acute and chronic illnesses?

What is the epidemiological transition?

What factors caused the decline in mortality between the nineteenth and early twentieth centuries?

What factors have caused the recent increases in infectious diseases, including tuberculosis and HIV disease?

How is globalization affecting rates of disease?

How have the "manufacturers of illness" increased deaths caused by tobacco? by alcohol? by toxic agents?

How have social forces and political decisions increased deaths caused by sexual behavior? caused by illicit drugs?

INTERNET EXERCISES

Try different Internet search strategies for finding information about writing a living will. First try one of the major search engines, such as Excite, Yahoo, or AltaVista, which you can probably access by clicking on the word "search" or a "search" icon on the main menu of your Internet browser. What kinds of information (quantity, quality, type) do you find? Then try using Metacrawler (www.metacrawler.com), which combines results from multiple engines.

If you searched for "living will," you probably found a great deal of irrelevant information. (Your search was probably more productive if you searched for "living wills." Can you figure out why?) To make your search more effective, you'll need to learn how to perform "advanced" or "power" searches. Instructions for doing so, or tips for searching, probably appear somewhere on the Web page for your browser. For example, in some browsers, to find Web pages on living wills (rather than on every document about living that includes the word "wills"), you must search for "living+wills," but in other browsers you would need to search for ("living wills"). Do your search again, using the proper syntax to specify your request. How does this affect the information you find?

Now try the same search, using Medline, the major online archive for medical and other health-related journals. You may be able to access *Medline* through your college library or its Web site. Otherwise, you will need to first search for and then connect to the Grateful Med Web site, the library of a major university, or the National Library of Medicine (a branch of the National Institute of Health). Check your screen, and see if it offers instructions for narrowing your search, power searches, or advanced searches. How does the information you get from *Medline* differ from the information you found using a Web browser?

Finally, try looking for articles on living wills in InfoTrac©, a large online archive of scholarly articles available through Wadsworth Publishing at http://www.infotrac-college.com/wadsworth. (You have free access to Infotrac this semester if your professor ordered it when ordering this textbook.) If you don't find anything after searching for "living wills," try searching for "right to die" or for "advanced directives" (a general term referring to legal documents specifying what types of medical care an individual would want in a given situation).

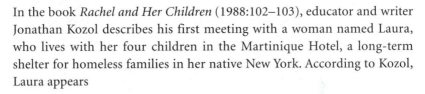

CHAPTER 3

The Social Distribution of Illness in the United States

—

In the book *Rachel and Her Children* (1988:102–103), educator and writer Jonathan Kozol describes his first meeting with a woman named Laura, who lives with her four children in the Martinique Hotel, a long-term shelter for homeless families in her native New York. According to Kozol, Laura appears

> *so fragile that I find it hard to start a conversation. Before I do, she asks if I will read to her a letter from the hospital. Her oldest son has been ill for several weeks. He was tested in November for lead poisoning. The letter tells her that the child has a dangerous lead level. She's told to bring him back for treatment. She received the letter some weeks ago. It's been buried in a pile of other documents she cannot understand.*
>
> *Although she cannot read, she knows enough to understand the darker implications of this information. The crumbling plaster in the Martinique Hotel is covered with sweet-tasting chips of paint that children eat or chew as it flakes off the walls. Infants may be paralyzed or undergo convulsions. Some grow blind. The consequences [including mental retardation and epilepsy] may be temporary or long lasting. They may appear at once or not for several years. This final point is what instills so much uneasiness; even months of observation cannot still a parent's fears.*
>
> *This, then, is her first concern, but there are others. The bathroom plumbing has overflowed and left a pool of sewage on the floor. A radiator valve is broken. It releases a spray of scalding steam at the eye level of a child. The crib provided by the hotel appears to be unstable. . . . When I test it with my hand, it starts to sway. The beds in the room are dangerous too. They are made of metal frames with unprotected corners; the mattresses do not fit the frames. At one corner or another, metal is exposed. If a child has the energy or playfulness to jump or do a somersault*

51

or wrestle with a friend, and if he falls and strikes his head against the metal ridge, the consequences can be serious. A child on the fourteenth floor, for instance, fell in just this way, cut his forehead, and required stitches just a week before. Most of these matters have been brought to the attention of the hotel management, [with] no visible results.

All of this would seem enough to make life difficult for an illiterate young woman in New York, but Laura has one other urgent matter on her hands. It appears that she has failed to answer a request for information from her welfare office. She's been cut from benefits for reasons that she doesn't understand. The timing is bad: It's a weekend. The crisis center isn't open, so there's nobody around to tide her over with emergency supplies. Her children have been eating cheese and bread and peanut butter for two days.

As Laura's story suggests, the causes of illness and disability are not randomly distributed among the U.S. population but rather fall more heavily on some than on others. For the same reason, the nature of illnesses varies dramatically from group to group, with poor people such as Laura and her children experiencing health problems—poor nutrition, lead poisoning, injuries from environmental hazards—unknown to more affluent Americans. This chapter looks at how **mortality** and **morbidity** are distributed among the U.S. population, examining the impact of four social factors: age, sex, social class, and race or ethnicity.

AGE

Overview

Not surprisingly, age is the single most important predictor of mortality and morbidity. As the last chapter described, until the twentieth century, deaths during the first year of life were common. Although far less common now, infant mortality remains an important issue because so many years of productive life are lost when an infant dies and because infant mortality so often is caused by preventable social and environmental conditions. Because infant mortality is closely connected to social class and race or ethnicity, however, it is discussed in more detail later in this chapter.

Once individuals pass the danger zone during and immediately after birth, mortality **rates** drop precipitously. Those rates begin to rise significantly beginning at about age 40 and escalate with age. For those who survive past age 65, **chronic illnesses** (such as cardiovascular disease, diabetes, and arthritis) rather than **acute illnesses** comprise the major health problems, often bringing years of disability in their wake.

Because age and illness are so closely linked, when the average age of a population changes, so does the overall health of that population. Throughout the twentieth century, the American population aged steadily. The proportion of Americans age 65 or older increased from 4 percent in 1900 to 13 percent in 1991, and is expected to near 20 percent by the year 2050 (U.S. Bureau of the Census, 1995). Similarly, the number of persons over age 85 is growing faster than any other group in the U.S. population (U.S. Bureau of the Census, 1998).

As the population ages—and even though most middle-aged and older persons are relatively healthy—the country will experience higher rates of illness, disability, and mortality. Similarly, both the total costs for health care and the percentage of health care dollars spent on the elderly—already greatly disproportionate to the size of that population—will increase. At the same time, as young persons become a smaller proportion of the population, the pool of persons who can provide or pay for care for the elderly will shrink. Consequently, providing services for elderly persons who need health care or assistance with daily tasks such as shopping or cooking will become more difficult.

These problems will be amplified by the **feminization of aging**—the increasing proportion of the elderly who are female. Between 1900 and 1996 the percentage of women among those over 65 increased from 49.5 percent to 58.9 percent, a trend that is still continuing (U.S. Bureau of the Census, 1998). Because elderly women more often than elderly men are poor and lack a spouse who can or will care for them, the feminization of aging has significant implications for health and health care. Moreover, as the next section will describe, women in general experience more illness than men. Consequently, as the proportion of the elderly who are female increases, so will the proportion needing health and social services and the proportion unable to pay for such services.

Taken together, the aging of the population and the feminization of aging will increase sharply the demand for services such as physical therapy, home health assistance, and nursing homes; one author estimates that "a new [nursing] home with over 150 beds would have to open every day for 40 years to keep pace with the current bed-to-elderly ratio" (Lindbloom, 1993:675). Meeting these demands will require a considerable social as well as financial commitment from younger Americans.

Case Study: Prostate Cancer and Aging in Men

Among men, one almost inevitable consequence of aging is cancer of the prostate, a poorly understood bodily organ that produces chemicals believed necessary for reproduction. By age 50, about 10 percent of men have prostate cancer, while by age 90, almost all do. Members of all racial and ethnic groups can get prostate cancer, but for some still-unknown reason African Americans are especially susceptible.

Because prostate cancer typically grows extremely slowly, most men who have it are killed by something else before the cancer can grow large enough to threaten their health. Because prostate cancer is so common, however, the small percentage who do develop health problems account for about 35,000 deaths per year, slightly fewer than the number of deaths per year caused by breast cancer. Moreover, when prostate cancer does grow, it often leads to excruciatingly painful bone cancer.

Before doctors can treat prostate cancer, they first must identify it. To do so, doctors, since the 1970s, have tested their male patients at periodic intervals for prostate-specific antigen (PSA), a chemical produced by the prostate (Russell, 1994). If a patient's PSA level has increased significantly, doctors then perform a biopsy—inserting a needle into the prostate to remove a few cells, which they then check for cancer. Unfortunately, PSA tests are highly inaccurate: about 30 percent of those who have cancer are not identified by the test and about two-thirds of those identified by the test as having cancer in fact do not have it (Russell, 1994:37). The test brings no benefits to those whose cancers are missed, while those who are falsely identified as having cancer suffer emotional trauma, financial costs, and uncomfortable physical procedures before they learn that the test results were incorrect.

If the biopsy suggests cancer, doctors usually perform a prostatectomy (that is, surgical removal of the prostate). The surgery succeeds in removing the cancer in about 80 percent of cases. Even in these cases, however, the risks of surgery can outweigh the benefits. Between 0.5 percent and 2 percent of patients die within a month of surgery, and another 5 percent experience serious and potentially deadly complications (Lu-Yao et al., 1993). In addition, more than 30 percent become impotent and 7 percent develop urinary incontinence, with many more experiencing periodic sexual or urinary problems. Perhaps most important, studies conducted during the 1980s and early 1990s that used **random samples** and **controlled** for other variables found no significant differences in five-year survival rates between men who do and do not receive prostatectomies, apparently because the risks of surgery counterbalance the benefits and because untreated prostate cancer rarely reduces life expectancy (Russell, 1994:28; Woolf, 1995). Recent studies, too, cast doubt on the utility of prostatectomy (Litwin et al., 1998).

Despite the limitations of current screening techniques and treatments, as of 1999 both the American Urological Association and the American Cancer Society recommend routine PSA screening for black men over age 40 and other men over age 50. If these recommendations are followed and a nationwide screening and treatment program is implemented, the cost of detecting and treating prostate cancer will rise more than tenfold to about $28 billion, as any money saved by identifying and treating persons with prostate cancer at earlier stages of the disease will be more than counterbalanced by money spent on screening and treating persons who probably would never have experienced any health problems related to prostate cancer (Mann, 1993).

In sum, at least among older men, the financial, emotional, and physical costs of identifying and treating prostate cancer seem to outweigh the benefits. Consequently, the rapid adoption of these strategies seems "a case study in one of the American medical system's worst shortcomings—its propensity to embrace expensive treatments without considering their long-term social or medical impact" (Mann, 1993:104). This **technological imperative,** which drives doctors to use all available technology, will be discussed in more detail in Chapter 11.

SEX

Overview

Sex, too, strongly affects health status. Prior to the twentieth century, complications of pregnancy and childbirth often cut short women's lives. Currently, however, American women in each racial and ethnic group and at each age have longer life expectancies than men (Table 3.1).

Despite these differences, most deaths in men and women are caused by the same diseases, including heart disease, cancer, and cerebrovascular disease (strokes). Men simply develop these diseases at younger ages.

If we look only at mortality rates, we might easily conclude that women are biologically hardier than men. When we look at *morbidity* as well as *mortality*, however, the relationship between sex and health becomes more ambiguous: at each age men have higher rates of *fatal* diseases, and, consequently, higher rates of mortality, but women experience higher rates of *nonfatal chronic* conditions and, consequently, higher rates of morbidity (Waldron, 1994). Arthritis, for example, is the most common chronic nonfatal condition among both men and women after age 45. Yet arthritis strikes women about 50 percent more often than it does men. In addition, at each age, women experience a 20 to 30 percent greater incidence of *acute* conditions (not including health problems related to their reproductive systems). In sum, women live longer than men but experience more illness, whereas men experience relatively little illness but die more quickly when illness strikes.

How can we explain these paradoxical findings? Some researchers have hypothesized that women's higher rates of illness are more apparent than real—that women do not actually experience more illness than men but simply *label* themselves ill and seek health care more often. Most researchers, however, have concluded based on various measures of health status that the health differences between men and women are real. These researchers trace these health differences to three factors: risk taking, use of

Table 3.1	Life Expectancy at Birth and at Age 65, by Race/Ethnicity and Sex					
	WHITE		BLACK		TOTAL	
	MALES	FEMALES	MALES	FEMALES	MALES	FEMALES
At birth:	73.9	79.7	66.1	74.2	73.1	79.1
At age 65:	80.8	84.1	78.9	82.2	80.7	84.0

Source: U.S. Department of Health and Human Services, 1998:200.

health care, and biology (Waldron, 1994). First, because of differences in male and female gender roles, women less often engage in potentially disabling or deadly activities. Men are more likely than women to use legal and illegal drugs, drive dangerously, participate in dangerous sports, or engage in violence. Work, too, more often endangers men, who more often work in dangerous occupations like agriculture or commercial fishing. Second, because women more often than men obtain routine health examinations, women's health problems are more often identified early enough for medical interventions to make a difference. Third, biological differences seem to favor females. Perhaps in natural compensation for those females who die from childbearing, in societies where females receive sufficient nourishment, more females than males survive at every stage of life from fetus to old age.

Case Study: Woman Battering as a Health Problem

One health issue in which sex plays an especially critical role is woman battering. Although neither health care workers nor the general public typically thinks of battering as a health problem, woman battering is a major cause of injury, disability, and death among American women, as among women worldwide.

The best data currently available on the extent of woman battering come from a national, random survey of 16,000 women and men, conducted during 1995–1996 by researchers co-sponsored by the U.S. Centers for Disease Control and Prevention and the U.S. National Institute of Justice (Tjaden and Thoennes, 1998). Half of the surveyed women (51.9 percent) had been physically assaulted during their lives and 17.6 percent had experienced rape or attempted rape. Three-quarters of those who were raped or assaulted as adults had been attacked by a current or former husband, lover, or date. Women were about twice as likely as men to report that they were seriously injured during an attack, and about one-third of the seriously injured women needed emergency health care. Extrapolating from these data, the researchers estimate that more than a half million women per year seek care at hospital emergency rooms for injuries resulting from assault by an intimate partner. Other studies have concluded that about 35 percent of women patients in hospital emergency rooms go there to seek treatment for injuries caused by battering (Council on Scientific Affairs, 1992; Novello et al., 1992).

That assaults by men should far surpass battering by women should not surprise us. Before 1962, U.S. courts consistently ruled that women could not sue their husbands for violence against them—in essence declaring wife battering a man's legal and even moral right. Even after that date, most police refused to arrest men for wife battering and most courts refused to prosecute, a situation that did not begin to change for more than a decade.

Woman battering continues to exist because it reflects basic cultural and political forces in our society and, indeed, around the world (Dobash and Dobash, 1998). Through religion, schools, families, the media, and so on, women often are taught to consider themselves responsible for making sure that their personal relationships run smoothly. When problems occur in relationships, women are taught to blame themselves, even if their husbands respond to those problems with violence. Moreover, once violence occurs, women's typically inferior economic position can leave them trapped in these relationships. Men, meanwhile, often receive the message—from sources ranging from pornographic magazines to religious teachings that give husbands the responsibility to "discipline" their wives—that violence is an acceptable response to stress and that women are acceptable targets for that violence. Although most men resist these messages, enough men absorb these messages to make woman battering a major social problem.

Battering occurs most often among men who believe that their power within the family is threatened. For example, men are significantly more likely to batter their wives if they are unemployed or in economic trouble, if their wives have higher educational or occupational levels than they do, or if their wives in some way appear to challenge their power (Lips, 1993:311–314). In addition, battering occurs most often among men who have a high need for power and who support traditional gender roles. Taken together, these data tell us that woman battering is not only an individual response to social stress, but, at a broader and largely unconscious level, a form of **social control** (that is, a way social norms and power relationships are reinforced—in this case, reinforcing men's power over women and women's inferior position within society). Consequently, as long as gender inequality remains the norm, so will woman battering.

Recognition of battering as a health risk has led various health-related organizations to enter the fight against woman battering. During the last decade, the U.S. Centers for Disease Control and Prevention has begun funding research on battering, while the U.S. Public Health Service has evaluated and helped develop violence prevention programs, trained health professionals and others in violence prevention, and encouraged health care workers to learn how to identify battered women in emergency rooms. Similarly, the American College of Obstetricians and Gynecologists now requires medical schools to teach how to identify and respond to battered women and publishes materials designed to aid health professionals in doing so.

SOCIAL CLASS

The Impact of Social Class on Health

Social class refers to individuals' education, income, and occupational status or prestige; some researchers use only one of these indicators to measure social class whereas others combine two or more indicators. Within the United States as elsewhere, at each age and within each racial or ethnic group, those with higher education, income, or occupational status have lower rates of morbidity and mortality (Feinstein, 1993; Marmot, Kogevinas, and Elston, 1987; Marmot and Shipley, 1996; Navarro, 1990; Williams and Collins, 1995). This relationship holds true for all major and most minor causes of death and illness and regardless of how researchers measure social class (Wilkinson, 1996). Heart disease, for example, occurs three times more often among low-income persons than among more affluent persons, while arthritis occurs twice as often. Moreover, these health differences appear not only when the poorest and the wealthiest are compared but also across the entire income scale, with each position on the scale having greater health than the group just below it (Wilkinson, 1996). Controlling for all known individual risk factors (such as obesity and smoking) only slightly reduces the impact of social class on mortality and morbidity rates (Wilkinson, 1996).

The relationship between social class and ill health begins at birth, with infant mortality significantly higher among those born to poor women (Nersesian, 1988). Similarly, poor children are more likely than other children to become ill or to die (Federal Interagency Forum on Child and Family Statistics, 1999). Only 65 percent of poor children are described by their parents as having very good or excellent health, compared with 84 percent of other children. Similarly, 12 percent of poor children compared with 7 percent of other children are described as having activity limitations caused by chronic health problems.

The overwhelming toll that poverty can take on a family's health is described vividly by journalist Laurie Kaye Abraham (1993) in the book *Mama Might Be Better Off Dead.* The book traces the health history of Jackie Banes and her family, who live in Chicago's predominantly African-American

North Lawndale neighborhood, where unemployment is the norm and almost half of all residents are on welfare. According to Abraham,

> accompanying this kind of poverty is a shocking level of illness and disability that Jackie and her neighbors merely take for granted. Her husband's kidneys failed before he was thirty; her alcoholic father had a stroke because of uncontrolled high blood pressure at forty-eight; her Aunt Nancy, who helped her grandmother raise her, died from kidney failure complicated by cirrhosis when she was forty-three. Diabetes took her grandmother's legs, and blinded her great-aunt Eldora, who lives down the block. . . .
>
> For the most part, the diseases that Jackie and her family live with are not characterized by sudden outbreaks but long, slow burns. As deadly infectious diseases have largely been eliminated or are easily cured—with the glaring exceptions of AIDS and drug-resistant tuberculosis—chronic diseases have stepped into their wake, accounting for much of the death and disability among both rich and poor. Among affluent whites, however, diabetes, high blood pressure, heart disease, and the like are diseases of *aging,* while among poor blacks, they are more accurately called diseases of *middle-aging.* In poor black neighborhoods on the West Side of Chicago, including North Lawndale, well over half of the population dies before the age of sixty-five, compared to a quarter of the residents of middle-class white Chicago neighborhoods (Abraham, 1993:17–18).

The Sources of Class Differences in Health

How can we explain the link between poverty and illness? One possible explanation is that illness causes poverty: as people become disabled or fall ill, their ability to earn a living or attract an employed spouse declines and they fall to a lower social status than that of their parents. Studies that have tracked cohorts of Americans over time, however, have demonstrated that this "**social drift**" only explains a small proportion of the poor ill population (Williams and Collins, 1995). Instead, far more often poverty causes illness. The work available to poorly educated lower-class persons—when they can find it—can cause ill health or death by exposing workers to physical hazards. A coal miner, for example, is considerably more likely than a mine owner to die of either accidental injuries or lung disease caused by coal dust. In addition, lower-status workers typically experience both demanding work conditions and low control over those conditions. For example, factory workers must keep pace with the production line but cannot control either the speed of the line or the timing of their personal breaks. Numerous studies have found that the combination of high demand and low control is particularly likely to cause both physical and psychological illness (North et al., 1996).

Second, environmental conditions can increase rates of morbidity and mortality among poorer populations. Chemical, air, and noise pollution all occur more commonly in poor neighborhoods than in wealthier neighborhoods both because the cheap rents in neighborhoods blighted by pollution attract poor people and because poor people lack the money, votes, and social influence needed to keep polluting industries, waste dumps, and freeways out of their neighborhoods (Bullard, 1993; Camacho, 1998). Pollution fosters cancer, leukemia, high blood pressure, and other health problems, as well as emotional stress. Because of this, both poor and middle class persons who live in poor neighborhoods have higher mortality rates than persons with similar incomes who live in more affluent neighborhoods (Haan et al., 1987).

Third, and as this chapter's opening story illustrated, inadequate, overcrowded, and unsafe housing increases the risk of injuries, infections, and illnesses, including lead poisoning when children eat peeling paint, deadly gas poisoning when families must rely on ovens for heat, and severe asthma when the lack of functioning elevators force already-ill persons to climb many stairs to reach their apartments (Reading, 1997). For example, Dr. Arthur Jones, who runs a clinic in Lawndale, told author Laura Abraham

> of one woman who was suffering a severe case of hives caused by an allergic reaction to her cat, yet repeatedly refused to get rid of the animal. "I really got kind of angry," Dr. Jones remembered, "and then she told me that if she got rid of the cat, there was nothing to protect her kids against rats." Another woman brought her 2-year-old to the clinic with frostbite, so Dr. Jones dispatched his nurse . . . to visit her home. . . . The nurse discovered icicles in the woman's apartment because the landlord had stopped providing heat (Abraham, 1993:18).

Fourth, the diet of poverty increases lifetime risks of illness (James et al., 1997). This diet relies heavily on fast foods children can prepare for themselves while their parents work and inexpensive fatty or sweet foods that provide energy and satisfy hunger but offer little nutrition. Such a diet also saps children's concentration and intellectual abilities, making it difficult for them to succeed in school and continuing the cycle of poverty.

Social class differences in diet begin at birth, for poor children are less likely than others to be breastfed. Yet according to a research review commissioned by the U.S. Department of Health and Human Services, breastmilk is "distinct and irreplaceable, . . . specific for the needs of the human infant [with a] unique composition [that] provides the ideal nutrients for human brain growth in the first year of life" (Lawrence, 1997:3). As a result, breastfed infants are less likely to develop infections, diabetes, allergies, and a host of other health problems and are less likely to die either as infants or in early childhood (Lawrence, 1997; Raisler et al., 1999). Breastfeeding also improves the health of mothers by reducing their risks of obesity, ovarian cancer, and aggressive breast cancers (Lawrence, 1997). Yet as of 1997, only

26 percent of American women breastfed for at least six months after giving birth, with rates of breastfeeding the lowest among women who were poor, young, or had not attended high school (U.S. Department of Agriculture, personal communication, August 1999).

Lower rates of breastfeeding among the poor partly reflect class differences in cultural attitudes toward breastfeeding. In addition, even when poorer women would like to breastfeed, their work situations are less likely to accommodate it: some women professionals have the option of bringing their babies to work with them or changing their schedules to work around the babies' needs, but few waitresses or factory workers do. Also, until the early 1990s, the federal "Women, Infants and Children" program, charged with improving the nutrition of poor pregnant women and young children, spent more of its funds helping participants to purchase infant formula than helping them to breastfeed (Lawrence, 1995). Since then, however, the program has placed considerably more emphasis on encouraging breastfeeding. Moreover, the supplemental foods (such as cheese and milk) available free through the program have significantly reduced morbidity and mortality among poor infants and children (Moss and Carver, 1998; Owen and Owen, 1997).

Fifth, the daily struggle to provide food, clothing, and shelter for oneself and one's family can leave little time or energy for potentially health-preserving activities like regular exercise. Meanwhile, the psychological stresses of living in poverty, coupled with the early deaths of many friends, neighbors, and relatives and the knowledge that one may meet the same fate, can encourage individuals to adopt behaviors (such as unprotected sexual activity and both legal and illegal drug use) that further endanger their health and lives. Even those who do not engage in high-risk activities can find themselves at risk if they live in neighborhoods where others do so—a bullet fired by a young gang member seeking excitement and status can kill a bystander as easily as it can kill another gang member.

Finally, poverty limits individuals' access to both preventive and therapeutic health care. In the United States, the poorest individuals can receive free health care under the **Medicaid** health insurance program (described in more detail in Chapter 8). However, these individuals still can find it difficult to obtain care if they cannot afford to take time off from work to see a doctor, to pay for transportation to the doctor, or to pay for child care while there. Individuals also sometimes are unable to obtain care even if they have Medicaid because many doctors refuse to work for the limited fees Medicaid pays. Still other individuals, referred to as the **medically indigent,** earn too much to qualify for Medicaid but too little to purchase either health insurance or health care; according to the U.S. Department of Health and Human Services (1998), 88 percent of those who lack health insurance are poor or near poor. (The problems of those who lack health insurance will be discussed in more detail in Chapter 8.) Not surprisingly, poor persons are less likely than others to have a personal physician or a regular source of care, and they are more likely to receive care in hospital clinics or emergency rooms, where quality of care is necessarily lower than in less-rushed and less-crowded settings. Similarly, poor children (regardless of race or ethnicity) are less likely to have received all necessary vaccinations by the recommended ages (Centers for Disease Control and Prevention, 1997). They are also more likely to have no usual source of health care, even if they have health insurance, as Table 3.2 shows (U.S. Department of Health and Human Services, 1998:292).

Access to health care cannot eliminate class differences in mortality and morbidity—differences that exist even in countries where access to care is universal—because it cannot eliminate the other factors that cause poor people to become ill more often than their more-affluent peers in the first place (Marmot et al., 1987). For this reason, access to health care plays less of a role in the relationship between poverty and ill health than do the other factors discussed so far (Feinstein, 1993; Williams and Collins, 1995). Nevertheless, adequate access to health care can protect against some problems, such as debilitating dental disease preventable through routine cleaning and treatment and disabling childhood illnesses preventable through immunization. In addition, access to health care can improve quality of life dramatically through such simple interventions as providing eyeglasses, hearing aids, and comfortable crutches or wheelchairs.

Conversely, lack of access can have deadly consequences (as Chapter 8 will describe in more detail). One large-scale study (Franks et al., 1993) found that by the end of a ten-year period, 18.4 percent of those who lacked health insurance had died, compared with only 9.6 percent of those

Table 3.2	*Percentage of Children Ages 6 to 17 with No Usual Source of Health Care, 1994–1995, by Ethnicity, Insurance Status, and Income*		
ETHNICITY	POOR	NEAR POOR	NOT POOR
White non-Hispanic	9.3	8.9	3.7
Black non-Hispanic	8.7	8.4	4.5
Hispanic	19.4	16.5	6.6
INSURANCE STATUS			
Insured	6.8	6.1	3.2
Uninsured	27.1	21.7	15.6

Source: U.S. Department of Health and Human Services (1998:292).

who had insurance. Even when the researchers statistically controlled for sex, age, race, education, preexisting illnesses, or use of tobacco, they still found 25 percent more deaths among uninsured persons than among insured persons.

In sum, poverty and illness are linked by underlying social conditions. Unfortunately, these social conditions have worsened over the last few decades, and social class differences in morbidity and mortality rates have continued to grow (Williams and Collins, 1995).

Importantly, health is affected more by social class than by race or ethnicity—which, in the United States, is highly correlated with class (Baquet et al., 1991; Navarro, 1990; Nersesian, 1988; Otten et al., 1990; Williams and Collins, 1995). For example, data from one national random sample found that apparent race differences in mortality rates between Mexican Americans, Asian Americans, and white Americans disappeared once social class was controlled for, and differences between African Americans and white Americans diminished substantially (Rogers et al., 1996). Looking at the same issue from a different angle, another study also using a national random sample found that class differences in rates of mortality and morbidity were almost twice as great as race differences (Navarro, 1990). For example, morbidity was 4.6 times more common among those making $10,000 or less per year compared with those making more than $35,000, but only 1.9 times more common among African Americans than among whites. These numbers suggest that social class is a more powerful predictor of mortality and morbidity than is race or ethnicity. This does not, however, reduce the importance of race or ethnicity, for both contemporary and historical racial discrimination remain at the root of minority poverty. Rather, it suggests that the racial gaps in health status can be overcome if minorities can raise their incomes and social positions (Farmer, 1999; Williams and Collins, 1995).

Case Study: Health Among the Homeless

The impact of social class on health falls heaviest on the homeless. Homelessness has been a major problem for the United States since the early 1980s, when the federal government slashed funds for low-income housing while increasing subsidies for "gentrifying" good-quality older buildings in inner-city neighborhoods. Although the latter policy was intended to improve quality of life in these neighborhoods, its unintended consequence was to raise rents. Meanwhile, the value of the minimum wage (adjusted for inflation) declined, and public assistance became harder to get and lower in value. As a result, in all fifty states, a person working 40 hours per week at minimum wage can no longer afford the average rent of a decent, modest, one-bedroom apartment, according to the nonprofit National Low Income Housing Coalition (1998).

Not surprisingly, given the physical and emotional strains of life on the streets, homeless persons experience a disproportionate share of chronic and acute illnesses, as well as greatly increased mortality rates. Researchers estimate that 35 percent of homeless people in Los Angeles have active tuberculosis and more than 30 percent have some other chronic health conditions (Cousineau, 1997; Kleinman et al., 1996). Homeless women face additional risks from rape and violence: One study of 53 long-term homeless women found that 15 percent had been raped and 42 percent battered in the last year (Fisher et al., 1995). Finally, a random survey of residents of New York City homeless shelters found age-adjusted death rates for both men and women four times higher than among other New Yorkers, with rates highest for those who had been homeless the longest (Barrow et al., 1999).

Homeless children, who make up about 25 percent of the homeless population, also face increased health risks (U.S. Conference of Mayors, 1998). A study conducted by the nonprofit Children's Health Fund found that 38 percent of children in New York City's homeless shelters had asthma—which can be life-threatening in children—and more than 60 percent had not received basic immunizations by the recommended ages (Herbert, 1999). Similarly, a Massachusetts study found that homeless children experience more ear infections, diarrheas, fevers, and severe asthma than do other children and are more likely to be in fair or poor health overall (Weinreb et al., 1998).

All the factors that explain the high rates of morbidity and mortality among poor persons also apply to homeless persons. However, maintaining health is even more difficult for homeless persons than for other poor persons. For example, because poverty, malnutrition, and cold weaken their bodies, and because they often can find shelter only in crowded dormitories where infections spread easily, homeless persons are more likely than others to develop upper respiratory infections. Moreover, those who do so cannot rest in bed until they have recovered for the simple reason that they have no beds to call their own. Similarly, homeless persons often suffer skin problems such as psoriasis, impetigo, scabies, and lice, which, if left untreated, can cause deadly infections. Even if homeless persons receive prompt treatment for these skin problems, their living conditions make it impossible for them to keep their linens and clothing clean enough to prevent reinfection. Finally, homeless persons, regardless of age or sex, often can support themselves only through prostitution, which dramatically increases their risks of rape, battering, and sexually transmitted diseases, including HIV disease.

Access to health care is also particularly difficult for homeless persons. The struggles necessary to meet basic needs for food, clothing, and shelter can leave individuals with little time, energy, or money for arranging transportation to health care facilities or for purchasing health care or prescription drugs. In addition, both substance abuse and mental illness—which affect more than 40 percent of homeless persons and can either cause or result from homelessness—can make it harder for individuals to recognize they need health care, to seek care promptly when they recognize it is needed, to follow the instructions of health care workers, and to return for needed follow-up visits (Cousineau, 1997).

In the book *Under the Safety Net,* Brickner and his colleagues (1990:10) describe the true costs homeless people pay and the limited benefits they receive when they seek health care:

> A homeless man with severe cellulitis [diffuse inflammation under the skin] of the legs, skin breakdown, and bilateral leg ulcers makes his way to the local hospital emergency room. Because he is not a genuine emergency, he waits for five hours. He loses his opportunity for lunch at a soup kitchen. He loses a bed for the night because he wasn't standing in line at the right time. He finally is examined by a physician and given a prescription for antibiotics, told to stay supine [on his back] for a week with his legs elevated and soaked in warm dressings, and given a return appointment for clinic. The realities of his life prohibit him from carrying out any portion of this treatment plan.

In sum, until the underlying conditions causing homelessness are alleviated, health care workers can offer homeless persons only the most temporary of help.

RACE AND ETHNICITY

The concept of "race" is a social construction, with almost no biological basis. For example, a century ago Jews and Irish people were considered separate and inferior races by many "white" Americans. Similarly, contemporary Americans typically label individuals "African American" if they have any known African ancestors, even if most of their ancestors were European. For this reason, from this point on this textbook will use the term *ethnicity,* which suggests cultural rather than biological differences, rather than the less accurate term *race.*

As the previous section explained, social class affects morbidity and mortality more than does ethnicity. Yet ethnicity remains an important and independent factor in predicting health status. This section looks at health and illness among African Americans (who, in 1997, made up 12.7 percent of the U.S. population), Hispanics (11.0 percent), Asian Americans (3.7 percent), and Native Americans (0.9 percent). As we will see, life expectancy is substantially curtailed for African Americans compared with the other groups, whereas Asian Americans experience the longest average life spans.

Ethnic differences are also apparent in "active life expectancy," the number of years individuals can expect to live in good health and free of disability, and in "inactive life expectancy," the number of years individuals can expect to live in poor health or with disabilities (Hayward and Heron, 1999).

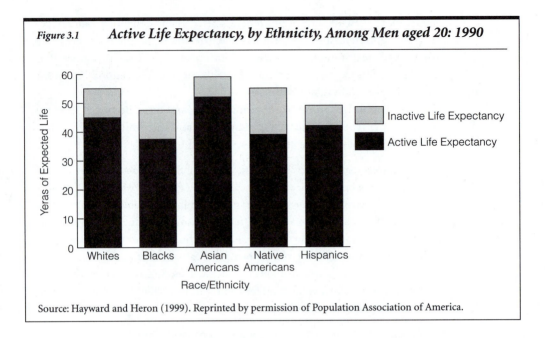

Figure 3.1   *Active Life Expectancy, by Ethnicity, Among Men aged 20: 1990*

Source: Hayward and Heron (1999). Reprinted by permission of Population Association of America.

As Figures 3.1 and 3.2 illustrate, on average, Asian Americans both live longer and have a higher active life expectancy than members of other ethnic groups. Both life expectancy and years in good health are greater for white non-Hispanics than for Hispanics, and greater for Hispanics than for African Americans. Finally, although Native Americans on average live as long as white non-Hispanics, Native Americans live more years in poor health than any other group. The remainder of this section explores in more detail some reasons for these ethnic differences in morbidity and mortality.

African Americans

The impact of ethnicity on health stands out vividly in studies of infant mortality. For all causes of infant deaths, African Americans have higher mortality rates than whites (National Center for Health Statistics, 1998b). African Americans have an infant mortality rate considerably higher than that found in such poor countries as Cuba, Portugal, and Greece and similar to—if somewhat worse than—the rates found in Costa Rica and Poland (Table 3.3). Moreover, whereas in 1950 African American infants were 1.6 times more likely than white infants to die, by 1994 African American infants were 2.4 times more likely to die (National Center for Health Statistics, 1998b; Schoendorf et al., 1992).

One possible explanation for the high rate of infant mortality among African Americans is their relatively low income levels. To determine whether ethnicity affects infant mortality among African Americans inde-

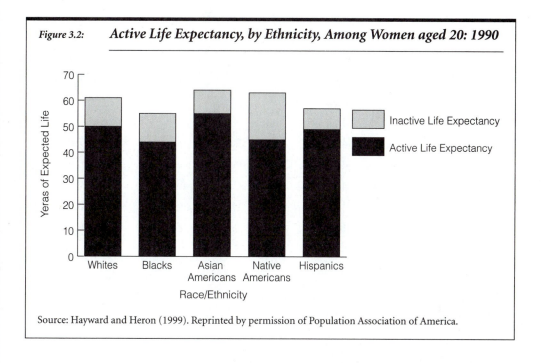

Figure 3.2: **Active Life Expectancy, by Ethnicity, Among Women aged 20: 1990**

Source: Hayward and Heron (1999). Reprinted by permission of Population Association of America.

pendent of income, Schoendorf and his colleagues (1992) looked at mortality rates among a national random sample of African-American and white infants whose parents were at least 20 years old and college graduates. Even within this relatively well-off sample, and after controlling for age, number of previous births, use of prenatal care, and marital status, African-American infants were almost twice as likely to die as white infants, largely because of higher rates of prematurity and low birthweight.

These differences, the authors theorize, reflect a constellation of factors stemming from racism, which, although far less common than in the past, remains deeply embedded in American culture (Feagin and Sikes, 1994; Williams, 1998). For example, data collected in the widely used national, random, General Social Survey (cited in Williams, 1998:308) found that in 1990, about half of all white respondents said that they believed African Americans are lazy (45 percent), prone to violence (51 percent), prefer to live off welfare (56 percent), and lack the motivation or willpower to pull themselves up out of poverty (60 percent); one might reasonably assume that even more held these views but did not admit it. Because of racism, even middle-class African Americans like those studied by Schoendorf and his colleagues, who could afford decent housing in neighborhoods free from pollution and violence, sometimes find it impossible to obtain such housing when landlords, realtors, or bankers flout laws banning discrimination and refuse to rent, sell, or give mortgages to African Americans (Williams, 1998). Other African Americans prefer to live in poorer, more segregated neighborhoods than to face the

Table 3.3	*Infant Mortality Rates per 1,000 Live Births*		
COUNTRY	RATE	COUNTRY	RATE
Japan	4.25	Spain	6.69
Singapore	4.34	New Zealand	7.24
Hong Kong	4.43	Israel	7.80
Sweden	4.45	Greece	7.93
Finland	4.72	Czech Republic	7.95
Switzerland	5.12	**U.S., all races**	**8.02**
Norway	5.23	Portugal	8.06
Denmark	5.45	Belgium	8.20
Germany	5.60	Cuba	9.40
Netherlands	5.61	Slovakia	11.19
Ireland	5.93	Puerto Rico	11.47
Australia	6.05	Hungary	11.55
Northern Ireland	6.05	Chile	11.99
England and Wales	6.20	Kuwait	12.68
Scotland	6.20	Costa Rica	13.71
Austria	6.25	Poland	15.13
Canada	6.30	**U.S. blacks**	**15.80**
France	6.47	Bulgaria	16.31
U.S. whites	**6.60**	Russian Federation	18.06
Italy	6.63	Romania	23.89

Source: U.S. Department of Health and Human Services (1998:193, 197).

daily hostility or, simply, social discomfort of white neighbors. Consequently, more-affluent African Americans sometimes live in conditions similar to those experienced by poorer African Americans. This hypothesis gains support from studies suggesting (if inconclusively) that African-American infant mortality rates are highest among those living in the most segregated cities (LaVeist, 1993; Polednak, 1996).

In addition, the psychosocial stresses of racism can harm health among African Americans (as well as among other minority groups). Several recent studies have found that as the number of incidents of ethnic discrimination that individuals have experienced increases, their physical and mental health deteriorates (Williams et al., 1997; Williams, 1998). (On the other hand, it is also important to note that African-Americans have lower rates of suicide, substance use, and mental illness than whites, possibly because of the higher levels of social support many African Americans receive from their families and churches; see Williams, 1998, for a review).

The disparities in health status between African Americans and whites do not end in infancy. At each age, and for 13 of the 15 leading causes of death, death rates are higher, and life expectancies are lower for African Americans than for whites. Moreover, the differences in life expectancies between whites and African Americans have remained essentially the same for the last 40 years (Williams, 1998).

These stark differences in life expectancy were demonstrated vividly by Colin McCord and Harold P. Freeman (1990) in a much-cited article in the *New England Journal of Medicine.* The article compared the chances of surviving to old age in Bangladesh, one of the poorest countries in the world, to the chances in Harlem, an overwhelmingly poor, African American, New York City neighborhood. As Figure 3.3 shows, although before age 5 both males and females have higher death rates in Bangladesh than in Harlem, after that age the death rate levels off in Bangladesh but rises in Harlem. Consequently, for females, the chances of surviving are lower in Bangladesh than in Harlem, but only because of the differences in the first five years of life. For males, the chances of surviving are lower in Bangladesh only until age 40 and almost solely because of deaths in the first five years of life. Among those who survive to age 5, both males and females have a greater chance of surviving to age 65 in Bangladesh than in Harlem.

Unfortunately, these high death rates among African Americans extend far beyond the borders of Harlem. Table 3.4 shows the death rates (controlled for age of the populations) for the ten leading causes of death in 1996. (These were the latest age-adjusted figures available by ethnicity when this book went to press.) This table shows that HIV disease causes about six times more deaths among African Americans than among whites. The table also highlights the disproportionately large role diabetes plays in African-American mortality. Diabetes, which is caused by both genetic factors and a diet of poverty, affects 10.8 percent of African Americans over age 20 and kills African Americans almost two and a half times more often than it kills whites, mostly by causing kidney disease (Centers for Disease Control and Prevention, 1998b).

Yet kidney disease need not kill, if transplants or dialysis can substitute for failing kidneys. However, African Americans are significantly less likely than whites to receive transplants or dialysis because standard procedures for selecting patients for these therapies unintentionally discriminate against them (Council on Ethical and Judicial Affairs, 1990; Gaston et al., 1993). Transplant programs generally require near-perfect biological matches between donor and potential recipient before they will perform a transplant, although

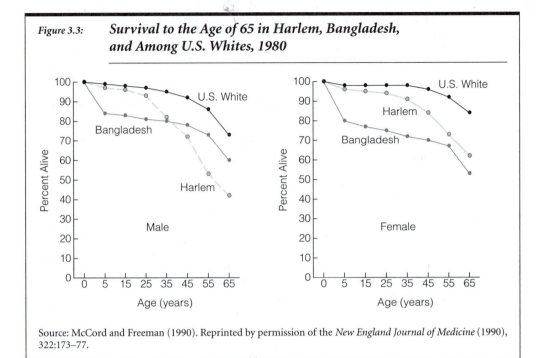

Figure 3.3: **Survival to the Age of 65 in Harlem, Bangladesh, and Among U.S. Whites, 1980**

Source: McCord and Freeman (1990). Reprinted by permission of the *New England Journal of Medicine* (1990), 322:173–77.

the difference in survival rates when kidneys are less well matched is small. Because African Americans donate kidneys far less often than whites do, African Americans who need kidneys less often match the available kidneys perfectly and, thus, less often receive transplants. African Americans also receive transplants less often because doctors less often refer them to transplant programs. This pattern mirrors behavior in other areas of health care: controlling for symptoms and insurance coverage, doctors less often refer African Americans than whites for angioplasty, bypass surgery, stroke-preventing procedures, and numerous other diagnostic and therapeutic procedures (Oddone et al., 1999; Williams, 1998). Moreover, even when African Americans are referred to transplant programs, they are more often rejected as patients because they lack transportation to care facilities and funds to pay for aftercare, which can costs thousands of dollars per year (Council on Ethical and Judicial Affairs, 1990). (The broader problem of how to allocate scarce health resources is discussed in Box 3.1.)

Hispanics

A similar pattern of health and illness occurs among Hispanics—although this is more true for some Hispanic groups than others. In general, Cubans (4.2 percent of U.S. Hispanics) have fared considerably better than have either Puerto Ricans (10.6 percent) or Mexican Americans (63.3 percent).

Table 3.4   *Age-Adjusted Death Rates per 100,000 for Leading Causes of Death, by Race, 1996*

CAUSES OF DEATH	ALL RACES	WHITE	AFRICAN AMERICAN	NATIVE AMERICAN
All causes	491.6	466.8	738.3	456.7
Heart disease	134.5	129.8	191.5	100.8
Cancers	127.9	125.2	167.8	84.9
Cerebrovascular disease (strokes)	26.4	24.5	44.2	21.1
Chronic obstructive pulmonary disease	21.0	21.5	17.8	12.6
Unintentional injuries	30.4	29.9	36.7	57.6
Pneumonia and influenza	12.8	12.2	17.8	14.0
Diabetes	13.6	12.0	28.8	27.8
HIV disease	11.1	7.2	41.4	4.2
Suicide	10.8	11.6	6.6	13.0
Chronic liver disease and cirrhosis	7.5	7.3	9.2	20.7

Source: U.S. Department of Health and Human Services (1998:203–204)

Relatively little is known regarding the health status of the newer immigrant groups from Central and South America.

As among African Americans, health problems among Hispanics largely reflect their generally lower social class status (Rogers et al., 1996). Hispanics are two and one-half times more likely than non-Hispanic whites to live in poverty and, except for Cubans, are half as likely to have completed college. In addition, cultural and language barriers as well as social discrimination can make it difficult for them to take advantage of health care resources even when they can afford them. In part as a result, Hispanic children are less likely than non-Hispanic white children to receive all necessary vaccinations by age three and, regardless of income, are almost twice as likely to have no regular source of health care, as Table 3.2 shows (Centers for Disease Control and Prevention, 1997; U.S. Department of Health and Human Services, 1998).

For reasons that remain unclear, rates of infant mortality among Hispanics (other than Puerto Ricans) are comparable with those of non-Hispanic white Americans (Table 3.5). On other measures of health, however, Hispanics fare less well. Life expectancy is lower for Hispanics than for non-Hispanic

---

Box 3.1   ***Ethical Debate: How Should We Allocate Scarce Health Resources?***

You are the chair of a regional organ bank charged with allocating one donated kidney. This kidney will mean the difference between life and death to whoever receives it. Would you give it to

James Russell, a world-famous pediatrician who is 60 years old, unmarried, and childless?

Julie Brown, a 35-year-old, unmarried mother and sole supporter of four young children, who is a high school drop-out and lives on government assistance?

or Sally Michaels, a 45-year-old housewife with children in college, who is married to a lawyer and is active in various local charities?

How to allocate scarce resources has animated public debate since the early 1960s. In 1961, the Seattle Kidney Center purchased three kidney dialysis machines—a then-new technology that offered life to those who would die otherwise. Because demand for dialysis far exceeded supply, the hospital established a committee of seven lay persons and two medical advisors—all affluent whites—to select which of the many medically qualified patients should receive treatment. Not surprisingly, the committee based their decisions on social criteria, such as sex, age, apparent emotional stability, social class, and marital status. When *Life* magazine in November 1962 published an article describing the committee's work, a public outcry arose, which eventually led Congress to pass a law making kidney dialysis the only medical procedure available virtually cost-free to all Americans. Demand for dialysis, however, continues to exceed supply, so doctors and others still must decide who should receive this treatment. The same question faces all those who must allocate expensive and scarce

treatments, for no national policies regulate how these decisions should be made.

Probably all observers would agree that medical factors must be considered in allocating scarce resources. For example, it makes little sense to give transplants to persons who are likely to die during or shortly after a transplant operation, such as persons whose tissue does not adequately match that of the prospective organ donor and whose body is therefore likely to reject the donated organ. In other circumstances, however, it is far from clear what role medical factors should play in these decisions. For example, some argue that we should give highest priority to those who are healthiest because they are most likely to survive a transplant and to have a good quality of life afterward. Others, however, argue that these individuals can live the longest *without* a transplant and so should have lowest priority.

Although it might seem fairest, relying on medical factors is also problematic because doing so may unintentionally discriminate against minorities and the poor. For example, for a variety of reasons, including generalized mistrust stemming from a history of poor treatment by the medical establishment, African Americans are less likely to donate organs than are whites. As a result, African Americans more often die while waiting for a closely matched donor kidney. Similarly, selecting the healthiest persons first discriminates against poorer persons, who on average are in worse health.

Using other "objective" criteria for selection also can unintentionally discriminate. Individuals are most likely to benefit from a procedure if they have family members who can take care of them while they recover; can afford to pay all necessary costs of receiving care, including costs for drugs, any special diet, and transportation to and from the health care delivery site; have the intellectual and emotional ability to follow the prescribed treatment and follow-up regimen; and have a stable life that allows them to do so. Yet all these factors encourage the selection of middle- and upper-class persons who share not only social status but also cultural values with those who control access to health care.

Perhaps this is not a problem. In fact, some consider it perfectly reasonable to use social characteristics overtly in making decisions, and probably most would agree that it makes more sense to allot scarce health resources to a 40-year-old than to an equally healthy 60-year-old because more years of productive life would be lost should the 40-year-old die.

Implicit in such a decision is a notion of social worth—that a younger person is automatically worth more than an older one. Similarly, many would argue, based on utilitarian ethical principles that emphasize achieving the greatest good for the greatest number, that scarce resources should be allocated to those most likely to benefit the community. This generally translates into those who are married and have children, those with more education and higher professional positions, and those who have engaged in service to the community. Such decision rules, of course, reflect the values of the middle- and upper-class persons who sit on hospital selection committees and are likely to work against minorities and the poor.

The difficulties with establishing equitable decision rules have led some to propose mechanisms that, in essence, eliminate the need to make decisions, such as holding a lottery open to anyone judged medically qualified. These proposals assume that all persons have equal social worth. Yet most people *do* consider some more morally worthy than others. Consequently, the hostility generated by a lottery could produce more social harm than the evils it is designed to avoid.

Another way to avoid making these difficult decisions is to allot scarce resources on a "first come, first served" basis. Such a policy, however, unintentionally privileges affluent persons because such persons typically receive accurate diagnoses and learn how to join waiting lists earlier in the course of their diseases. Consequently, this system would be inequitable in practice.

A third way to avoid having to make decisions is to give priority to those who are sickest. Many, however, object that such a policy would be wasteful because it would result in more unsuccessful procedures. Moreover, whether consciously or unconsciously, under such a system doctors likely would define as sickest those whom they consider most worth saving.

Finally, rather than establish equitable decision rules, we could allocate scarce resources simply on the basis of ability to pay. Such a policy, as ethicist George Annas explains, "puts a very high value on individual rights, and a very low value on equality and fairness. It has properly been criticized on a number of bases, including that the transplant technologies have been developed and are supported with public funds . . . and that financial success alone is an insufficient justification for demanding a medical procedure" (1992:550).

In sum, decisions regarding how to allocate scarce health resources seem destined to rely on social and cultural as well as medical factors. Perhaps the best we can hope for is that decision makers will recognize how these factors affect their decisions and use that recognition to work for more equitable policies.

Table 3.5	**Infant Mortality Rate by Ethnicity, United States**		

MOTHERS' ETHNICITY	RATE
African American	14.6
Native American	9.0
White	6.3
Hispanic origin	6.3
Puerto Rican	8.9
Mexican American	6.0
Central and S. American	5.5
Cuban	5.3
Other Hispanic origin	7.4
Asian or Pacific Islander	5.3
All mothers	7.6

Source: U.S. Department of Health and Human Services (1998:190).

whites, even though the main causes of death are the same for both groups. Like African Americans, Hispanics are at greater risk than non-Hispanic whites for diabetes and for its more serious complications. Hispanics also die at higher rates from violence and from legal and illegal drug use. In addition, largely because of their higher rates of cigarette smoking, Hispanics have higher rates of morbidity from hypertension and lung cancer. Finally, Hispanics are twice as likely as whites to die from HIV disease. Conversely, Hispanics have lower rates of heart disease and most forms of cancer simply because on average they do not live long enough to develop these diseases.

Health status is particularly poor among those who are migrant workers. Of course, most Hispanics are not migrant workers, but the majority of migrant workers are Hispanic, and most other migrant workers belong to other minority communities. Consequently, issues of minority status and social class are tightly interwoven in understanding why these individuals are so vulnerable to health problems.

About five million migrant laborers—most foreign-born, many illegal aliens—work in agricultural fields in the United States (Sandhaus, 1998). The work itself is physically hazardous, with long days of repetitive stooping and bending, heavy lifting, and exposure to toxic pesticides (Gwyther and Jenkins, 1998; Sandhaus, 1998). Access to clean water and sanitary toilets is often minimal, and workers are often exposed to weather extremes. Living conditions, too, are often poor, with many individuals crowded together in poorly heated or cooled rooms with insufficient water and toilets and low wages that make it difficult to obtain nutritious foods. Yet because so many migrant laborers are illegal aliens, they cannot protest these conditions without risking deportation. Finally, lack of transportation, cultural differences, and communication problems make it difficult for laborers and their families to obtain good health care. As a result, life expectancy is substantially reduced among migrant workers and their families, and chronic health problems, infectious diseases (including tuberculosis, typhoid, and hepatitis), miscarriages, and infant mortality are several times more common than among the rest of the population (Gwyther and Jenkins, 1998; Sandhaus, 1998).

Native Americans

As is true with any ethnic group, Native Americans are highly diverse. Native Americans in the United States belong to more than 500 different tribes, each with a distinct language and culture. Slightly more than half of Native Americans live off reservations, often in large urban areas.

Native American life expectancy has improved substantially since the 1950s and now almost equals that of white Americans. However, the particular patterns of disease experienced by Native and white Americans continue to differ (Kunitz, 1996; Rhoades et al., 1987; U.S. Department of Health and Human Services, 1990).

These differences begin at birth. Although lower than among African Americans and lower than in the past, infant mortality (Table 3.5) remains considerably higher than among whites (Kunitz, 1996; U.S. Department of Health and Human Services, 1998). Moreover, research suggests that official rates underestimate all forms of mortality, including infant mortality, among Native Americans because death certificates often erroneously report the deceased as white (U.S. Department of Health and Human Services, 1998).

The differences between whites and Native Americans become clearer when we divide infant mortality into **neonatal infant mortality** (deaths occurring during the first twenty-seven days after birth) and **postneonatal infant mortality** (deaths occurring between twenty-eight days and eleven months after birth). The neonatal infant mortality rate among Native Americans is essentially the same as the white rate (4.0 per 1,000 live births versus 4.1). However, the postneonatal infant mortality rate is more than twice as high among Native Americans (5.1 per 1,000 live births versus 2.2). These differences reflect differences in rates of pneumonia and gastritis. Although less common than in the past, these easily preventable diseases—precipitated by poverty, malnutrition, and poor living conditions and normally controllable through prompt medical attention—still occur more often among Native Americans than among others (Rhoades et al., 1987). Box 3.2 describes the benefits and limitations of the Indian Health Service, the federally funded program charged with providing health care to Native Americans.

Among those who survive past infancy, cancer actually kills fewer Native Americans than whites (see Table 3.4). However, Native Americans are almost three times more likely to die from liver disease and cirrhosis, the end-results of heavy alcohol use. They are also twice as likely to die from homicide or unintentional injuries, with alcohol use often contributing to these deaths. Although the rate of alcohol-related deaths among Native Americans has decreased in recent years, it remains four times more common than in the U.S. population as a whole and more common than in any other ethnic group (Kunitz, 1996).

During adolescence, too, Native Americans face unusual health risks. A national survey of Native-American teenagers attending reservation schools found that 19.1 percent had been knocked unconscious in a fight, 16.9 percent had attempted suicide, and 10 percent could be considered potentially problem drinkers, with the last figure rising steadily during the teen years. Moreover, these numbers underestimate the health risks faced by Native-American teens because those most at risk are least likely to attend school and hence least likely to have participated in this survey. Because of these and other risks, death rates among Native-American teenagers are almost twice those of American teenagers as a whole, with rates among Native-American males three times the national average (Blum et al., 1992).

Native Americans differ from other Americans in their pattern of diseases as well as their pattern of deaths. The rate of respiratory disease is 31 percent higher than in the U.S. population as a whole, partly because of high rates of tobacco use: Native Americans have higher rates of tobacco use than any other ethnic group in the United States, are the only group in which rates have not declined since the 1970s, and are the only group in which women are as likely to smoke as men (*Morbidity and Mortality Weekly Report*, 1998b). Native Americans also have mortality rates from infectious diseases twice as high as those found among white Americans, primarily because of inadequate sanitation, lack of access to clean water, and the general physical debilitation associated with poverty. In addition, diabetes now affects approximately 9 percent of Native American adults, who are three times more likely than whites to die because of it (Centers for Disease Control and Prevention, 1998b; Claiborne, 1999).

Asian Americans

Overall, Asian Americans enjoy far better health than other American minority groups. The largest Asian-American groups (Chinese, Japanese, and Filipino) have life expectancies and infant mortality rates equal or superior to those of white Americans (Table 3.5, Figures 3.1 and 3.2). As a group, Asian Americans experience the same causes of death as whites and in similar percentages. The only exception is the rate of heart disease, which is

---

Box 3.2   ***The Indian Health Service***

Under the provisions of various treaties, the U.S. government has since the 1830s provided health care to Native Americans (Kunitz, 1996). Today, more than one million Native Americans, living in urban and rural areas, both on and off reservations, receive comprehensive health services from the Indian Health Service (IHS).

The IHS offers both "direct" health care and "contract" care. Direct care programs, generally located on Indian reservations, provide access to generalist medical care from internists, family doctors, and pediatricians and are open to all Native Americans. In addition, the IHS contracts with private health care providers to offer specialty care. This contract health program, however, is only open to Native Americans who either live on a reservation or live in the contract area affiliated with their tribe. For example, a Navajo who moves to Flagstaff, where the IHS's contract health program includes Navajos, can obtain care through that program. The same individual could not receive services in Phoenix, where the IHS's contract health program does not include Navajos, or in Minneapolis, where the IHS has no contract health program.

Since the 1970s, the IHS increasingly has moved toward local control (Kunitz, 1996). Tribes now can sign agreements to take over some services offered by the IHS or to provide additional services. Unfortunately, per capita funding for the IHS is about one-third lower than per capita health care expenditures for the U.S. population as a whole, with serious consequences for health care. For example, only 15 of the IHS's 515 health care facilities can provide tertiary-level care, and funds for these 15 facilities usually run out early in each fiscal year. Similarly, in 1994 the ratio of doctors to patients in the IHS was less than 90 per 100,000, compared with 229 per 100,000 in the United States as a whole (Claiborne, 1999). Because of problems like these, the move toward tribal control of health care has pitted tribes against each other in the fight for limited federal dollars—a battle that has particularly hurt smaller, poorer tribes and tribes located in isolated regions where finding qualified health care providers is difficult and expensive. The need for additional funds to pay for tribal health care costs partly explains why tribes have pursued casino gambling aggressively in recent years (Kunitz, 1996).

---

lower among Asian Americans than among other Americans, possibly because of dietary differences.

These statistics, however, tell only part of the story. Since 1975, a substantial portion of Asian immigration has come from the war-torn countries of southeast Asia. These immigrants typically have far lower income and education levels than established Asian Americans. In addition to the health problems that always accompany poverty, these individuals often also suffer from unavoidable dietary changes, cultural shock, tropical diseases for which diagnosis and treatment can prove elusive, and the long-lasting traumas of warfare and refugee life.

The limited available data on the health status of southeast Asians in the United States suggest that they have significantly higher mortality and morbidity rates than have whites or other Asians (Association of Asian Pacific Community Health Organizations, 1997). For example, only 22.7 percent of Vietnamese Americans report that their health is excellent compared with just over 40 percent of Americans who are white non-Hispanic, Japanese, or Asian Indian (Kuo and Porter, 1998). Tuberculosis is 13 times more common and hepatitis B is 25 times more common among southeast Asian immigrants than among white Americans, with the latter causing much higher rates of liver cancer. Lung cancer is also more common among male southeast Asian immigrants, reflecting rates of tobacco use two to three times higher than among other American men.

At the same time, southeast Asians typically have more limited access to health care (Association of Asian Pacific Community Health Organizations, 1997). Rates of health insurance coverage are low, and even those who have insurance sometimes find that linguistic or cultural barriers make it nearly impossible to communicate with health care workers and to obtain quality health care. As a result, southeast Asians are less likely than are other Americans to use western health care (although some continue to use traditional Asian healers and therapies). For example, 54 percent of Vietnamese American women have never had a Pap smear, and 34 percent have never had a mammogram (Kuo and Porter, 1998).

The communication barriers between new immigrants and their doctors and the problems these barriers create for both groups are described poignantly by writer Anne Fadiman in her prize-winning book, *The Spirit Catches You and You Fall Down: A Hmong Child, Her American Doctors, and the Collision of Two Cultures*. The book describes the completely divergent worldviews of American doctors and Hmong patients in Merced, California, where many Hmong refugees from Laos have settled:

> Most Hmong believe that the body contains a finite amount of blood that it is unable to replenish, so repeated blood sampling [for lab tests] . . . may be fatal. When people are unconscious, their souls are at large, so anesthesia may lead to illness or death. If the body is cut or disfigured, or if it loses any of its parts, it will remain in a condition of perpetual imbalance, and the damaged person not only will become frequently ill but may be physically incomplete during the next reincarnation; so surgery is taboo. If people lose their vital organs after death, their souls cannot be reborn into new bodies and may take revenge on living relatives; so autopsies and embalming are also taboo. . . .
>
> Not realizing that when a man named Xiong or Lee or Moua walked into the Family Practice Center with a stomachache he was actually complaining that the entire universe was out of balance, the young doctors of Merced frequently failed to satisfy their Hmong patients. How could they succeed? . . . They could hardly be expected to "respect" their patients' system of health beliefs (if indeed they ever had the time and the interpreters to find out what it was), since the medical schools they had attended had never informed them that diseases are caused by

fugitive souls and cured by [sacrificing] chickens. All of them had spent hundreds of hours dissecting cadavers, . . . but none of them had had a single hour of instruction in cross-cultural medicine. To most of them, the Hmong taboos against blood tests, spinal taps, surgery, anesthesia, and autopsies—the basic tools of modern medicine—seemed like self-defeating ignorance. They had no way of knowing that a Hmong might regard these taboos as the sacred guardians of his identity, indeed, quite literally, of his very soul. [Moreover], what the doctors viewed as clinical efficiency the Hmong viewed as frosty arrogance. And no matter what the doctors did, even if it never trespassed on taboo territory, the Hmong, freighted as they were with negative expectations accumulated [during years under military siege and in refugee camps] before they came to America, inevitably interpreted it in the worst possible light (Fadiman, 1997:33, 61).

Growing recognition of problems like these has spurred medical schools to incorporate training in working with culturally diverse populations in their programs, as Chapter 11 will discuss in more detail.

Case Study: Environmental Racism

One health issue that cuts across America's minority communities is environmental racism. Environmental racism refers to the disproportionate burden of environmental pollution experienced by ethnic minorities. According to Benjamin F. Chavis (1993:3),

Environmental racism is racial discrimination in environmental policymaking. It is racial discrimination in the enforcement of regulations and laws. It is racial discrimination in the deliberate targeting of communities of color for toxic waste disposal and the siting of polluting industries. It is racial discrimination in the official sanctioning of the life-threatening presence of poisons and pollutants in communities of color. And, it is racial discrimination in the history of excluding people of color from the mainstream environmental groups, decision-making boards, commissions, and regulatory boards.

Environmental racism first became a subject for widespread discussion following the 1983 publication of a groundbreaking study by sociologist Robert D. Bullard. Bullard documented how, since the 1920s, the city of Houston had located all of its landfills and 75 percent of its garbage incinerators in African American neighborhoods, even though those neighborhoods constituted only a tiny fraction of the city. Since Bullard's study appeared, federal agencies, social activists, and scholars around the country have collected evidence demonstrating that minority communities bear a disproportionate share of the nation's environmental hazards, from Hispanic farm workers exposed to dangerous pesticides to Navajo communities poisoned by deadly uranium mines and inner-city African Americans plagued by asthma-inducing air pollution (Bullard, 1997; Camacho et al., 1998).

The most important of these environmental hazards, because it is so widespread and devastating, is lead, found in polluted air, contaminated soil, and the paints and pipes of older residences. Among children under age 5, 2 percent of white non-Hispanics have high levels of lead in their blood, compared to 4 percent of Mexicans and 11 percent of African Americans (*Morbidity and Mortality Weekly Report,* 1997). As Table 3.6 shows, minorities also are exposed more often than whites to dust and soot, carbon monoxide, ozone, sulfur, and sulfur dioxide, as well as to pesticides, emissions from hazardous-waste dumps, and other hazardous substances (U.S. Environmental Protection Agency, 1992). Researchers have found that exposure to environmental pollution is more highly correlated with race than with any other factor, including poverty (Bullard, 1993; Stretesky and Hogan, 1998).

Environmental racism exemplifies the workings of internal colonialism (Bullard, 1993). The term **internal colonialism** was coined to highlight the similarities between the treatment of minority groups within a country and of native peoples by foreign colonizers, such as under the former apartheid system (Blauner, 1972). Those who use this concept argue that, in the same way colonizers exploit native labor power and lands and keep native peoples economically dependent for the benefit of the colonizing power, majority groups can exploit internal colonies of minority group members. In the case of environmental racism, racial discrimination enables industrialists, with the tacit approval of government bureaucrats and politicians, to place environmental hazards in these internal colonies without worrying that those communities will have sufficient political power or financial resources to resist. Poverty and lack of other job opportunities can even encourage minority communities to welcome polluting industries for the jobs they will bring. This does not mean, however, that those who make decisions about where to locate environmental hazards *intend* to discriminate against minorities—certainly those who make

Table 3.6			
**Percentage of U.S. Whites, African Americans, and Hispanics Living in Polluted Areas**			

TYPE OF POLLUTION	WHITES	AFRICAN AMERICANS	HISPANICS
Particulate matter	15%	17%	34%
Carbon monoxide	34	46	57
Ozone	53	62	71
Sulfur dioxide	7	12	6
Lead	6	9	19

Source: U.S. Environmental Protection Agency (1992).

these decisions would argue that they decide solely on economic and technical considerations—only that their actions have the *effect* of discriminating.

Currently, about 200 grassroots groups of African Americans, Hispanics, Asian Americans, and Native Americans are working to fight for environmental justice (Bullard, 1997; Sandweiss, 1998), as are numerous national civil rights and environmental organizations; Box 3.3 describes the work of one of these groups. Similarly, the Environmental Protection Agency recently began using the Civil Rights Act of 1964, which forbids racial discrimination in any federally funded programs, as grounds for investigating how companies and local governments decide where to locate environmental hazards. Its first study found that 90 percent of major industrial polluters in Louisiana were located in predominantly African-American areas, and resulted in the cancellation of a hazardous waste permit in that state (Sandweiss, 1998).

CONCLUSIONS

Far from being purely biological conditions reflecting purely biological factors, health and illness are intimately interwoven with social position. In the United

---

Box 3.3   *Making a Difference: The Center for Health, Environment and Justice*

During the 1970s, the families of Love Canal, New York, were plagued by a series of unexplained deaths of children from cancer and leukemia. Eventually, local community activists traced these deaths to a nearby toxic waste site and won federal funding to relocate their families to safer areas. Perhaps more important, their work led to passage of the federal Superfund program to clean up toxic waste sites around the country.

In 1981, some of these activists founded the Center for Health, Environment and Justice (CHEJ) to assist other grassroots groups in similar battles. Since then, CHEJ has served as an invaluable resource. Much of their work consists of responding to letters and phone calls from people seeking information about toxic threats and helping people access information in their extensive library. In addition, CHEJ publishes more than 100 guidebooks and information packages on issues related to chemical hazards

and to environmental justice more broadly, plus two magazines, *Everyone's Backyard* and *Environmental Health Monthly,* designed to inform community activists and health care professionals about issues in the field.

In addition to providing technical information, CHEJ puts interested individuals in touch with appropriate organizations and runs periodic workshops to help organizations work more effectively. CHEJ provides both strategic planning and technical training on the specific environmental issues relevant to their fight. As of 1999, CHEJ has worked with more than 8,000 local groups (www.essential.org/cchw/cchwinf.html).

Since its founding, CHEJ has had many important successes. These include working with local activists to obtain the passage of laws that established state Superfund programs and laws that prohibited corporations forced to clean up toxic waste in one state from setting up business or dumping wastes in another state.

States as elsewhere, those who are poor or are targets of racial discrimination die younger than others. Conversely and ironically, women's lower social status (coupled with their biological advantages) has in some ways contributed to preserving their health, while men have paid for their higher social position with shorter life spans. As a result, social trends can help us predict future health trends. For example, as women's rate of tobacco use has approached men's, so has their rate of lung cancer. Similarly, if economic and ethnic inequality decrease, we should expect to see decreased disparities in health status.

SUGGESTED READINGS

Boston Women's Health Book Collective. 1998. *Our Bodies, Ourselves for the New Century: A Book By and For Women.* New York: Touchstone. An excellent overview of women's health issues, emphasizing self-help but also discussing the political and social aspects of health and health care.

Bullard, Robert D. (ed.) 1997. *Unequal Protection: Environmental Justice and Communities of Color.* San Francisco: Sierra Club Books. Excellent collection of essays on the topic.

Russell, Louise B. 1994. *Educated Guesses: Making Policy About Medical Screening Tests.* Berkeley: University of California Press. A fascinating exploration of the dangers of routine screening for health problems.

GETTING INVOLVED

National Coalition Against Domestic Violence. P.O. Box 18749, Denver, CO 80218. (303) 839-1852. www.ncadv.org. A national organization that can refer you to organizations in your region.

National Women's Health Network. 514 10th St. NW, Suite 400, Washington, DC 20004. (202) 628-7814. www.womenshealthnetwork.org. Educational and lobbying group concerned with all issues affecting women's health.

Citizen's Clearinghouse for Hazardous Wastes. P.O. Box 6806, Falls Church, VA 22040. (703) 237-2249. www.essential.org/cchw. Central clearinghouse for the environmental justice movement, providing assistance for grassroots organizations located primarily in poor and minority communities.

Habitat for Humanity. 121 Habitat Street, Americus, GA 31709. (912) 924-6935. www.habitat.org. Ecumenical Christian organization that helps poor families build low-cost housing.

Association of Asian Pacific Community Health Organizations. 1440 Broadway, Ste. 510, Oakland, CA 94612-2025. (510) 272-9536. www.aapcho.org. Excellent source of information about health and health care among both new and old Asian American communities.

REVIEW QUESTIONS

What are the consequences of an aging population and of the feminization of aging for the health care system?

Why might sociologists and other observers argue *against* early detection and treatment of prostate cancer?

Why do men have higher mortality rates than women but lower morbidity rates?

What are the sources and consequences of woman battering? Why do some health care workers consider woman battering a serious health problem?

How and why does social class affect individuals' health?

What are the special health problems of homeless persons? of migrant farmworkers?

How does ethnicity affect health separately from social class? How does social class affect health separately from ethnicity? How can you tell which is the more powerful factor?

How and why do the particular health problems of African Americans, Hispanics, Native Americans, and Asian Americans differ from those of whites?

What is environmental racism?

INTERNET EXERCISES

Both the United Nation's World Health Organization (www.who.int) and the U.S. National Institute of Health (www.nih.gov) have Web sites devoted to health problems associated with aging. Find those sites, and compare the major health problems identified by the World Health Organization with the major problems identified by the National Institute of Health. How do you explain the differences?

The U.S. Census Bureau (www.census.gov) provides a wealth of information about the U.S. population. Can you find out what percentage of Americans currently live below the poverty line?

To find out how social class affects individuals' perceived health status, first locate the Web site for the University of California's Survey Documentation and Analysis (SDA) Archive. This archive contains data from several national random surveys. Click on the GSS Cumulative Datafile, 1972–1996. Find the Select an Action section, then click the button for Frequencies or Crosstabulations. Next click on Start. A form with several blank spaces will appear on your screen. For row variable, type health. For column variable, type class. Click on the boxes to the left of Column Percentaging, Statistics, and Question Text. Then click the button to Run the Table. What effect does social class have on people's perceptions of their health status?

CHAPTER 4

Illness in the Developing Nations

—

For almost twenty years, Paul Farmer, an American doctor and anthropologist, has worked among Haiti's rural poor. One of his patients is Jean Dubuisson, who

> lives in a small village in Haiti's Central Plateau, where he farms a tiny plot of land. He shares a two-room hut with his wife, Marie, and their three surviving children. All his life, recounts Jean, he's "known nothing but trouble." His parents lost their land [when] the Péligre hydroelectric dam [was built and flooded their village]—a loss that plunged their large family into misery. Long before he became ill, Jean and Marie were having a hard time feeding their own children: two of them died before their fifth birthdays, and that was before the cost of living became so intolerable.
>
> And so it was a bad day when, some time in 1990, Jean began coughing. For a couple of weeks, he simply ignored his persistent hack, which was followed by an intermittent fever. There was no clinic or dispensary in his home village, and the costs of going to the closest clinic . . . are prohibitive enough to keep men like Jean shivering on the dirt floors of their huts. But then he began having night sweats. Night sweats are bad under any conditions, but they are particularly burdensome when you have only one sheet and often sleep in your clothes (Farmer, 1999:187–188).

Although Jean and Marie both recognized that he needed to seek medical care, doing so seemed economically impossible. Over the next few months, however, Jean's health continued to decline and his weight to drop. Even more frightening, in December 1990 Jean began to cough up blood, which, given how common tuberculosis is in Haiti, they easily recognized as a symptom of the disease. At that point, Jean agreed to go to a clinic:

> At the clinic, he paid $2 for multivitamins and the following advice: eat well, drink clean water, sleep in an open room and away from others, and go to a hospital. Jean and Marie recounted this counsel without a hint of sarcasm, but they nonetheless evinced a keen appreciation of its

total lack of relevance. In order to follow these instructions, the family would have been forced to sell off its chickens and its pig, and perhaps even what little land they had left. They hesitated, understandably.

Two months later, however, a second, massive episode of [coughing up blood] sent them to a church-affiliated hospital [some distance away, where Jean] was charged $4 per day for his bed; at the time, the per capita income in rural Haiti was about $200 a year. When the hospital's staff wrote prescriptions for him, he was required to pay for each medication before it was administered. Thus . . . he actually received less than half of the medicine prescribed. . . . [Jean] discharged himself from the hospital when the family ran out of money and livestock (Farmer, 1999:188–189).

Some months later, Jean learned of a nonprofit clinic Farmer had founded in a nearby village, and sought care there. As Farmer describes,

Jean was cured of his tuberculosis, but this cure, in many respects, came too late. Although he is now free of active disease, his left lung was almost completely destroyed, . . . forever compromising his ability to feed his family—a precarious enough enterprise in contemporary Haiti, even for the hardy (Farmer, 1999:197).

As Jean and Marie's story suggests, the sources and patterns of illness and health care in poorer countries differ dramatically from those found in more affluent countries. This chapter begins by comparing some of these differences and then focuses on health problems in the poorer countries.

DISEASE PATTERNS AROUND THE WORLD

In making international comparisons, politicians, social scientists, medical researchers, and others typically divide the world into two broad groups, the **industrialized nations** and the **developing nations.** Essentially, this division reflects the economic status of the various nations. The industrialized nations are primarily defined by their relatively high gross national product (GNP) per capita compared with developing nations. In addition, the industrialized nations are characterized by diverse economies made up of many different industries, while the developing nations have far simpler economies, in some cases still relying heavily on a few agricultural products such as rubber or bananas. Because of these economic differences, the developing nations as a group have higher infant mortality, lower life expectancies, and a greater burden of infectious and parasitic diseases than do the industrialized nations. Table 4.1 shows life expectancies for various nations.

Table 4.1   *Life Expectancy at Birth*

COUNTRY	LIFE EXPECTANCY
Japan	80
France	78
Singapore	77
Costa Rica	76
United States	76
Cuba	75
Mexico	72
China	71
Thailand	69
Brazil	67
Egypt	67
Philippines	66
Bolivia	60
India	59
Haiti	51
Somalia	47
Ethiopia	42
Zimbabwe	40
Sierra Leone	34

Source: Population Reference Bureau, 1998.

Although dividing the globe into industrialized versus developing nations is a useful analytic tool, it is important to remember that development level is a scale, not a dichotomy. So, for example, the most rapidly developing nations like Mexico and Thailand have many complex industries in addition to traditional agricultural crops and enjoy infant mortality rates and life expectancies approaching those found in the United States and Europe. In addition, although infectious and parasitic diseases remain more common in the rapidly developing nations than in the industrialized nations, chronic diseases are now the most common sources of mortality in both sets of nations (Murray and Lopez, 1996). In contrast, in the 31 *least developed* nations, life expectancy remains less than 50 years and infectious and parasitic diseases still claim most lives (Population Reference Bureau, 1998).

The division between developing and industrialized nations also should not keep us from recognizing that social conditions and hence health patterns vary from community to community and from social group to social group within each nation. Thus conditions in Harlem in some ways resemble those in Bangladesh, as the previous chapter discussed, whereas conditions in wealthy sections of Bangkok resemble those in many U.S. cities. Similarly, in the same way that differences in health and access to health care within the United States have grown wider in recent years because of federal economic policies that have encouraged a growing income gap between rich and poor, differences also have grown in many developing nations because of "structural adjustment" policies adopted by the International Monetary Fund, under which developing nations could not get economic aid unless they agreed to cut back social programs such as food subsidies and health care for the poor (Kolko, 1999; Peabody, 1996).

Finally, although the terms *developing* and *industrialized* nations imply linear progression from one status to the other, this is not necessarily the case. For example, economic and health conditions have worsened in Eastern Europe following the collapse of the Soviet Union and in parts of Africa as a result of the AIDS epidemic.

With these caveats, the remainder of this chapter explores the sources and nature of disease in the developing nations. Keep in mind, though, that *diseases respect no national borders.* With rising **globalization,** diseases are spreading rapidly from developing to industrialized nations and vice versa. This is especially likely between the United States and Mexico, where only an imaginary line divides two territories that share the same water, air, and, to a large extent, economies. For example, the city of Juarez, Mexico, has no sewage treatment facilities for its population of more than one million, while neighboring El Paso, Texas, has thousands of homes that lack sanitary septic systems (Skolnick, 1995). As a result, from both sides of the border human wastes drain into the Rio Grande River, which provides water for drinking and for agriculture in these two cities and in downstream communities, including Laredo and Brownsville, Texas. Diseases like cholera or hepatitis could easily take root in these areas and spread into the interiors of both countries. Thus, as this suggests, those who live in the industrialized nations have a vested interest in understanding health and illness in the developing nations.

THE SOURCES AND NATURE OF DISEASE IN THE DEVELOPING NATIONS

Poverty, Malnutrition, and Disease

The primary factor underlying the high death rates and low life expectancies in the developing nations is poverty. The previous chapter showed

Table 4.2	Gross National Product (GNP) per Capita by Life Expectancy	

GNP PER CAPITA	LIFE EXPECTANCY
Low-income countries (average per capita GNP=US $320)	60 years
Middle-income countries (average per capita GNP=US $1,930)	66 years
High-income countries (average per capita GNP=US $17,470)	76 years

Source: World Health Organization, 1993a:27.

how, in the United States, individuals with higher incomes experience less illness and live longer than individuals with lower incomes. In the same way, nations with higher median incomes have lower rates of illness and mortality than nations with lower median incomes. As Table 4.2 shows, residents of poor nations (where the per capita GNP averages $320) die an average of sixteen years earlier than residents of rich nations (where the per capita GNP averages $17,470). Similarly, as the median income of a nation rises, life expectancy rises, at least until the threshold point described in Chapter 2. Between 1965 and 1989, for example, the GNP per capita in the poor nations rose 2.9 percent annually, while life expectancy increased by fourteen years (World Health Organization, 1993a: 27).

In large part, poverty causes disease and death by causing chronic malnutrition. According to the **World Health Organization (WHO),** malnutrition accounts for 49 percent of deaths before age five in the developing nations (World Health Organization, Programme of Nutrition, 1998). Malnutrition indirectly causes disease and death by damaging the body's immune system, leaving individuals more susceptible to all forms of illness and contributing to both infant and maternal mortality. In addition, malnutrition directly causes numerous health problems, including brain damage due to iodine deficiency, blindness due to vitamin A deficiency, and mental retardation due to anemia (World Health Organization, Programme of Nutrition, 1998).

The Roots of Chronic Malnutrition

Given the link between malnutrition, illness, and death, the importance of investigating the roots of chronic malnutrition is clear. At first thought, one might easily assume that malnutrition in developing nations that have not yet experienced the **epidemiological transition** results naturally from overpopulation combined with insufficient natural and technological resources. Yet food production has surpassed population growth in most countries, including most of those where hunger is common (Lappé et al., 1998). In fact, most of the "hungry" countries export more food than they import, and almost every country has access to sufficient food to feed its entire population.

Nor can malnutrition be blamed on population density (Lappé et al., 1998). The Netherlands, for example, is one of the most densely populated countries in the world, yet chronic malnutrition no longer occurs there. Similarly, Honduras has twice the cropland per person as Costa Rica, yet malnutrition remains common only in the former.

If overpopulation, lack of food, population density, and lack of cropland do not explain chronic malnutrition, what does? The answer lies in the social distribution of food and other resources: *malnutrition occurs most often in those countries where resources are most concentrated.* In other words, malnutrition occurs not in countries where resources are scarce, but in countries where a few people control many resources while many people have access to very few resources (Dreze and Sen, 1989; Lappé et al., 1998; Turshen, 1989). Similarly, within each country, malnutrition occurs most often among those groups—typically females and the poor—with the least access to resources (Messer, 1997). In essence, then, malnutrition is a disease of powerlessness.

If powerlessness causes malnutrition, then eliminating inequities in power should eliminate malnutrition. Evidence from China and Costa Rica supports this thesis. These two nations—one essentially communist and one essentially capitalist—have both adopted socialistic strategies for redistributing resources somewhat more equitably. By giving farmland to formerly landless peasants, extending agricultural assistance to owners of small farms, working to raise the status of women, and so on, they have made chronic malnutrition almost unknown within their borders. On the other hand, China has not proved immune to *acute* malnutrition caused by famines. According to Nobel Prize-winning economist Amartya Sen, famines occur when natural events reduce harvests in countries whose nondemocratic governments need not fear being voted out of office if they do not meet their citizens' needs (Sen, 1999).

The Role of International Aid

Similarly, international aid—both food aid and development projects—has helped improve the standard of living and health status in developing nations run by democratic governments, but often has had the opposite effect in nondemocratic nations (World Bank, 1998). Most international food aid comes from the United States, under the 1954 Food for Peace Act, or PL480. The primary purpose of this law is to protect U.S. economic and military interests (Lappé et al., 1998). By sending U.S. farm surpluses overseas as food aid, agricultural producers can maintain prices for their goods at home while opening new markets to U.S. agricultural commodities. In addition, because the United States sells food aid on credit rather than giving it away, food aid helps offset U.S. trade deficits. Food aid also helps to protect U.S. military interests by bolstering the governments of nations with strategic military importance for this country. This explains why U.S. food aid primarily goes not to the hungriest countries, but to countries where the U.S. has military interests, such as Egypt, Israel, El Salvador, Pakistan, and Turkey.

Once food aid reaches the developing nations, its distribution can unintentionally reinforce inequities in access to resources and thus malnutrition (Lappé et al., 1998). Food aid goes directly to foreign governments, which can distribute it as they choose. In countries run by democratic governments committed to social equality, aid is likely to benefit those who need it most. Unfortunately, many developing nations are run by small, economically powerful elites, which sometimes instead sell on the open market any food their governments receive and pocket the profits, thus accentuating social inequities.

Because the hungriest people cannot afford to buy food aid sold in the marketplace, food aid does not improve their nutritional status. In fact, food aid *contributes* to the malnutrition of the landless tenants, sharecroppers, and day laborers who form the overwhelming bulk of those who suffer from malnutrition (Lappé et al., 1998). When the United States sells its surplus agricultural commodities in the developing nations, the prices of those commodities within the developing nations plummet. As a result, owners of small farms may no longer be able to earn a living and must sell their land to larger landowners who can take advantage of economies of scale. Thus land ownership and power become further concentrated, as do the inequities that underlie malnutrition and illness in the developing nations.

Like international food aid, internationally–sponsored development projects have had mixed impacts on malnutrition and on health in general (World Bank, 1998). According to the politically conservative World Bank, carefully designed projects, sensitive to local conditions and culture and located in countries with democratic governments, open trade, social safety nets, and conservative economic policies can reduce malnutrition and its root causes. In Pakistan, for example, enrollment of girls in schools soared in 1995 when development money was given to local communities to open new schools on the condition that the enrollment rate of girls increased (World Bank, 1998). In the long run, this should increase the status of women, which, as we will see, is directly linked to malnutrition, infant mortality, and maternal mortality.

On the other hand, projects like the Péligre dam in Haiti, the Akosombo Dam in Ghana, and the Aswan Dam in Egypt have brought electricity to urban elites and industrial sites run by multinational corporations while flooding and destroying agricultural fields and rural villages and bringing plagues of waterborne diseases to rural dwellers (Basch, 1990; Farmer, 1999). Agricultural development projects have been particularly likely to contribute to malnutrition among women and children (Lappé et al., 1998). These projects often start from the assumption, based on Western ideas about the family and the economy, that raising cash crops will benefit families more than raising food crops and that men rather than women should be responsible for agricultural efforts. However, cultural traditions in many developing nations hold women responsible for growing food and feeding the family (Lappé et al., 1998). When development projects encourage men to grow cash crops, the men sometimes take over land women had used to grow food and, because men consider feeding the family a woman's responsibility, use their profits not to purchase food but, rather, to purchase high-status goods for themselves such as tobacco or Western clothes. As a result, malnutrition increases among both women and children.

Infectious and Parasitic Diseases

One result of malnutrition and, more broadly, of poverty is a high rate of infectious and parasitic disease. As stated earlier, such diseases cause the majority of deaths in the least developed nations and still account for considerably more deaths in the rapidly developing nations than in the industrialized nations. Table 4.3 documents the central role these diseases played in 1997 in the developing nations.

As in Europe and the United States before the twentieth century, the high rates of infectious and parasitic diseases reflect the dismal circumstances in which many people live. In addition to malnutrition, overcrowding promotes the spread of airborne diseases like tuberculosis, while contamination of the water supply with sewage spreads waterborne diseases such as cholera and intestinal infections. Similarly, poor housing and lack of clean water for bathing result in frequent contact with disease-spreading rats, fleas, and lice.

The infectious and parasitic diseases that most commonly cause deaths in the developing nations are respiratory infections, diarrheal diseases, measles, and malaria. In addition, **HIV disease** now causes more deaths in central Africa than any other infectious disease. The next section discusses these diseases in more detail.

Respiratory Infections

The most important respiratory infection in the developing world is tuberculosis, which kills between two and three million people yearly (World Health Organization, 1998a). The disease is most common in Asia, followed by Africa and the Middle East, and then by Latin America and the Caribbean.

As described in Chapter 2, the **incidence** of tuberculosis is increasing around the world due to the HIV epidemic and the development of drug-

Table 4.3: **Percentage of Deaths by Causes, Industrialized and Developing Nations, 1997**

DISEASE	DEVELOPING NATIONS	INDUSTRIALIZED NATIONS
Infectious and parasitic disease	43%	1%
Circulatory disease (e.g., heart problems)	24	46
Infant and maternal mortality	10	1
Cancers	9	21
Respiratory disease	5	8
All other and unknown	9	23
Total	100	100

Source: World Health Organization, 1998b.

resistant tuberculosis strains which have made treating that disease economically unfeasible in many developing nations. WHO estimates that HIV causes 1.5 million new tuberculosis cases annually, and that up to 22 percent of persons with tuberculosis in several "hot zones" around the world, including India, Russia, the Dominican Republic, and the Ivory Coast, are infected with multiple drug-resistant strains (World Health Organization, 1997a; 1998a).

Diarrheal Diseases

In industrialized nations, diarrhea is generally a source of passing discomfort. In developing nations, diarrheal diseases can be fatal, especially among children under age two (World Health Organization, Division of Child Health and Development, 1992). The World Health Organization estimates that diarrheal diseases kill 4 million children under age five yearly and that children suffering from diarrheal diseases fill more than a third of pediatric hospital beds in some developing nations.

Diarrhea is a symptom, not a disease, and can be caused by infection with any of a number of bacteria, viruses, or parasites. Diarrhea kills by causing dehydration and electrolytic imbalance. It also leads to malnutrition, as affected children both eat less and absorb fewer nutrients from the foods they eat. In turn, malnutrition leaves children susceptible to other fatal illnesses. Conversely, other illnesses can leave children susceptible to both diarrheal diseases and malnutrition.

Diarrheal diseases (which include dysentery, cholera, and infection with *E. coli*) occur when individuals ingest contaminated water or foods.

Consequently, the likelihood of severe diarrhea is greatest when families lack refrigerators, sanitary toilets, sufficient fuel to cook foods thoroughly, or safe water for cooking and cleaning. As of 1997, statistics reported to WHO by government officials indicate that 25 percent of people living in the developing nations lack access to clean water, down from 39 percent a few years earlier (Kristof, 1997). However, these figures may substantially overestimate access, as governments may report to WHO that citizens of their countries have access to clean water even if the only water source is a single, sporadically working, faucet, a mile or more away and shared by many families.

Survival rates for children with diarrheal diseases in developing nations have improved rapidly in recent years. Before the 1960s, those suffering from diarrheal diseases could only be treated using expensive intravenous fluids, making treatment unfeasible for many in the developing nations. Since then, however, scientists have demonstrated that a simple and inexpensive solution of dried salts and water is just as effective, and the World Health Organization has actively and successfully promoted this "oral rehydration therapy."

Measles

To persons living in the industrialized nations, where measles is considered a minor childhood illness, it might seem odd to see measles listed as a major cause of death. Yet measles kills more children under age five in the developing nations than any other infectious disease—about 1 million children per year (*Morbidity and Mortality Weekly Report,* 1998a). These deaths occur when children, already weakened by malnutrition and poor living conditions, become further weakened by measles. As a result, their bodies' ability to fight disease diminishes, leaving them susceptible to potentially deadly pneumonia, respiratory infections, and diarrhea. Unlike tuberculosis and malaria, however, rates of measles declined almost by half during the 1990s, as a result of a worldwide measles vaccination campaign run by WHO. Immunization rates have remained unchanged in Africa, however, as a result of ongoing and severe economic problems on that continent.

Unfortunately, even if vaccination becomes more widespread and rates of measles continue to decline, the overall health of children in the developing nations will not improve unless social conditions also improve. As long as conditions in the developing nations continue to foster diseases of all kinds, those children who do not die from measles are still likely to die at equally young ages from other diseases; at least one study has found that reducing a country's death rate from measles has no effect on its rate of childhood mortality (Turshen, 1989). Only when the basic inequities in living conditions that underlie death and disease are eliminated or at least substantially reduced will more children survive.

Malaria

More than 40 percent of the world's population lives in areas where malaria is **endemic** (World Health Organization, Division of Control of Tropical Diseases, 1998). Each year, approximately 300 to 500 million people (more than 90 percent of whom live in tropical Africa) develop malaria, and between 1.5 and 2.7 million people die from the resulting anemia, general debility, or brain infections. Most, however, will survive, but will experience intermittent disabilities from the intermittent chills, fevers, and sweats that malaria brings. As a result, looking solely at deaths from malaria understates the true toll malaria imposes on a population's health.

Malaria poses the greatest threat to pregnant women, infants, and young children. Among pregnant women, malaria increases the risks of miscarriage, anemia, and premature labor, each of which increases the risk of potentially fatal hemorrhaging. Infants born to these women typically have lower than average birthweights and, hence, higher chances of death or disability. Malaria is often fatal among young children, whose immune systems have not yet developed sufficiently to fight infection; as many as two-thirds of those who die from malaria are under age five (World Health Organization, Division of Control of Tropical Diseases, 1998).

Malaria is caused by protozoan parasites belonging to the genus *Plasmodium*. Malaria is transmitted only by *Anopheles* mosquitoes and, consequently, exists only where those mosquitoes live. (*Anopheles* mosquitoes and malaria used to exist throughout the United States and appear to be making a comeback; in 1999, for the first time in decades, malaria was diagnosed in a U.S. resident who had neither lived nor traveled in another country.) The disease cycle begins when a mosquito bites an infected individual and ingests the parasite from the individual's blood. The parasite reproduces in the mosquito's stomach and then migrates to the mosquito's salivary glands. The next time the mosquito bites someone, it transmits the parasite to that person.

Because of this transmission cycle, eliminating *Anopheles* mosquitoes will eliminate malaria. Since the 1940s, antimalaria campaigns have depended heavily on using pesticides to kill mosquitoes. Although these campaigns initially work well, over time the pesticides lose their potency as pesticide-resistant mosquitoes evolve. As a result, nations must constantly search for new and ever more toxic pesticides, each of which can endanger birds, fish, and insects that benefit humans. Because of these problems, some recent campaigns have instead focused on encouraging the use of insect repellents, mosquito netting, and screens to prevent infection. These campaigns also have focused on encouraging the use of drugs, such as chloroquine and mefloquine, which can both prevent and treat malaria. Unfortunately, because these drugs can cause debilitating side effects and cost more than many residents of developing nations can afford, infected individuals often stop taking the drugs before they are cured. This continual undertreatment of malaria, like the undertreatment of tuberculosis, has encouraged the evolution of drug-resistant strains of the disease around the globe. Consequently, although malaria has been eliminated in some regions, the situation has worsened during the last decade in the rest of the world.

HIV Disease

More than 90 percent of HIV-infected persons in the world live in developing nations (Stine, 1998:361). HIV infection is now endemic in parts of the Caribbean and in much of Africa, and is spreading especially rapidly in Asia (especially in India and Southeast Asia).

From the beginning of the epidemic, heterosexual intercourse has been the major mode of HIV transmission in the developing nations. Consequently, women account for half or more of all cases in these nations. Because many of these countries lack the funds needed to test blood for HIV, transmission via blood transfusions remains common. Similarly, in part because they lack the funds to supply infected women with the drug AZT, which can prevent transmission from mother to fetus, such transmission remains common. (However, the recent development of new, inexpensive drug regimens to prevent transmission could make it less common in future.) Infection is most common in urban areas, but is spreading rapidly in the coun-

---

Box 4.1 **War and Health**
Lisa Comer

In addition to poverty, malnutrition, germs, and parasites, another important, although often overlooked, source of death and disease in the developing nations is war (Geiger and Cook-Deegan, 1993; Toole and Waldman, 1993). War is an ongoing fact of life in many developing nations. Civilians may face anything from malnutrition, forced labor, and deportations to mass executions, starvation, chemical warfare, torture, and rape. These physical traumas typically are magnified by the psychological traumas of losing one's family, community, and, frequently, dignity.

Human rights violations committed during wars often lead to a rise in illness among civil-

ians. Forcing refugees into overcrowded, unsanitary relocation camps frequently results in epidemics of communicable diseases, which often go untreated because health care workers and medical facilities—insufficient in the best of times—are targeted for destruction by the military (Geiger and Cook-Deegan, 1993; Toole and Waldman, 1993). For the same reason, wars often disrupt public health services, including vaccination programs for children.

Given the profound impact of war on public health, the medical community can and sometimes does play a crucial role in documenting and preventing war crimes and other human rights violations. Health care workers' docu-

tryside, especially in areas where war has disrupted families and increased both consensual and nonconsensual sexual intercourse between soldiers and villagers. (Box 4.1 discusses in more detail how war and militarism affect health.) In the hardest hit countries (most located in sub-Saharan Africa), at least 20 percent of the entire adult population is infected, while life expectancies have plummeted by 20 or more years (Olshansky et al., 1997).

As stunning as these numbers might appear, they understate the impact of HIV disease. Unlike most illnesses, HIV disease most commonly strikes at mid-life, normally the most economically productive years. Moreover, HIV disease disproportionately has affected the most educated segments of the population in the developing nations. Consequently, HIV disease has crippled the economy in the hardest hit countries. The resulting increase in unemployment and poverty is sending ripples of illness and death throughout these countries. In addition, HIV disease typically strikes during the childrearing years. As a result, about 16 percent of all children in the hardest hit countries have been orphaned, resulting in a second wave of deaths among orphans who have no surviving relatives to care for them (Olshansky et al., 1997).

Several theories have been proposed to explain why HIV disease has hit Africa so hard. The two theories that have gained the most supporters are the cultural theory used by demographer John Caldwell and his colleagues and the materialist theory used by sociologist Charles Hunt.

---

mentation of these horrors is especially important because politicians and others can less easily dismiss testimony from health care workers as biased than testimony from other civilians (Geiger and Cook-Deegan, 1993; Swiss and Giller, 1993). Consequently, health care workers can help to awaken public awareness of war crimes and human rights violations, speeding health care and other assistance to the survivors of war and bringing war criminals to justice.

Over the years, individual health care workers and nonprofit groups, such as Physicians for Human Rights and *Medécins Sans Frontières* (Doctors Without Borders)—which won the Nobel Peace Prize in 1999—have eased the burdens of war victims substantially. In addition, beginning in the early 1990s, the American medical community moved toward officially asserting a commitment to war survivors. For example, the *Journal of the American Medical Association* in 1993 devoted part of an issue to this topic, including an editorial that called on doctors to support the United Nations in its drive to provide health care to civilians affected by warfare (Cobey et al., 1993). If this advice is heeded, doctors may play a growing role in the future in documenting, treating, and perhaps even preventing this significant source of death and disease in the developing nations.

Caldwell and his colleagues (1989, 1991, 1992) have argued that because soils are poor throughout much of Africa, farming there always has been highly labor-intensive, and farming families have needed to have many children to help them in the fields. Consequently, a cultural system developed that valued fertility over chastity or monogamy and valued ties between parents and children more than ties between spouses. As a result, individuals tended to have relatively high numbers of sexual partners over their lifetimes. In past centuries, Africans typically obtained these sexual partners within small social and geographic circles. Since the rise of European colonization, however, and the attendant growth of towns, bars, transportation networks, and a commercial sex industry, both the size and geographic spread of Africans' social circles have broadened, causing dramatic increases in both average numbers of sexual partners and the geographic diversity of those partners. As a result, Africans are particularly likely to be exposed to sexually transmitted diseases, including HIV.

Whereas Caldwell and his colleagues give primary emphasis to cultural factors in explaining the devastating rates of HIV disease in Africa, Charles Hunt (1989, 1996) emphasizes the impact of material conditions. Hunt's argument is based on "**world systems theory**," which divides the world's nations into **core nations, peripheral nations,** and a few **semiperipheral nations** (Chase-Dunn, 1989; Wallerstein, 1974). The core nations, such as France and the United States, are, in effect, an upper class of nations—enjoying a highly diversified, industrialized economy that provides a high standard of living for most citizens. Conversely, the peripheral nations form a lower class of nations, where modernization and industrialization have developed slowly if at all and the standard of living is low for all but a small elite.

World systems theory argues that the core nations have achieved and maintained their present economic position through exploiting the resources of the peripheral nations. Rather than establishing industries in peripheral nations that would allow those nations to modernize their economies, multinational corporations instead have established industries in the peripheral nations based on extracting raw goods, such as rubber, minerals, or specialized food crops, and have brought the profits from those industries back to the core nations. Lacking their own modern industries, peripheral nations must buy most manufactured goods and, sometimes, basic foods from the core nations. In this way, the core nations maintain a favorable trade balance with peripheral nations, force the peripheral nations to rely for their economic well-being on inherently unstable markets for raw materials, and perpetuate the underclass position of the peripheral nations.

Applying this theory to HIV in Africa, Hunt (1989, 1996) argues that the African nations remain largely under the economic control of corporations based in the former colonial powers. To increase their profits, those corporations have concentrated industries in a few sites, rather than distributing manufacturing, mining, and corporate agriculture around the continent. Attracted by the prospects of cash income and faced with few means of earning a living in their home villages, native men leave the countryside to seek employment at these sites, often living apart from their wives and families for weeks, months, or even years at a time. These conditions foster the use of prostitutes and, in turn, the spread of sexually transmitted diseases, including HIV disease. Once workers become ill, their employers fire them and send them back to their villages, where they spread infection still further.

Meanwhile, health conditions also deteriorate among women and children left in rural villages. The loss of men's labor makes it more difficult for women to grow sufficient crops to feed themselves and their children. As a result, women typically adopt agricultural practices and crops that are less labor-intensive, even though they also deplete the soil and provide less nutrition. Consequently, those left in rural villages grow progressively more malnourished and susceptible to disease. Faced with these conditions, women's only option is to seek employment in cities, where many find prostitution the only available job. This completes the cycle through which multinational corporations indirectly encourage HIV infection among both men and women, in rural and urban areas.

Support for this theory comes from data suggesting that HIV was most common and appeared earliest in areas where migrant laborers worked, was next most common in the rural areas from which migrant laborers were recruited, and was least common in areas without links to migrant labor (Hunt, 1989). Other studies similarly have found that economic and structural factors better explain the explosive spread of HIV in Africa than do cultural factors (Simmons et al., 1996). At this point, however, the poor quality of data on HIV rates in Africa makes it difficult to test any theory with confidence. Moreover, neither the cultural theory used by Caldwell and his colleagues nor the materialist theories used by Hunt and others can account fully for the geographic distribution of HIV infection in Africa (Hunt, 1996). Thus, neither theory can be considered fully supported.

Infant Mortality

Like infectious and parasitic diseases, infant mortality is far more common in the developing nations than in the industrialized nations. As of 1998, the average infant mortality rate in the developing nations was 64 per 1,000 live births, down from 119 per 1,000 in 1991 but still eight times higher than the rate in the industrialized nations (Population Reference Bureau, 1998). These averages, however, hide the great range in infant mortality rates within the developing nations, which are illustrated in Table 4.4.

To understand the sources of infant mortality, we need to differentiate between **neonatal infant mortality** (during the first twenty-seven days after birth) and **postneonatal infant mortality** (between twenty-eight days

Table 4.4   *Infant Mortality per 1000 Live Births*

COUNTRY	INFANT MORTALITY RATE
Afghanistan	150
Ethiopia	128
Somalia	122
Bolivia	75
Haiti	74
India	72
Egypt	63
Zimbabwe	53
Brazil	43
Philippines	34
China	31
Mexico	28
Thailand	25
Costa Rica	12
Cuba	7
United States	7
France	5
Singapore	4
Japan	4

Source: Population Reference Bureau, 1998.

and eleven months after birth). Rates of *neonatal* infant mortality, typically caused by genetic defects, do not differ substantially between developing and industrialized nations. Rates of *postneonatal* mortality, on the other hand, are far higher in the former countries, especially in the least developing nations, producing significantly higher rates of infant mortality overall. The most common killers of infants in the developing nations are malnutrition and infections (particularly respiratory infections and diarrheal diseases). Because these factors were described earlier in this chapter, in this section, I will focus on two other important sources of postneonatal infant mortality: women's status and infant formula manufacturers.

The Role of Women's Status

The low status of women plays a critical role in infant mortality in developing nations. That status "can be readily perceived from the observation that, although women constitute one-third of the world's official labor force, they are responsible for nearly two-thirds of the total hours worked, receive only one-tenth of the world's income, and own less than 1 percent of its property" (Basch, 1990:24).

Infant mortality occurs most often among babies born with low birthweights. Whereas in the industrialized nations, low birthweight typically occurs when babies are born prematurely, in the developing nations, low birthweight typically occurs among babies born after a normal gestation period to mothers who have malaria, are underfed, routinely perform heavy labor, or suffer from anemia, which affects more than 50 percent of pregnant women in developing nations (World Health Organization, 1998b). These conditions reflect women's typically low status. Throughout the developing nations, girls and women often spend long hours in heavy labor and, in many nations, receive less nutrition than boys and men (Messer, 1997). In addition, girls are less likely than boys to be immunized against disease, to receive health care when ill, and to receive health care promptly (Messer, 1997). Consequently, girls are more likely to become ill and less likely to survive their illnesses. In Bangladesh, for example, the death rate between the ages of one and four is 58 percent greater among girls than among boys (Smyke, 1991:32). Consequently, women often enter their childbearing years already ill and malnourished—a situation that worsens as pregnancies further stress their bodies and drain their energy.

Similarly, infant mortality is highest among infants born to very young or very old mothers and to infants born less than eighteen months after a sibling. This occurs most commonly in cultures that expect women to marry at young ages and that judge women's worth by the number of sons they produce. In part, these cultural values reflect the economic realities of agricultural life: in agricultural societies, children produce more economic resources than they consume, so a family with many children is more likely to succeed than a family with few children. In addition, in the absence of any formal provisions for social security, individuals can only guarantee their security in old age by having sons. (Having daughters usually does not help because daughters in most cultures are expected to take care of their husbands' parents rather than their own.)

Nevertheless, even in these societies many women would like to limit their fertility at least somewhat. This desire is so great that throughout the world, women often choose illegal abortion over childbearing: 44 percent of all abortions performed worldwide (and 54 percent in developing nations) are illegal (Henshaw et al., 1999). In fact, statistics from Romania,

---

Box 4.2   ***Ethical Debate: The Ethics of Sex Preselection***

Zhang Zhiquan and his wife Mei live in a rural village in the People's Republic of China. Growing up in rural China, they learned early that couples needed sons to prosper and to care for them in their old age. They also learned that sons were essential for passing on the family name, that wives who produced no sons deserved mockery and abuse, and that girls were so useless that in the past many rural families did not even bother to name them. When Mei became pregnant, therefore, they had to decide what they would do if the baby were female. In the past, should they have felt unable or unwilling to raise a daughter, their only options would have been to kill the baby or give it up for adoption—choices that some families still make. Now, however, they had one additional option: having a health care worker identify the fetus' sex through ultrasound or amniocentesis and perform an abortion if the fetus were female.

Half a world away, the same issues of sex preselection and selective abortion arise, although in a different form:

Sharon and James Black live in Denver, Colorado, with their two young daughters. Because they both believe that children need a parent home at the end of the school day, Sharon works only part-time as a secretary, while James works two jobs so they can make ends meet. Sharon has just learned she is pregnant again. Although they had only planned on having two children, James always wanted a son with whom he could share his interests in sports and automobiles. Having another child, however, will further strain their finances and make it difficult for Sharon to return to full-time work for several more years. Consequently, continuing the pregnancy does not seem worthwhile unless they know the fetus is male.

Is sex preselection ethically justified in these cases? Although the circumstances differ enormously, for both families the birth of a daughter would bring substantial economic hardship. For both families, too, a daughter would enter life unwanted and already having failed to meet her parents' expectations. In addition, for the Chinese family and possibly (although to a lesser extent) for the American family, the birth of another daughter might

---

where abortion was outlawed between 1966 and 1989, suggest that making abortion illegal has almost no long-term impact on either the abortion rate or the birth rate—although it dramatically increases the number of women who die or become infertile following unsafe abortions (World Health Organization, Division of Reproductive Health, 1998). Meanwhile, the slums of Bombay and Rio de Janeiro, like the orphanages of Romania after 1966, are filled with abandoned children whose families could not support them. Similarly, in parts of Asia, infanticide of girl babies continues to occur among families that want babies only if those babies are male, and abortions now often occur when women learn through genetic testing that they

lower the wife's status and strain the marriage. Given these circumstances, wouldn't it be best for all concerned if the families use the available medical technology to test their fetuses' sex and to abort them if they are female?

For hundreds of thousands of couples in Asia and a growing number in the West, the answer, resoundingly, is yes. A recent international survey of geneticists and genetic counselors found that 47 percent had received requests from couples desiring fetal sex selection (Wertz and Fletcher, 1998). Twenty-nine percent of the respondents reported that they would test fetal sex for a couple with four daughters who intend to abort if their fetus is female, and another 20 percent would refer the couple to someone they knew would do so.

Those who support prenatal sex selection argue that selective abortion causes little harm, whereas the birth of unwanted girls financially strains families, leaves mothers open to ridicule or even physical abuse, and results in child neglect, abuse, or abandonment.

Conversely, those who oppose sex preselection argue that it does more harm than good because it reinforces the low status of females. Although in rare circumstances families use medical technologies to ensure that their babies are female (such as families with a history of hemophilia, a disease that affects males but not females), most often in the United States and almost always in the developing nations, sex preselection means selecting males. In the United States, both women and men prefer boys as their first child and prefer two boys and a girl to two girls and a boy; families are most likely to have three children if their first two are female (Rothman, 1986).

When families consistently select male fetuses over female fetuses, their actions proclaim male babies inherently preferable. Moreover, when health care workers help families to select male babies, the workers in essence validate this preference. Finally, when health care workers assist in sex preselection—whether helping families to select males or females—they reinforce the idea that males and females are inherently different. After all, if male and female personalities, interests, and aptitudes were more similar than different, why would families need to choose one over the other?

In sum, to assess the ethics of sex preselection, we need to weigh the potential benefits and costs for families and for society as a whole.

---

are carrying a female fetus (Smyke, 1991:39). Box 4.2 discusses some of the ethical quandaries posed by using abortion for sex selection.

In sum, research suggests that if women's social status were higher, they would enter their childbearing years with healthier bodies, wait longer to begin having babies, wait longer between babies, and have fewer babies in total, with each of these factors lowering the infant mortality rate. For all these reasons, many researchers and public health workers have suggested that the most effective way to reduce infant mortality is to improve the status of women, thereby increasing their power to make decisions for themselves. This in fact seems to explain at least partly why infant mortality is so

---

Box 4.3   **Making a Difference: Freedom from Hunger**

Freedom from Hunger (FFH) began in the 1940s as a traditional food aid program, providing food relief to the hungry in the developing nations of Africa, Asia, and Latin America. By the 1980s, however, the organization had concluded that the only way to reduce hunger in the long run was to help poor women in the developing nations to become economically self-sufficient. As a result, in 1989 FFH committed all its resources to providing **microcredit** to women in developing nation through its Credit with Education program. Microcredit refers to the practice of loaning very small, short-term loans (typically between $10 and $300 for four to six months) to poor women who lack meaningful assets or affordable access to cash credit. FFH distributes these loans through community-based credit associations, each made of 20 to 30 women neighbors. The association is then responsible for allocating credit to individual women and collecting debt payments from them. To date,

the FFH credit associations have had exceptional success in repaying their loans and the system has become largely self-sustaining.

Although the amounts given in these loans might seem too small to be meaningful, they can make an enormous difference in women's lives. Women who receive micro-credit loans no longer have to purchase supplies or raw materials from local vendors on credit at usurious rates and, instead, can start investing in their own businesses, such as raising chickens or making clothing. In addition, FFH links microcredit to health and nutrition education, using its credit associations to provide basic information about such topics as breastfeeding and treating infant diarrhea. As a result, FFH provides women and, in the long run, their children and families, with both information on how to improve their health and the resources necessary to do so. As of March 31, 1999, more than 112,000 women had participated in FFH's Credit with Education program.

---

much lower in Costa Rica, China, and Zimbabwe than in other countries at similar levels of development. Box 4.3 describes the actions of one nonprofit agency that is working to improve women's status and thus health status overall in the developing nations.

The Role of Infant Formula Manufacturers

A final cause of infant mortality in the developing nations is the use of artificial foods instead of or in addition to breastfeeding. The preceding chapter explained the basic biological benefits of breastfeeding for both infants and mothers. The benefits are even greater in the developing nations, where babies who receive artificial foods (whether infant formula, juice, water, or any other substances) are twenty-five times more likely than breastfed babies to die from infections (*Lancet,* 1990). The World Health Organization (1993b) estimates that about one and a half million babies die unnecessarily each year because they are fed artificial foods.

Several factors contribute to the especially high rates of death and disease among artificially fed infants in the developing nations. First, in addition to the inherent limitations of artificial foods, the process of bottle-feeding itself can expose infants to tremendous risks. Infant formula is typically sold as a powder that must be mixed with water and then transferred to a bottle before it can be used. In most developing nations, this water contains dangerous infectious organisms. Those organisms can be killed if the water, bottle, and nipple are boiled. However, families do not necessarily understand how or why they should do so. Moreover, throughout the developing nations, women and children already may spend hours each day getting water and firewood and often lack the time and energy to get the extra supplies needed for proper sterilization.

Second, artificial foods cost far more than breastmilk (which is not actually free, as it reduces mothers' nutritional stores and can prevent their return to paid employment). Consequently, families often stretch the formula by diluting it with water. Babies fed this diluted formula in essence starve to death even while filling their stomachs.

Finally, by altering the hormonal levels in a woman's body, breastfeeding serves as a moderately effective contraceptive. Consequently, breastfeeding helps women to space out pregnancies and gives each baby a better chance for survival. For all these reasons, WHO (1995) recommends that children throughout the world, in both industrialized and developing nations, receive only breastmilk during the first four to six months of life and a combination of breastmilk and other foods until at least age two.

Given all the benefits of breastfeeding, why do so many women in developing nations instead offer their babies artificial foods? Part of the answer lies in traditional cultural beliefs. For example, some cultures believe that beyond a given age children require certain traditional foods for health. Other cultures indirectly encourage artificial feeding by forbidding sexual relations as long as a woman breastfeeds (Dettwyler, 1995). Second, some women choose to use artificial foods because they cannot mesh breastfeeding with their work, whether paid or unpaid. Finally, campaigns by multinational food corporations based in the industrialized nations have convinced many women in the developing nations that infant formula is superior to breastfeeding.

To create a market in the developing nations, corporations have provided free or subsidized formula to patients in maternity hospitals (Gerber, 1990; *Lancet*, 1990). If these women use the formula instead of breastfeeding while in the hospital, they may find it physiologically impossible to switch to breastfeeding later. Corporations also have mounted massive advertising campaigns throughout the developing nations to convince women that bottle-feeding produces healthier babies and even lightens babies' skin—a status symbol in many developing nations. One particularly

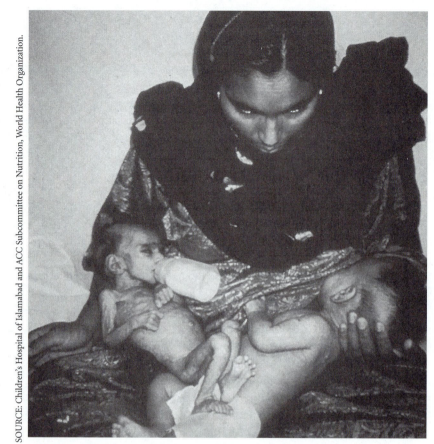

SOURCE: Children's Hospital of Islamabad and ACC Subcommittee on Nutrition, World Health Organization.

This woman knew breast milk was healthier, but, fearing she would not have enough breast milk, she breast-fed only her son and bottle-fed his twin sister.

pernicious strategy is to dress saleswomen as nurses and send them to villages and maternity hospitals to encourage women to bottle-feed.

During the 1970s, recognition of bottle-feeding's role in infant mortality led to the rise of an international, consumer-led campaign, based in the United States and Europe, against the multinational corporations that produce infant formula (Gerber, 1990). The campaign focused especially on Nestlé, the most aggressive marketer of infant formula in the developing nations. The campaign's main tools were an international awareness campaign and a consumer boycott of infant formula and other products made by Nestlé.

In 1981, and partly in response to this campaign, WHO's Assembly adopted an International Code of Marketing of Breastmilk Substitutes aimed at sharply limiting the promotion and sale of formula in the developing nations. (The United States was the sole nation to vote against the Code, finally ending its opposition only in 1996.) Among its provisions, the

Code calls for manufacturers to refrain from advertising infant formula, providing free samples to mothers, promoting infant formula through health care facilities, hiring nurses or women dressed as nurses to promote infant formula, providing gifts or personal samples to health care workers, and providing free or low-cost supplies to hospitals.

By 1984, all the major formula producers had agreed to accept WHO's Code, bringing an end to the boycott. Within the developing nations, however, the mistaken notion that bottle-feeding was more "modern" and healthier already had taken root. Moreover, it soon became obvious that the manufacturers had reneged on their promise to abide by the Code. Consequently, in 1988, the International Baby Food Action Network and Action for Corporate Accountability, its U.S. affiliate, began a new boycott. Partly as a result of this consumer pressure, billboards and other advertisements for infant formula have become less common and health care workers in developing nations now more often actively support women's efforts to breastfeed. Manufacturers continue to break the code, however, although they now focus more on encouraging mothers to stop breastfeeding early rather than encouraging mothers never to begin (Wise, 1998). A recent study conducted through random sampling in four developing nations found that 10 percent of all interviewed mothers with children under six months old and 25 percent of all surveyed health care facilities had received free samples of bottle-feeding supplies from manufacturers, in direct violation of the WHO Code (Taylor, 1998). Moreover, this survey probably underestimates the problem as it only studied developing nations known for reasonably good compliance with the Code (Costello and Sachdev, 1998).

Maternal Mortality

Although maternal mortality is now rare in the industrialized nations, it remains the primary cause of death among women of reproductive age in the developing nations. For example, the lifetime risk of dying from childbirth complications is one in 1400 in Europe, one in 65 in Asia, and one in 16 in Africa (World Health Organization, 1998b).

How can we account for the tremendous toll maternal mortality takes in the developing nations? Patricia Smyke, writing for the United Nations (1991:61–62) explains:

> If you ask, "Why do these women die?" the technical response is: "The main causes of maternal death are hemorrhage, sepsis (infection), toxemia, obstructed labor and the complications of abortion." But looking beneath those immediate causes, one must ask why they occurred or why they were fatal. The answer to that is: lack of prenatal care; lack of trained personnel, equipment, blood or transport at the moment the obstetrical emergency arose, or earlier, when it might have been foreseen and avoided; lack of family planning to help women

avoid unwanted pregnancies, too many or too closely spaced births, or giving birth when they were too young or too old; pre-existing conditions like malaria, anemia, fatigue and malnutrition that predispose to obstetrical complications; problems arising from female circumcision. From that list of intermediary causes one must go deeper still to identify the cultural and socioeconomic factors that put young girls, almost from birth, on this road to maternal death: ... low status of women and discrimination against them; poverty; lack of education; local customs; and government policies that give low priority to the needs of women.

Like infant mortality, maternal mortality occurs most often among women who suffer from malnutrition or from illness (most commonly, malaria). Hemorrhage more often occurs during abortion or childbirth in women who develop anemia because of malaria or inadequate diets. Maternal mortality is also most common among women who give birth before age 20 or after age 35. In Bangladesh, for example, where half of all women marry by age 15, maternal mortality is five times higher among those ages 10 to 14 than among those ages 20 to 24 (Basch, 1990). Maternal mortality also rises with each birth after the third. Finally, maternal mortality is more common among women who give birth in unsanitary conditions and

---

Box 4.4   ***Female Circumcision***

According to the World Health Organization, approximately 130 million girls and women across Africa as well as in Malaysia, Indonesia, Yemen, and elsewhere have experienced the ordeal of female circumcision, and about two million additional girls are circumcised each year (World Health Organization, 1997b). Female circumcision is a brutal and sometimes fatal procedure, in no way analogous to male circumcision. In clitoridectomy, the first and least common of the three types of female circumcision, either the tip of the clitoris or the skin over the clitoris is cut off. In excision, which comprises about 80 percent of cases, the entire clitoris and labia minora are removed but the vulva is left untouched. In infibulation, which comprises about 15 percent of cases, the clitoris, labia minora, and parts of the labia majora are removed and the sides of the vulva are stitched together, leaving only a small opening for urine and menstrual fluid to escape. Most commonly, a midwife or other lay healer performs the circumcision using a razor blade, knife, or piece of broken glass.

Those who support circumcision believe it makes women more docile and reduces their sex drive, making them better wives and less likely to disgrace their families by engaging in premarital or extramarital sexual relationships. In addition, supporters of circumcision believe that circumcised women are cleaner, healthier, more fertile, and prettier. In countries where circumcision is the norm, these beliefs leave uncircumcised women with few marriage prospects and pressure parents to

among those who have been circumcised; Box 4.4 provides further details on this dangerous practice.

Another cause of maternal mortality throughout the developing nations is unsafe abortion (Dixon-Mueller, 1990; World Health Organization, Division of Reproductive Health, 1998). Abortion is a technically simple procedure, far safer than childbirth when performed by trained professionals working in sterile conditions with proper tools (World Health Organization, Division of Reproductive Health, 1998). However, most developing nations have criminalized or legally restricted abortion because of cultural traditions, religious beliefs, a desire by political elites to increase population, or financial and political pressures from the United States—which since 1973 has withheld family planning funding from any agencies that offer abortions (Dixon-Mueller, 1990). In other countries, abortion is legal, but many women cannot afford to obtain abortions from a trained health care worker. Consequently, almost 20 million women yearly—most of whom are married with several children—receive unsafe abortions (World Health Organization, Division of Reproductive Health, 1998). Unsafe abortion accounts for about 13 percent of maternal mortality in the developing nations, most commonly because of infections caused by unsterile instruments, hemorrhage when those instruments pierce the uterus, or poisoning

have their daughters circumcised even if the parents disapprove of the practice.

Circumcision substantially impairs the health of young girls and women. Given the unsanitary conditions in which it is usually performed, the operation can cause life-threatening shock, hemorrhage, infections, or tetanus. Those who survive often experience pain during intercourse and chronic urinary, vaginal, or pelvic infections, sometimes resulting in infertility. If they do become pregnant, scar tissue and the narrowed vaginal opening can make it difficult for a baby to emerge, causing women to die from hemorrhage and babies to die from brain damage. These health problems have convinced some doctors and nurses to perform circumcisions to protect girls who would otherwise be circumcised under more dangerous conditions.

Currently, most nations where circumcision occurs officially oppose female genital mutilation, and Senegal and Egypt have outlawed the practice. These actions, however, have had little impact on its prevalence (World Health Organization, 1997b). Western opposition has proven similarly ineffective, as it is difficult, if not impossible, for Westerners to condemn circumcision without appearing to condemn the cultures in which it is embedded. As a result, the most effective opponents of female circumcision are those who come from within these cultures. With this in mind, feminists and health care workers native to these cultures have formed alliances aimed at stopping this practice, such as the Inter-African Committee on Traditional Practices Affecting the Health of Women and Children.

when women try to abort themselves by swallowing toxic chemicals. Unsafe abortion can also cause illness or permanent disability. As a result, hospitals in the developing nations spend as much as 50 percent of their resources on treating the aftereffects of unsafe abortion (World Health Organization, Division of Reproductive Health, 1998).

Respiratory Diseases

Finally, respiratory disases, such as emphysema, are also major killers in the developing nations. The percentage of deaths caused by respiratory disease is almost twice as high in developing nations as in industrialized nations.

As with all disease in the developing nations, poverty and malnutrition increase individual susceptibility to illness. In addition, long periods spent cooking over open fires in closed rooms expose millions of women to cancer-causing toxins that are equivalent to smoking several packs of cigarettes daily. Meanwhile, those who live in cities like Caracas, Mexico City, or Calcutta risk their health daily because of pollution from automobiles and industries. Unfortunately, in some developing nations, government officials lack the political or economic power to control polluting industries and in other nations officials are unwilling to do so because they benefit economically from these industries. Equally important, officials in developing nations sometimes believe that pollution and the attendant morbidity and mortality are short-term costs they must pay to industrialize and to improve their nation's health in the long run.

To these factors must be added the growing role of tobacco, which, in the developing nations as in the industrialized nations, is a major cause of chronic obstructive pulmonary disease. In addition, tobacco serves as a catalyst that increases the risks of other diseases (World Health Organization, 1998c). For example, compared with nonsmokers, smokers who have parasitic bladder infections are more likely to get bladder cancer and smokers who work in uranium mines are more likely to develop leukemia. In addition, tobacco use promotes disease by taking a large bite out of small incomes. Smokers spend as much as 15 percent of family income in Brazil and as much as 10 percent in India on tobacco; in Egypt wives name their husband's smoking as the main reason their children go hungry (Nichter and Cartwright, 1991). WHO (1998c) estimates that by 2020, tobacco use will cause 11 percent of all deaths in developing nations (and 18 percent of deaths in industrialized nations).

Tobacco use has grown steadily in the developing nations since 1964, when the U.S. Surgeon General declared tobacco a cause of lung cancer. In response, sales of cigarettes plummeted in North America and tobacco manufacturers turned to the developing nations for new markets (Hammond, 1998; Nichter and Cartwright, 1991). This tactic proved highly successful: today 73 percent of all tobacco users live in the developing nations.

Tobacco manufacturers (most of which are based in the United States) have relied on three strategies to increase their sales in the developing nations: heavy advertising, the distribution of tobacco as "food aid," and trade sanctions (Hammond, 1998; Nichter and Cartwright, 1991). Manufacturers have devoted enormous sums to advertising tobacco in the developing nations (Hammond, 1998). In countries where direct advertising of tobacco on television or radio is restricted, manufacturers have marketed their products by sponsoring cultural and athletic events, especially those oriented toward youths. For example, the Chinese national soccer league is now named the "Marlboro Professional Soccer League."

U.S. manufacturers also fostered a market for their products in the developing nations by lobbying successfully for inclusion of cigarettes in U.S. food aid projects from the 1940s through the 1980s. More recently, responding to efforts by rapidly developing nations like Thailand to restrict advertising or sales of tobacco, manufacturers have convinced the U.S. government to threaten to take sanctions against these nations under the federal 1974 Trade Act and to keep them out of the World Trade Organization (Hammond, 1998). So far, these threats have proved successful.

CONCLUSIONS

In this chapter, I have described how poverty and inequality—rather than over-population, tropical environments, lack of natural resources, or other biological factors—underlie the high rates of illness and death found in the developing nations. Consequently, reducing poverty and inequality in the developing nations should raise them to the health levels found in the industrialized nations. Conversely, the situation in the former Soviet Union demonstrates how an industrialized nation can slide toward health levels lower than that found in some developing nations (Feshbach, 1999; Feshbach and Friendly, 1992).

With the political and economic upheaval of the last fifteen years, poverty has spread across the former Soviet Union and living conditions have deteriorated. The decline in income in these countries during the early 1990s exceeded that in the United States during the Great Depression and seems to have become permanent (Little, 1998). Increasingly across this vast territory, people live in inadequately heated, overcrowded, and ramshackle housing. Almost three-fourths of the water supply is polluted, with one-fourth completely untreated. At the same time, the growing realization that the government can no longer guarantee citizens a minimum standard of living has demoralized people, encouraging many to find solace in alcohol. Partly as a result, more than three times as many Russians die each year of acute alcohol poisoning as die from poisoning of any sort in the United States (Wines, 1999).

To these problems must be added those caused by environmental degradation. In past decades, the Soviet Union expanded its economic base as rapidly as possible, with little regard for the human or environmental toll. The Soviet government rarely established and almost never enforced regulations designed to protect the environment from industrial pollution. As a result, industries wreaked far greater environmental havoc in the Soviet Union than in other industrialized nations, polluting farmlands and waterways beyond repair and leaving radioactivity, lead, and other dangerous toxins behind. Similarly, the emphasis on increasing agricultural yields as quickly as possible led to overplowing, which has caused perhaps permanent soil erosion, and to overuse of herbicides, chemical fertilizers, and pesticides, which have poisoned the water, the land, and the food grown on it.

This environmental damage and downturn in living conditions is now taking its toll in human lives. As summarized by *New York Times* reporter Michael Specter (1997:A1):

> There is almost no current demographic fact about Russia that would fail to shock: Per capita alcohol consumption is the highest in the world, nearly double the danger level drawn by the World Health Organization; a wider gap has developed in life expectancy between men (59) and women (73) than in any other country; the mortality rate of 15.1 deaths per 1,000 people puts Russia ahead of only Afghanistan and Cambodia among the countries of Europe, Asia and America (the rate for the United States is 8.8); the death rate among working-age Russians today is higher than a century ago.

Although government officials claim that infant mortality is now 17.5 per 1,000, informed observers believe that it is almost twice that, putting Russia on a par with China (Feshbach and Friendly, 1992). Infant mortality is twice as high in agricultural areas where pesticides were used heavily compared with less-poisoned nearby regions. Meanwhile, **incidence** rates for numerous infectious diseases have increased. For example, in 1998, the former Soviet Union experienced the first large diphtheria epidemic in an industrialized nation in thirty years (Vitek and Wharton, 1998), while tuberculosis—which has a mortality rate 34 times higher in Russia than in the United States—is quickly becoming a more common cause of death than cancer and heart disease combined (Feshbach, 1999). In addition, the collapse of the social structure and economy has contributed to a proliferation of sexually transmitted diseases. The incidence of syphilis among girls under age 15 is thirty times greater now than it was only five years ago, while the incidence of HIV is skyrocketing among all groups. Only 54 percent of 16-year-old Russian boys are expected to survive to age 60, compared with 83 percent of American boys (Feshbach, 1999).

In sum, no natural progression leads countries toward an increasingly healthy citizenry. Rather, as the political and economic fortunes of a country shift, and as the natural environment improves or declines, so too will the health of its population. Only by continued commitment to eliminating poverty and inequality and to protecting the environment can a nation guarantee that it will keep whatever health gains it has achieved.

SUGGESTED READINGS

Desowitz, Robert S. 1991. *The Malaria Capers.* New York: Norton. An ironic, first-hand account of why efforts to eliminate malaria in the developing nations have failed.

Farmer, Paul. 1999. *Infections and Inequalities: The Modern Plagues.* Berkeley: University of California Press. A brilliant analysis of the link between disease and social inequality, written by a physician/anthropologist who for many years has divided his time between a clinic in inner-city Boston and one in rural Haiti.

Lappé, Frances Moore, Joseph Collins, and Peter Rosset. 1998. *World Hunger: Twelve Myths.* New York: Grove Press. Excellent summary of the issues. The first author is one of the most important figures in this field.

Smyke, Patricia. 1991. *Women and Health.* London: Zed Books. A well-written overview, produced for the United Nations, of women's health issues in the developing nations.

GETTING INVOLVED

Action for Corporate Accountability. 910 17th Street, NW Suite 413, Washington, DC 20006. www.action4corpacct.org. Educational, lobbying, and activist organization concerned with the sale of infant formula in both the developing and industrialized nations.

Amnesty International USA. 322 8th Avenue, New York, NY 10001. (212) 807-8400. www.amnesty-usa.org. Powerful international organization working to end torture and the death penalty and to obtain fair trials and freedom for persons jailed solely because of their beliefs, color, sex, ethnicity, language, or religion.

Freedom from Hunger. 1644 DaVinci Court, Davis, CA 95617. www.freefromhunger.org. Provides small loans to women in developing nations to enable them to become economically self-sufficient and, in the long run, to reduce the chances that they or their families will experience hunger.

The Institute for Food and Development Policy. 398 60th Street, Oakland, CA 94618. (510) 654-4400. www.foodfirst.org/index.html. Popularly known as Food First, this nonprofit organization was founded in 1975 by Frances Moore Lappé and Joseph Collins to promote awareness of the social causes of hunger and poverty around the world.

REVIEW QUESTIONS

How do social conditions limit the effectiveness of modern medicine in developing nations?

How do social factors contribute to illness in developing nations?

How do international politics and multinational corporations contribute to illness in developing nations?

How do the role and status of women contribute to illness in developing nations?

INTERNET EXERCISES

One way to identify the range of opinions on a given topic is to browse listserves. Listserves are online discussion groups in which any eligible individual can post a question or an answer to someone else's question. (Some discussion groups are open to everyone, but some are only open to certain groups of individuals, such as members of an organization.) For example, there are a wide variety of opinions regarding female genital mutilation, and regarding what, if anything, westerners should do about it. To identify the range of opinion on this topic, search for posts on this topic using deja.com (www.deja.com). Deja.com is divided into several categories: make sure you are searching discussions or discussion groups, not ratings or communities. (You might have to use the "power search" feature.) Identify and summarize three different views.

Obtain current information from the nonprofit Population Reference Bureau's Web site (www.prb.org) regarding life expectancy at birth by country. Compare that information with the information contained in your textbook. Are there any countries in which life expectancy has changed markedly since this textbook was printed? If so, what might explain those changes?

The Meaning and Experience of Illness

—

CHAPTER FIVE
The Social Meanings of Illness

CHAPTER SIX
The Experience of Chronic Illness and Disability

CHAPTER SEVEN
The Sociology of Mental Illness

—

Our common-sense understandings of the world tell us that illness is a purely biological condition, definable by objectively measured biological traits. As I will show in Part Two, however, definitions of illness vary considerably over time and space and across social groups. In Chapter 5, I explore the social meanings of illness, and trace how ideas about the nature and causes of illness have changed historically, from biblical explanations that attributed illness to punishment for sin to modern New Age explanations that attribute illness to lack of self-love. I also describe how defining something as an illness can act as a form of social control.

Whereas Chapter 5 discusses the meaning of illness in the abstract, Chapter 6 looks at the consequences of chronic illness and disability for individuals. I begin with a discussion of how Western society historically has treated those who have chronic illnesses and disabilities. I then describe the modern experience of illness from the response to initial symptoms to the search for mainstream or alternative therapies to the process of coming to terms with a changed body and self-image.

In Chapter 7, I raise parallel questions regarding mental illness. This chapter explores what we mean when we say something is a mental illness. I then describe how and why mental illness is distributed among social groups, how Western society historically has treated persons with mental illnesses, and how individuals experience mental illness, from initial symptoms, to treatment, to social status following treatment.

The AIDS quilt has helpted change Americans' image of AIDS from a "deserved" illness to a shared national tragedy.

CHAPTER 5

The Social Meanings of Illness

—

All Marco Oriti has ever wanted, ever imagined, is to be taller. At his fifth birthday party at a McDonald's in Los Angeles, he became sullen and withdrawn because he had not suddenly grown as big as his friends who were already 5: in his simple child's calculus, age equaled height, and Marco had awakened that morning still small. In the six years since then, he has grown, but slowly, achingly, unlike other children.

"Everybody at school calls me shrimp and stuff like that," he says. "They think they're so rad. I feel like a loser. I feel like I'm nothing." At age 11, Marco stands 4 feet 1 inch—4 inches below average—and weighs 49 pounds. And he dreams, as all aggrieved kids do, of a sudden, miraculous turnaround: "One day I want to, like, surprise them. Just come in and be taller than them."

Marco, a serious student and standout soccer player, more than imagines redress. Every night but Sunday, after a dinner he seldom has any appetite for, his mother injects him with a hormone known to stimulate bone growth. The drug, a synthetic form of naturally occurring human growth hormone (HGH) produced by the pituitary, has been credited with adding up to 18 inches to the predicted adult height of children who produce insufficient quantities of the hormone on their own—pituitary dwarfs. But there is no clinical proof that it works for children like Marco, with no such deficiency. Marco's rate of growth has improved since he began taking the drug, but his doctor has no way of knowing if his adult height will be affected. Without HGH, Marco's predicted height was 5 feet 4 inches, about the same as the Nobel Prize-winning economist Milton Friedman and this year's Masters golf champion, Ian Woosnam, and an inch taller than the basketball guard Muggsy Bogues of the Charlotte Hornets. Marco has been taking the shots for six years, at a cost to his family and their insurance company of more than $15,000 a year. . . .

A Cleveland Browns cap splays Marco Oriti's ears and shadows his sparrowish face. Like many boys his age, Marco imagines himself some-day in the NFL. He also says he'd like to be a jockey—making a painful incongruity that mirrors the wild uncertainty over his eventual size. But he is unequivocal about his shots, which his mother rotates nightly be-tween his thighs and upper arms. "I hate them," he says.

He hates being short far more. Concord, the small Northern Califor-nia city where the Oriti family now lives, is a high-achievement commu-nity where competition begins early. So Luisa Oriti and her husband, Anthony, a bank vice president, rationalize the harshness of his treat-ment. "You want to give your child that edge no matter what," she says, "I think you'd do just about anything" (Werth, 1991).

Does Marco have an illness? According to his doctors, who have recom-mended that he take an extremely expensive, essentially experimental, and potentially dangerous drug, it would seem that he does. To most people, however, Marco simply seems short.

In the first part of this chapter, I look at what we mean by illness, and contrast the sociological model of illness with the medical model. I also ex-plore how people explain why illness occurs and some of the consequences of these explanations. In the second part of this chapter, I look at how med-icine can act as an institution of social control, highlighting the process through which behaviors or conditions become defined as illnesses and the consequences of these definitions.

MODELS OF ILLNESS

The Sociological Model of Illness

What makes something an illness? As Marco's story suggests, the answer is far from obvious. Most Americans are fairly confident that someone who has a cold or cancer is ill. But what about someone who has trouble learn-ing, drinks to excess, or enjoys hitting others? What about a 60-year-old woman who no longer menstruates or whose bones have become more brittle with age? Or a 65-year-old man who, like many his age, is balding or has an enlarged prostate gland and urinary problems?

As these examples suggest, defining something as an illness reflects more than simply the objective nature of the condition or behavior. Rather, it re-flects a *subjective* judgment regarding its meaning (Weitz, 1991). For exam-ple, defining menopause, as many American doctors now do, as a "hormonal deficiency disease" means defining it as abnormal and undesirable, rather than as the natural result of aging. Yet many women experience menopause not as a sign of a declining body but as freedom from the constraints of re-production (Martin, 1987). Moreover, in some parts of the world, surviving to menopause signals not decline but rather maturity and wisdom and brings a higher social status. In the same manner, when we define cancer, polio, or diabetes as illnesses, we judge the bodily changes these conditions produce as abnormal and undesirable, rather than as normal variations in functioning, abilities, and life expectancies. (Conversely, when we define a condition as healthy, we judge it normal and desirable.)

The subjective and even political elements involved in defining illness are sometimes explicit. For example, in 1989, the American Society for Plastic and Reconstructive Surgery, the foremost professional organization for plastic surgeons, petitioned the federal government to loosen its restric-tions on the use of breast implants, on the grounds that implants are med-ically necessary to cure the "disease" of small breasts. In their words: "these deformities [i.e., small breasts] are really a disease which in most patients results in feelings of inadequacy, lack of self-confidence, distortion of body image and a total lack of well-being due to a lack of self-perceived feminin-ity" (American Society of Plastic and Reconstructive Surgeons, 1989:4–5). Conversely, in Brazil, where large breasts are considered a sign of African heritage and, consequently, lower-class status, breast *reduction* is the most popular form of cosmetic surgery and is a popular sixteenth birthday gift (Gilman, 1999). These two opposite examples strongly suggest that defin-ing breast size as a disease requiring medical intervention reflects cultural values and economic interests at least as much as biological reality.

Illness, then, is a **social construction,** something that exists in the world not as an objective condition but *because we have defined it as existing.* This does not mean that the virus that causes measles does not exist or that it does not cause a fever and rash. It does mean, though, that when we talk about measles as an illness, we have organized our ideas about that virus, fever, and rash in only one of the many possible ways. In another place or time, people might conceptualize those same conditions as manifestations of witchcraft, as a healthy response to the presence of microbes, or under a different illness rubric. In sum, "illness," like "crime" or "sin," refers to biological, psychologi-cal, or social conditions subjectively defined as undesirable by those within a given culture who have the power to create such definitions.

If labeling a condition an illness reflects the perceived undesirability of that *condition,* then labeling someone ill suggests, if unintentionally, that there is something undesirable about that *person.* By definition, an ill per-son is one whose actions, ability, or appearance do not meet social **norms,** or expectations within a given culture regarding proper behavior or ap-pearance. Such a person will typically be considered less whole and less so-cially worthy than those deemed healthy. Illness, then, like virginity or laziness, is a **moral status**—a social condition that we believe indicates the goodness or badness, worthiness or unworthiness, of a person.

From a sociological standpoint, illness is not only a moral status but (again, like crime or sin) a form of **deviance** (Parsons, 1951). To sociologists, labeling something deviant does not necessarily mean it is immoral. Rather, deviance refers to behaviors or conditions that socially powerful persons within a given culture *perceive,* whether accurately or inaccurately, as immoral or as violating social norms.

We can tell whether behavior violates norms (and, therefore, whether it is deviant) by seeing if it results in **negative social sanctions.** This term refers to any punishment, from ridicule to execution. (Conversely, the term **positive social sanctions** refers to rewards, ranging from token gifts to knighthood.) These social sanctions are enforced by **social control agents** including parents, police, teachers, peers, and doctors. In the next section, I look at how societies explain illness and describe some of the negative social sanctions imposed against those who are ill.

Explaining Illness

The moral status of illness is reinforced by the tendency to explain illness by blaming it on those who are ill (Brandt and Rozin, 1997; Weitz, 1991). Be-

---

| Box 5.1 | *The Social Construction of HIV Disease* |
| | Karl Bryant |

As is true for all illnesses, our ideas about HIV disease are a social construction, reflecting the particular history of HIV disease as well as ideas about illness in general.

Medical researchers and, subsequently, the media first identified HIV disease as an illness affecting "fast-lane" gay men who lived in major cities, had many sexual partners, and sometimes used drugs to heighten their sexual pleasure. As a result, from the start the public linked HIV disease to homosexual activities and recreational sex—both stigmatized behaviors (Weitz, 1991). When researchers later linked HIV disease to illegal intravenous drug users (many of them African-American or Hispanic) and Haitians (many of them illegal immigrants and all people of color), the stigma of HIV disease increased even further.

The apparent links, emphasized in media portrayals, between HIV disease and homosexuality, Haitians, and intravenous drug users encouraged much of the public to see HIV disease as "someone else's" disease. This tendency was reinforced by public health officials' emphasis on risk groups rather than on risk behavior. In public announcements, prevention campaigns, and media reports, officials emphasized that certain groups were at high risk of infection, rather than emphasizing that anyone who engaged in certain behaviors, regardless of what groups they belonged to, was at risk. As a result, high school athletes who share needles to inject steroids, for example, do not typically think they are at risk of HIV infection because they do not think of themselves as belonging to the risk group of intravenous drug

cause people consider illness undesirable and because it can strike anyone at any time, people often react to illness with fear and confusion. As a result, people typically need to explain why illness occurs and, especially, why illness strikes some persons and not others. Having an explanation helps to relieve anxiety by making the world seem less capricious and frightening.

Most often—and as the example of **HIV disease** has vividly demonstrated—these explanations relieve anxiety by defining illness as a deserved punishment, blaming the individual for his or her illness (Box 5.1). Whether accurate or inaccurate, these explanations provide psychological reassurance by reinforcing people's belief in a "just world," in which punishment falls only on the guilty (Burger, 1981; Lerner and Miller, 1978; Meyerowitz, Williams, and Gessner, 1987). Even those who are themselves ill may prefer to blame themselves rather than to believe that random, impersonal events control their fates.

Prescientific Explanations

Throughout history and across cultures, people typically have explained illness by blaming those who are ill. According to George Foster (1976), all traditional theories of illness causation around the world divide into only

users, even though they have the same basic risks as anyone who shares needles.

Ironically, while defining HIV disease as "someone else's disease," many people have exhibited extreme and unwarranted fears of contracting the disease, resulting in periodic calls for quarantining or incarcerating HIV-infected persons, banishing HIV-infected children from public schools, and so on. Yet medical researchers quickly had identified the few means by which HIV could be transmitted and had demonstrated that no one need fear infection if he or she follows a few simple precautions. The continued fears regarding persons with HIV disease, therefore, seem to reflect something other than straightforward medical concerns. Most basically, these fears seem to reflect Americans' horror at discovering that medicine has not conquered infectious disease and

a resulting need to formulate strategies—no matter how irrational—to protect themselves and their families from those diseases.

The excessive fears of HIV disease also reflect the strong link in the public imagination between HIV disease and death. Until the development of protease inhibitors in the 1990s, the media typically portrayed HIV disease as quickly fatal, adding to its stigma. Yet, while full-blown AIDS is terminal and often horrifying, most people live for more than a decade following infection with HIV and remain healthy for more than half of that time.

These social constructions have affected the lives of individuals with HIV disease as well as society as a whole. Only by changing these social constructions can we improve the lives of those who live with this illness and develop more effective strategies for combating the illness.

two, somewhat overlapping, categories: "personalistic" and "naturalistic." **Personalistic theories,** the more common type (Murdock, 1980), hold that illness occurs when a god, witch, spirit, or other supernatural power lashes out at an individual, either deservedly or maliciously. **Naturalistic theories** assert that illness occurs when heat, cold, wind, damp, or other natural forces upset the body's equilibrium. Both personalistic and naturalistic theories blame ill persons for causing their illness, whether by displeasing supernatural beings or by exposing themselves to harmful natural elements. And both define ill persons as less morally worthy than others, whether as sinners or as fools.

Personalistic theories have played an especially important role in the Western world, which in the past often equated illness with divine punishment for sin (Murdock, 1980:42–52). The history of leprosy provides an obvious example. Both the Jewish and Christian Bibles consistently describe leprosy as punishment for an individual's sin. Biblical explanations for leprosy, coupled perhaps with some awareness that leprosy was contagious, led early Western societies to isolate affected individuals, an approach that continued essentially unchanged for centuries. Throughout the Middle Ages and until the Reformation, Christian society required anyone diagnosed with leprosy to participate in a special mass for the dead known as the "lepers' Mass." Following the mass, a priest would shovel dirt on the individual's feet to symbolize his or her civil and religious death. From then on, the individual was legally prohibited from entering public gathering places, washing in springs or streams, drinking from another's cup, wearing anything other than the special "leper's dress," touching anything before buying it, talking to anyone without first moving downwind, and so on (Richards, 1977:123–124). This social banishment continued even after death: like those who committed suicide or other mortal sins, persons with leprosy could not be buried in church graveyards.

The Rise of Scientific Explanations

By the early nineteenth century, prescientific ideas about illness had begun to erode, as the idea grew, especially among the elite, that scientific principles controlled the natural order. According to the new scientific thinking, illness occurred when biological forces combined with personal susceptibility. Doctors (still lacking a concept of germs) argued that illness occurred when persons whose constitutions were naturally weak or had been weakened by unhealthy behaviors came in contact with dangerous "**miasma,**" or air "corrupted" by foul odors and fumes. According to this theory, therefore, individuals became ill because of unhealthy rather than immoral behavior.

As the history of cholera shows, however, these new ideas still allowed the healthy to blame the ill for their illnesses. Cholera first appeared in the Western world in about 1830, killing its victims suddenly and horrifyingly, through overwhelming dehydration brought on by uncontrollable diarrhea and vomiting. Cholera is caused by waterborne bacteria, generally transmitted when human wastes contaminate food or drinking water. Because of the link to sanitation, cholera most often strikes poor persons who lack clean water and are weakened by insufficient food, clothing, or shelter.

To explain why cholera had struck, and why it struck the poor especially hard, early nineteenth century doctors asserted that cholera could only attack individuals who had weakened their bodies through improper living (Risse, 1988; Rosenberg, 1987). According to this theory, the poor caused their own illnesses, first by lacking the initiative required to escape poverty and then by choosing to eat an unhealthy diet, live in dirty conditions, or drink too much alcohol. Thus, for example, the New York City Medical Council could conclude in 1832 that "the disease in the city is confined to the imprudent, the intemperate, and to those who injure themselves by taking improper medicines" (Risse, 1988:45). Conversely, doctors assumed that wealthy persons would become ill only through gluttony, greed, or "innocently" inhaling some particularly noxious air.

Using this theory, doctors, foreshadowing what would happen with HIV disease, divided patients into the "guilty" (the overwhelming majority), the "innocent," and the "suspect," and hospitals provided or refused care accordingly (Risse, 1988; Rosenberg, 1987). This theory of illness allowed the upper classes to adopt the new, scientific explanations for illness while retaining older, moralistic assumptions about ill people and avoiding any sense of responsibility for aiding the poor or the ill. In sum, instead of believing that immorality directly *caused* illness, people now believed that immorality *left one susceptible* to illness.

Modern Explanations for Illness

Despite the tremendous growth in medical knowledge about illness during the last century, popular explanations for illness have remained remarkably stable. Theories connecting illness to sin continue to appear, as do theories that conceptualize illness as a direct consequence of poorly chosen and hence irresponsible (although not necessarily sinful) behavior (Blaxter, 1983; Brandt and Rosin, 1997; Helman, 1986; Pill and Scott, 1982; Zola, 1972). For example, although most Americans know that viruses cause influenza and the common cold, most continue to hold essentially naturalistic theories regarding these illnesses—warning their children to eat warm foods, wear hats and gloves, and cover up against the rain to avoid infection.

Similarly, the mass media, public health authorities, and the general public now often blame illness on individual life-styles (Brandt and Rozin, 1997; Tesh, 1988). Magazines regularly print articles such as "Beat Your Risk Factors" (Libov, 1999) and "24 Hours to a Healthier You" (Colino, 1998), exhorting individuals to protect or restore their health through diet, exercise, stress reduction, and the like. Simultaneously, the U.S. government—even while, as described in Chapter 2, continuing to subsidize the tobacco and beef industries—spends millions on education campaigns to encourage the public to stop smoking and to eat a healthier diet.

Another popular ideology ties illness not to individual actions but to individual personalities (Sontag, 1978). For example, a newspaper account of comedian Gilda Radner's death from ovarian cancer quoted her "therapist" explaining how

> Gilda always had this wonderful will to live. Yet she also exhibited the same preconditioning virtually all [cancer patients] have. Fear. Hopelessness. Negativity. What . . . Gilda came to appreciate [in her therapy], is that a positive outlook can improve the quality of life—up to and including the immune system" (Kahn, 1989).

Similarly, the media continue to warn that aggressive and competitive Type A personalities can cause heart problems (Siegman and Dembroski, 1989), despite a string of studies refuting this concept (Aronowitz, 1998).

In its most extreme form, this sort of theorizing has led some to claim that illness occurs not because individuals ignore their bodies or have illness-producing personalities but because they *choose* to become ill. The most influential statement of this theory appears in the best-selling book *Love, Medicine and Miracles* by surgeon Bernie Siegel (1990). Siegel postulates that people become ill because they "need" their illness—to escape a stressful work situation, receive sympathy from their spouses, punish themselves for misdeeds, and so on—and because they do not love themselves enough to take care of their emotional needs. Consequently, Siegel advises ill persons that they will find lasting cures only when they truly desire a healthy, long life.

Theories such as Siegel's draw on research suggesting that stress, personality, and life-style can increase personal susceptibility to illness. Such factors may indeed affect the distribution of illness in society. Yet by focusing on these factors as the primary source of illness, these theories encourage the healthy to devalue and reject those who are ill and promote depression and lowered self-esteem among those who blame themselves for their illnesses.

In addition, by emphasizing how individuals cause their own illnesses, these theories encourage policymakers to ignore how social and environmental factors can foster illness (Crawford, 1979; Tesh, 1988; Waitzkin, 1981; Zola, 1972). For example, magazines that emphasize how individuals make themselves ill rarely discuss how factors largely beyond individual control (such as poverty, malnutrition, pollution, or unsafe conditions in our houses, cars, or workplaces) can produce ill health. Nor do these magazines discuss how social factors (including the advertisements for alcohol and cigarettes in some of these same magazines) can pressure individuals to adopt unhealthy life-styles—how unemployed teenagers with poor job prospects sometimes smoke cigarettes to demonstrate their adulthood, how young mothers who lack assistance with child care probably also lack time for the recommended three sessions per week of aerobic exercise, or how

workers sometimes suffer injuries because of unsafe equipment rather than because of personal carelessness. As Barbara Katz Rothman (1989:21) notes,

> Think of the anti-smoking, anti-drinking "behave yourself" campaigns aimed increasingly at pregnant women. What are the causes [as identified in these campaigns] of prematurity, fetal defects, damaged newborns—flawed products? Bad mothers, of course—inept workers. One New York City subway ad series shows two newborn footprints, one from a full-term and one from a premature infant. The ads read, "Guess which baby's mother smoked while pregnant?" Another asks, "Guess which baby's mother drank while pregnant?" And yet another: "Guess which baby's mother didn't get prenatal care?" I look in vain for the ad that says "Guess which baby's mother tried to get by on welfare?"; "Guess which baby's mother had to live on the streets?"; or "Guess which baby's mother was beaten by her husband?"

In sum, whether or not they are accurate, theories of illness that focus on individual responsibility reinforce existing social arrangements and help us rationalize our tendency to reject, mistreat, or simply ignore those who suffer illness.

The Medical Model of Illness

According to the sociological model, then, illness is a moral status referring to conditions or behaviors deemed undesirable by powerful social groups. But what do doctors mean when they declare something an illness? To answer this question, we need to look at the **medical model of illness.** This model is not accepted in its entirety by all doctors—those in public health, pediatrics, and family practice are especially likely to question it—and is not rejected by all sociologists, but it is the dominant conception of illness in the medical world.

The medical model consists of five doctrines: that illness is (1) deviation from normal, (2) specific and universal, (3) caused by unique biological forces, (4) analogous to the breakdown of a machine, and (5) defined and treated medically through a neutral, scientific process.

First, the medical model defines illness narrowly as a deviation from normal biological functioning (Mishler, 1981). In contrast, other cultures define illness more broadly as an absence of balance or well-being or as undue pain, discomfort, or disability. For example, traditional Navajos define illness as a lack of harmony in and with the universe, and the **World Health Organization** defines illness as anything less than a complete state of physical, social, and mental well-being.

Defining illness as deviation from normality assumes that we know and can easily recognize what is normal. Yet the nature of normality varies from person to person and group to group. A height of 4 feet 6 inches would be normal for a Pygmy man but not for a man born in the United

States. Similarly, drinking three glasses of wine a day would be normal for an Italian woman but would raise eyebrows in many American social circles. In defining normality, therefore, we need to look not only at individual bodies but also at the broader social context. Moreover, even within a given group, "normality" is a range and not an absolute. The median height of men in the United States, for example, is about 5 feet 9 inches, but most people would consider someone several inches taller or shorter than that still normal. Similarly, individual Italians routinely and without social difficulties drink more or less alcohol than the average Italian.

In addition, defining illness as deviation from normal encourages doctors to define those with measurable *risks* of illness as if they already *have* an illness. For example, doctors now typically label persons who have higher than average blood pressure—and, hence, based on statistical averages, an elevated risk of heart disease or strokes—as having the disease of hypertension.

Second, the medical model assumes illness is both specific and universal (Dubos, 1961; Mishler, 1981). The model assumes that each illness has specific features recognizable through clear, objective measures that differentiate it both from other illnesses and from health. As I will explain in the next section, however, this assumption is problematic, for deciding which biological or psychological conditions constitute illnesses is a subjective and political process. In addition, diagnosing whether an individual has an illness is also highly subjective. Two patients with the same symptoms may receive different diagnoses depending on their doctors, countries of residence, ages, genders, ethnicities, and so on. At the same time, doctors often differentiate illnesses from health according to arbitrary statistical cutoffs rather than according to any inherent or absolute differences. For example, doctors typically recommend dietary changes for anyone with a cholesterol level higher than average for the U.S. population and recommend both drug therapy and dietary changes for anyone with a level in the top 20 percent of the population (Sempos et al., 1993). These treatment regimens, which in essence define half of Americans as ill, are based on statistical data regarding cholesterol levels, rather than on any absolute differences between those defined as healthy and those defined as ill.

The assumption that illness is universal—that each illness manifests itself in the same way across time, peoples, and cultures—is also problematic. In practice, the model assumes that illnesses manifest themselves in other cultures in the same way as in Western culture and, by extension, that doctors can readily transfer their knowledge of illness developed in the West to the treatment and prevention of illness elsewhere. Yet other cultures recognize illnesses for which Western medical culture has no counterpart. For example, Latin American culture assigns the diagnosis of "susto" to individuals who experience restless sleep, listlessness, depression, and loss of appetite and attributes this illness to frights that cause the soul to wander from the body. Similarly, illnesses recognized by Western medical culture can have a different appearance and natural progression elsewhere. For example, as I described in the last chapter, the measles virus typically causes a mild childhood illness in industrialized nations but causes a devastating disease in developing nations, whereas the microorganism that in the developing nations causes yaws, a mild childhood disease, causes syphilis in the United States (Brothwell, 1993:1097). By the same token, illnesses can manifest themselves differently in different age groups: the polio germ, for example, typically causes paralysis in adults but only flu-like symptoms in very young children.

Third, the medical model assumes each illness has a unique **etiology,** or cause (Mishler, 1981). Modern medicine assumes, for example, that tuberculosis, polio, HIV disease, and so on, are each caused by a unique microorganism. Similarly, doctors continue to search for limited and unique causes of heart disease and cancer, such as high-cholesterol diets and exposure to asbestos. Yet even though illness-causing microorganisms exist everywhere and environmental health dangers are common, relatively few people become ill because of these conditions. By the same token, although cholesterol levels and heart disease are strongly correlated among middle-aged men, many men eat high-cholesterol diets without developing heart disease, while others eat low-cholesterol diets but fall ill anyway. The doctrine of unique etiology discourages medical researchers from asking why individuals respond in such different ways to the same potentially illness-causing factors and encourages researchers to search for **magic bullets**—a term first used by Paul Ehrlich, the discoverer of the first effective treatment for syphilis, to refer to drugs that almost miraculously prevent or cure illness by attacking one specific etiological factor.

Fourth, the medical model conceptualizes the body as a machine or factory and illness as a breakdown of those mechanisms (Martin, 1987; Mishler, 1981; Osherson and AmaraSingham, 1981; Waitzkin, 1993). For example, medical textbooks routinely describe the biochemistry of cells as a "production line" for converting energy into different products and describe the female reproductive system as a hierarchically organized factory of signaling machines, which "breaks down" at menopause (Martin, 1987). Similarly, medical writers typically describe HIV disease as a mechanical failure of the body's immune system (Sontag, 1978).

This mechanistic model encourages doctors to treat individuals in a reductionistic rather than holistic fashion. **Reductionistic treatment** refers to treatment in which doctors consider each bodily part separately from the whole, in the same way auto mechanics might replace an inefficient air filter without worrying whether the problem with the air filter reflected or caused problems in the car's fuel system. Similarly, doctors might perform wrist surgery to correct problems experienced by typists without first investigating whether simple behavioral changes, such as getting furniture that fits the typists better or using wrist rests to take the strain off their wrists while typing, might solve the problem. In contrast, **holistic treatment** assumes that all aspects of an individual's life and body are interconnected—that, for example, to treat an individual with cancer, health care workers must not only treat the tumor but also explore sources of illness elsewhere in the body and in the individual's psychological and social circumstances.

Like the doctrine of unique etiology, the mechanistic model encourages doctors to seek the source of problems within the individual body rather than within the broader social environment. By extension, this doctrine encourages doctors to develop increasingly sophisticated technological fixes, such as organ transplants for persons with lung cancer, rather than to work toward social interventions, such as banning tobacco advertisements to prevent lung cancer in the first place (Waitzkin, 1993:174).

Finally, the medical model assumes that the definition, diagnosis, and treatment of illness are neutral, scientific matters, unaffected by moral or subjective judgments or vested personal interests (Waitzkin, 1993). Yet, for example, rates of surgery for any given condition vary wildly across the United States—and usually are highest in areas where large numbers of surgeons compete for the same pool of patients (Center for the Evaluative Clinical Sciences, 1996). The medical model further assumes that, because medicine is an objective, scientific field requiring highly technical skills and knowledge, only doctors are qualified to define, diagnose, or treat illness. These assumptions reinforce the social power of medicine and ignore how social values affect our definitions of and responses to illnesses (Waitzkin, 1993).

MEDICINE AS SOCIAL CONTROL

Creating Illness: Medicalization

The process through which a condition or behavior becomes defined as a medical problem requiring a medical solution is known as **medicalization** (Conrad and Schneider, 1992). For example, as social conditions have changed, activities formerly considered sin or crime, such as masturbation, homosexual activity, or heavy drinking, have become defined as illnesses. The same has happened to various natural conditions and processes, such as uncircumcised penises, aging, and the entire life course of the female reproductive system, from menstruation, to pregnancy and childbirth, to menopause (for example, Figert, 1996; McCrea, 1983; Sullivan and Weitz, 1988).

Medicalization doesn't simply happen. Rather, it is the end stage in a series of events. For medicalization to occur, one or more organized social group must have both a vested interest in working toward that end and sufficient power to convince others (including doctors, the public, and insurance companies) to accept their new definition of the situation.

Not surprisingly, doctors often play a major role in medicalization, for medicalization can increase their power, the scope of their practices, and, as a result, their incomes. For example, during the first half of the twentieth century, improvements in the standard of living coupled with the adoption of numerous public health measures substantially reduced the number of seriously ill children. As a result, the market for pediatricians declined and their focus shifted from treating serious illnesses to treating minor childhood illnesses and offering well-baby care. Consequently, pediatrics became a less remunerative, interesting, and prestigious field. To increase their market while obtaining more satisfying and prestigious work, some pediatricians have expanded their practices to include children whose behavior concerns their parents or teachers and who are now defined as having medical conditions such as attention deficit disorder or antisocial personality disorder (Halpern, 1990; Pawluch, 1983). Doctors have played similar roles in medicalizing premenstrual syndrome (Figert, 1996), drinking during pregnancy (Armstrong, 1998), impotence (Tiefer, 1994), and numerous other conditions.

In other instances, however, doctors have proved indifferent or even opposed to medicalization. For example, although some doctors believe that woman battering is a medical problem and that doctors should accept responsibility for identifying it and intervening when it occurs, others believe that women provoke their own battering, that doctors can do little to help, or that woman battering is best dealt with by the police rather than by doctors (Kurz, 1987). As a result, many doctors oppose medicalizing woman battering and prefer to treat women's injuries without delving into their causes.

In circumstances such as these, pressure for medicalization can instead come from lay groups. Alcoholics Anonymous, for example, has fought to medicalize alcoholism partly as a means of relieving alcoholics of some of the stigma of their addiction. Other lay groups similarly have argued for medicalization in the hope that medical control will be more humanitarian than legal control, in such areas as compulsive gambling, erratic and violent behavior, and homosexuality. In addition, individuals might press for medicalization as a way of gaining validation for their experiences and stimulating research on treatments and cures (Ziporyn, 1992). For example, much of the pressure to define premenstrual syndrome and chronic fatigue syndrome as illnesses has come from persons who believe they suffer from these syndromes.

The third major force behind medicalization is the pharmaceutical industry. This industry has a vested economic interest in medicalization whenever it can provide a drug as treatment. The medicalization of shortness exemplifies this process (Werth, 1991). In 1985, the pharmaceutical company Genentech patented a genetically engineered and mass-produced form of human growth hormone (HGH). At that time, the available data suggested that HGH could increase final height in children whose pituitary glands did not naturally produce enough HGH, but could not increase final height in children without pituitary defects. Moreover, it was known that HGH could promote a drastic loss of body fat and increase in muscle, with unknown consequences for the bodies of growing children. Nevertheless, Genentech and, subsequently, Eli Lilly Pharmaceuticals (which patented a slightly different synthetic hormone) embarked on a major campaign to sell HGH. Together, they underwrote two-thirds of the budget of the Human Growth Foundation, a nonprofit advocacy group that works to increase public awareness of the problems experienced by short children. With the pharmaceutical companies' help, the Foundation began broadcasting news of HGH across the nation at health fairs, shopping malls, and the like. The pharmaceutical companies also began spending millions of dollars annually to underwrite medical research supporting HGH, to advertise the drug to doctors, and to sponsor in-school screening programs that identified children in the shortest 3 percent of the population and informed their parents that the children had a disease requiring medical treatment.

By 1999, about 30,000 children—20 percent of whom have no disease other than shortness—were being treated with HGH in the United States, at an average cost per child of about $25,000 yearly for an average of ten years (Greenberg, 1999). According to the only long-term study (partially funded by Genentech) of the drug's effectiveness on children with normal pituitary glands, these children can expect to add about two inches to their adult height (Hintz et al., 1999). Because of HGH's limited effectiveness and potential for long-term health problems (such as tumors and diabetes) and because identifying short children as "diseased" and treating them with daily injections over several years can lead to social stigma and lowered self-esteem, the American Academy of Pediatrics (1997) currently recommends against its use in short but otherwise healthy children.

Working Together to Medicalize Hyperkinesis

Neither doctors, lay groups, nor pharmaceutical companies individually have enough influence to medicalize a condition. Consequently, successful medicalization depends on the interwoven interests and activities of these three groups and sometimes others. The history of hyperkinesis illustrates this process.

As originally defined, hyperkinesis lacked any definitive biological markers and instead referred to children above age 5 who were overactive, impulsive, and easily distracted but who had no brain damage (Diller, 1998). Since the late 1930s, doctors have known that amphetamines (including methamphetamine or "speed") can reduce distraction in children and adults, regardless of their mental status. In addition, even though biologically amphetamines are stimulants, they provide an intense focus that can make users appear less active. These characteristics made amphetamines a natural choice for treating hyperkinesis. However, because amphetamines are highly addictive and have dangerous side effects, physicians avoided prescribing them.

In the absence of a viable treatment, physicians rarely made the diagnosis of hyperkinesis. This situation only changed in the 1960s, when the amphetamine Ritalin (methylphenidate) appeared on the market (Conrad and Schneider, 1992). Ritalin has fewer short-term side effects than other amphetamines have and, in the short term, improves the ability to concentrate, reduces the tendency to act on impulse, and increases willingness to accept discipline. Yet Ritalin is far from a panacea. The drug's immediate side effects can include addiction, loss of appetite, sleep deprivation, headache, and stomach ache. Its long-term side effects are unknown, and its long-term benefits seem minor at best: the little available research suggests that it does not improve users' chances of graduating high school, holding a job, refraining from illicit drugs, or avoiding trouble with the law (Diller, 1998).

Following the development of Ritalin, pharmaceutical companies embarked on a huge campaign to "sell" hyperkinesis. According to Peter Conrad and Joseph Schneider:

> After the middle 1960s it is nearly impossible to read a medical journal or the free "throw-away" magazines [mailed by pharmaceutical companies to doctors] without seeing some elaborate advertising for either Ritalin or Dexedrine [another amphetamine]. These advertisements explain the utility of treating hyperkinesis . . . and urge the physician to diagnose and treat hyperkinetic children. The advertisements may run from one to six pages. They often advise physicians that "the hyperkinetic syndrome" exists as "a distinct medical entity" and that the "syndrome is readily diagnosed through patient histories and psychometric testing" and "has been classified by an expert panel" of the Department of Health, Education and Welfare as MBD [Minimal Brain Dysfunction]. These same pharmaceutical firms also supply sophisticated packets of "diagnostic and treatment" information on hyperkinesis to physicians, pay for professional conferences on the subject, and support research in the identification and treatment of hyperkinesis (1992:159–160).

In addition, pharmaceutical companies promoted Ritalin to the public, spending $610 million on direct-to-consumer advertisements in 1996, up from $44 million in 1990 (Diller, 1998:139).

As noted earlier, pediatricians proved a ready audience for this marketing campaign, which promised a way to boost their flagging income and prestige. This market further increased in the late 1980s, when the diagnosis of hyperkinesis was replaced by "attention deficit disorder" (ADD). Unlike hyperkinesis, the definition of ADD sets no age limits and includes girls who daydream as well as boys who express boredom or dissatisfaction through physical activity.

Like pediatricians, many teachers readily adopted the concept of ADD, if for different reasons (Diller, 1998). Faced with cuts in staffing and larger classes at the same time that school boards began placing an increased emphasis on testing and competition at earlier and earlier ages, teachers can hardly be blamed for looking with favor on drugs that make their students more manageable. In addition, diagnosing a student with ADD shifts blame for poor student performance from teacher to student. Not surprisingly, the suggestion to place a child on Ritalin now often comes initially from a teacher (Diller, 1998).

Parents, too, are often relieved to find an explanation other than poor parenting for their child's behavioral or educational problems. In addition, like those who argue that alcoholism or compulsive gambling is a disease, these parents hope to remove blame from their children, reduce the chances of legal sanctions against their children, and stimulate research on treatment. Finally, recent legal changes have encouraged parents to push for a diagnosis of ADD as a means of gaining educational assistance for their child (Diller, 1998). Since 1991, federal guidelines have defined ADD as a disability covered under federal antidiscrimination statutes (which are discussed in the next chapter). Those statutes set aside funds for individualized educational services for disabled students, while making it extremely difficult for schools to discipline children for any problem behaviors that could be considered part of their disability. Thus, many parents find that having their child diagnosed with ADD increases the child's educational opportunities while it reduces the chances that the child will be suspended or expelled. For this reason, children are much more likely to be diagnosed with ADD if they are wealthy and white than if they are poor or nonwhite. Similarly, adults with ADD can legally request accommodations in the workplace, such as quiet space or extra time to finish tasks, as long as their disability does not substantially interfere with their job performance. As a result, adults increasingly seek diagnosis with ADD for themselves as a means of getting these accommodations (Diller, 1998).

Taken together, these factors have produced an astounding increase in the number of persons diagnosed with ADD, from about 150,000 U.S. children in 1970 to 900,000 in 1990 and almost 5 million in 1998 (Diller, 1998:2, 27). By 1996, according to the United Nations's International Narcotics Control Board, Ritalin was prescribed more often in the United States than anywhere else in the world, with between 3 to 5 percent of all U.S. schoolchildren— and between 10 and 12 percent of all boys ages 6 and 14—receiving the drug (Crossette, 1996). Recent reports have focused on the burgeoning numbers of children who now receive Ritalin in pre-school.

The Consequences of Medicalization

In some circumstances, medicalization can be a boon, leading to increased social awareness of a problem, sympathy toward its sufferers, and the development of beneficial therapies. Persons with epilepsy, for example, lead far happier and more productive lives now that drugs usually can control their seizures and that few people view epilepsy as a sign of demonic possession. Yet as the previously discussed examples suggest, it also can produce **unintended negative consequences** (Conrad and Schneider, 1992; Zola, 1972).

First, defining a condition as an illness does not necessarily improve the social status of those who have that condition. Those who use alcohol excessively, for example, continue to experience social rejection despite the medicalization of alcoholism.

Second, once a situation becomes medicalized, doctors become the only experts considered appropriate for responding to it, increasing the power of doctors at the expense of other social groups. With the medicalization of troublesome behavior in children, for example, the voices of parents, teachers, and the children themselves have lost credibility when they disagree with doctors' assessments.

Third, once a condition is medicalized, medical treatment becomes the only logical response to it. For example, if woman battering is considered a medical condition, then doctors need to treat women and the men who batter them. However, if woman battering is considered a social problem stemming from male power and female subordination, then it makes more sense to arrest the men, assist the women in developing financial and emotional independence, and work for broader structural changes that will increase all women's status and options.

Medicalization can justify not only voluntary but also involuntary treatment. Yet treatment does not always help and sometimes can harm. For example, around the country during the last two decades, U.S. courts have forced women to submit to cesarean deliveries, in which babies are surgically removed from their mothers' uteruses rather than delivered naturally through the vagina (Daniels, 1993). In these cases, doctors argued successfully that childbirth is a dangerous medical condition, not a natural process, and that therefore mothers lack the expertise to decide whether cesarean deliveries are in their and their babies' best interests. Yet doctors' judgment is not infallible. A 1987 study found that in six of the first fifteen cases in which doctors sought court orders to force cesarean deliveries, the mothers in the end delivered healthy babies vaginally (Kolder et al., 1987); the remaining nine women were forced to have cesareans, so we cannot know whether they might have safely delivered vaginally. Moreover, the rate of cesarean deliveries in the United States is about twice that recommended by the World Health Organization (1985:437), suggesting that U.S. doctors are far too ready to perform this potentially life-threatening surgery. Box 5.2 explores the ethical issues involved in forced obstetrical interventions, and the broader issue of "fetal rights."

Fourth, and as these examples suggest, medicalization significantly expands the range of life experiences under medical control. For example, the existence of "fetal alcohol syndrome"—a constellation of birth defects including mental retardation believed caused by drinking alcohol during

Box 5.2   *Ethical Debate: Medical Social Control and Fetal Rights*

In 1985, Pamela Rae Stewart became pregnant. Her doctor, knowing her history of drug use, warned her to stop using amphetamines. Later, when problems developed during her pregnancy, he advised her to stay off her feet, avoid sexual intercourse, and seek medical attention if she began to bleed heavily.

On November 23, 1985, Stewart gave birth to a severely brain-damaged baby. On the day her child was born, according to police reports, Stewart took amphetamines and had intercourse with her husband. She subsequently began bleeding but did not go to the hospital for several hours. Six weeks later, the baby died, and the District Attorney filed criminal charges against Stewart for child neglect.

Beginning in the mid 1980s, doctors and the courts increasingly have imprisoned pregnant women to keep them from using illicit drugs or have involuntarily hospitalized them to force them to follow medical advice. A 1995 survey of directors of substance abuse and child protective services found that 71 percent of the 50 states had criminally prosecuted a woman for using drugs while pregnant, 65 percent routinely reported pregnant women who used drugs to child protective services, and 24 percent legally mandated substance abuse treatment for pregnant women (Chavkin et al., 1998). Ironically, pregnant drug-users are most likely to face criminal sanctions if they are poor or minorities, even though such women are least likely to have access to substance abuse treatment (Chasnoff et al., 1990). Less commonly if more severely, doctors and the courts have forced women to have cesarean sections in the belief that these operations were in the babies' best interests. A 1987 study identified 21

cases nationally in which doctors sought court orders to force obstetrical interventions and found that the doctors succeeded in 86 percent of these cases (Kolder et al., 1987). In these successful suits, 81 percent of the women were African American or Hispanic, 44 percent were unmarried, 24 percent were not fluent in English, and all were poor.

These actions reflect a growing tendency among doctors, lawyers, and the general public to view mother and fetus as separate beings, with separable and sometimes conflicting rights, and to see the fetus rather than the mother as obstetricians' primary patient (Daniels, 1993; Rothman, 1989). This tendency reflects both the growth of technologies like ultrasound, electronic fetal monitoring, and fetal surgery that have allowed doctors to view and act on the fetus (Casper, 1998) and the successes of the anti-abortion movement in convincing many Americans to think of fetuses as children or "almost children" (even though more than three-quarters of Americans continue to believe that women should have access to abortion in some circumstances [Harris Poll, 1996]).

The state has a legal obligation to protect children from parents who abuse or otherwise endanger them. Similarly, both ethical and legal guidelines require doctors who learn of child abuse to report it to the state. Should doctors and the state have a similar obligation to protect the fetus even if it means superseding parents' wishes?

Those who argue in favor of medical intervention find it illogical to protect children from bodily harm *after* birth but to deny them protection that might ensure their health *before* birth. Children born prematurely, addicted to drugs, or with birth defects because their mothers did not follow medical advice may suffer short, painful lives or—perhaps worse—long and painful lives, with reduced mental or physical abilities. In addition, these children cost hospitals and taxpayers vast sums every year. Those costs alone, one could argue, give the medical and legal systems the right to intervene when women endanger their fetuses.

Others, however, have raised several objections to placing **fetal rights** above mothers' wishes. First, these individuals question whether doctors necessarily know better than do mothers what is in the fetus' best interest. Thirty years ago, for example, doctors routinely prescribed diethylstilbestrol (DES) for pregnant women, told women to limit their weight gain during pregnancy to 15 pounds, and x-rayed women's abdomens to check fetal growth. Yet doctors later discovered that limiting maternal weight gains increased the rate of birth defects and that DES and x-rays sometimes caused women to miscarry or their children to develop cancer (Rothman, 1989).

In addition, arresting or forcibly hospitalizing pregnant drug users to keep them from using drugs might not serve any useful medical purpose. Punitive responses to pregnant drug users can encourage such women to avoid health care altogether. Moreover, withdrawal from drugs can endanger the fetus more than continued drug use (Pollitt, 1990), because over time bodies adapt to the presence of drugs whereas withdrawal severely shocks the system; indeed, to keep addicts from dying during withdrawal, doctors often have to prescribe other highly toxic drugs. Similarly, to treat addiction doctors often must prescribe drugs such as methadone that can harm fetuses as much as illicit drugs can. Finally, some studies suggest that drug use during pregnancy has few if any long-term effects on children (Armstrong, 1998; Koren et al., 1989; Pollitt, 1990). This information has

had relatively little impact on public attitudes, partly because of the natural bias built into all publishing enterprises, including medical journals. Like other periodicals, medical journals are interested in "making news"—presenting new and interesting data that will make headlines and grip readers. As a result, scientific journals in general are more likely to accept for publication manuscripts that describe significant effects rather than articles that say that a given variable, experiment, legal drug, or illegal drug has no effect (Koren and Klein, 1991); one study found that medical journals more often reject manuscripts describing studies that find that drugs have little or no effect on fetuses than poorer-designed studies that find drugs do have an effect (Koren et al., 1989).

Opponents of forced intervention further argue that doctors cannot make better decisions than do mothers because they cannot understand fully the circumstances in which mothers make those decisions. For example, many women continue to use drugs during pregnancy only because they cannot obtain access to treatment programs, which usually have long waiting periods and often will not accept pregnant women. In addition, to enter a treatment program, women almost always have to leave their existing children with relatives or in foster care; for example, Arizona currently has an estimated 5,000 drug-addicted parents but only one treatment facility with a total of ten beds that allows parents to keep their children with them (Bland, 1999). Yet leaving children with relatives or in foster care can place children at greater risk than having a drug-using mother, given that women often begin drug use because of problems in their family and that foster care sometimes results in physical, sexual, or mental abuse.

(continues)

Box 5.2   ***(continued)***

Opponents of forced intervention also argue that the benefits of intervention do not justify the costs to women's civil liberties. Once we decide that women must put their fetuses' welfare above their own, where do we draw the line? Given that tobacco poses a far greater threat to fetuses than any illicit drug does, do we prosecute or hospitalize women who continue to smoke during pregnancy? What about women who continue to eat junk food rather than eating healthy meals? Or women who work two jobs and get insufficient rest? Already, some employers have tried to use the language of fetal rights to bar women (but not men) from work involving toxic chemicals (Nelkin and Tancredi, 1989).

Finally, the costs of fetal rights to women's rights lead to questions regarding the true purposes of the fetal rights movement. Although we require parents to guard their children's health and welfare, we do not require them to donate kidneys, bone marrow, or even blood for their children's sake. Why, then, should we require women—and only women—to protect their fetuses? After all, fathers' use of tobacco, alcohol, and other drugs may damage sperm and therefore fetuses, but no court yet has charged a man for fetal abuse. Similarly, working in toxic environments damages sperm as well as ova and fetuses, yet no employers have tried to "protect" men from holding such jobs. And during Pamela Stewart's pregnancy, her

---

pregnancy—was widely accepted by American doctors based on three articles: two case studies describing a total of 11 cases born to women who were chronic alcoholics and one retrospective study that found six cases out of 55,000 patient records in which doctors had noted that the mothers were chronic alcoholics and that their babies had symptoms now considered possible indicators of fetal alcohol syndrome (Armstrong, 1998). None of these studies **controlled** for other variables or used **random** samples, and none gave data suggesting a widespread problem, even among severe alcoholics. Yet based on these articles, and the very limited additional data that has been collected since then, doctors and others have campaigned to forbid restaurants and bars from serving alcohol to pregnant women; to require liquor manufacturers, restaurants, and bars to post warning labels and signs warning of the dangers of drinking during pregnancy; and to jail or hospitalize involuntarily women who drink during pregnancy. Similarly, through defining tobacco use and overweight as risk factors for illness and, at times, as illnesses in themselves, doctors have laid the ideological groundwork for employers to discriminate against smokers and overweight persons. Using this medical rhetoric, some corporations now justify firing or refusing to hire smokers and overweight persons on the grounds that they are more likely to perform poorly, become ill, or

husband not only used amphetamines and had sexual intercourse with her but also beat her periodically. Yet no District Attorney arrested him for wife abuse or fetal abuse. These facts have led some to conclude that the true, if perhaps unconscious, motive behind the rhetoric of fetal rights is not to protect fetuses or children but to restrict women's lives—especially the lives of those women who are most different from and, hence, considered most suspect by those who make laws and policy.

Because of the problems inherent in forced medical intervention, medical organizations have adopted oddly ambivalent positions toward this strategy. In 1990, the American Medical Association Board of Trustees issued a policy statement that strongly argued against forced medical intervention but nevertheless concluded that it could be justified in extraordinary circumstances in which the risks to the mother are low, the threat to the fetus in the absence of intervention is high, and the intervention is likely to be effective. (None of these terms was defined.) The American Academy of Pediatrics and the American College of Obstetricians and Gynecologists issued similar statements in 1999. On the other hand, the American Medical Association, the American Academy of Pediatrics, the American Nurses Association, and the American Public Health Association unambiguously oppose the use of legal sanctions against pregnant drug users.

drive up the company's health insurance rates (Brandt and Rozin, 1997;Nelkin and Tancredi, 1989).

This expansion of medical control over broad areas of life diminishes the power of other social authorities, including judges, the police, religious leaders, legislators, and teachers. Through medicalization, questions such as who should receive abortions or organ transplants, how society should respond to drug use, and whether severely disabled infants should receive experimental but potentially life-saving surgery become defined as strictly medical, rather than social, religious, economic, or ethical issues.

Fifth, medicalization can hide the political nature of social actions. For example, both China and the former Soviet Union removed many political dissidents from the public eye by committing them to mental hospitals. By so doing, these governments discredited and silenced individuals who might otherwise have offered powerful dissenting voices. In other words, medicalization allowed these governments to **depoliticize** the situation—to define it as a medical rather than a political problem. Similarly, medicalizing woman battering encourages a limited focus on the behavior and personality of individual men and women rather than on the social and political forces that give men the power to beat women and deny women the power to leave their abusers.

The Rise of Demedicalization

The dangers of medicalization have fostered a countermovement of **demed-icalization** (Fox, 1977). A quick look at medical textbooks from the late 1800s reveals many "diseases" that no longer exist. For example, nineteenth century medical textbooks often included several pages on the health risks of masturbation. One popular textbook from the late nineteenth century asserted that masturbation caused "extreme emaciation, sallow or blotched skin, sunken eyes, . . . general weakness, dullness, weak back, stupidity, laziness, . . . wandering and illy defined pains," as well as infertility, impotence, consumption, epilepsy, heart disease, blindness, paralysis, and insanity (Kellogg, 1880:365). Today, however, medical textbooks describe masturbation as a normal part of human sexuality.

Like medicalization, demedicalization often begins with lobbying by lay groups. For example, medical ideology now defines childbirth as an inherently dangerous process, requiring intensive technological, medical assistance. Since the 1940s, however, growing numbers of American women have attempted to redefine childbirth as a generally safe, simple, and natural process and have promoted alternatives ranging from natural childbirth classes, to hospital birthing centers, to home births assisted only by midwives (Sullivan and Weitz, 1988). Similarly, and as will be described in Chapter 7, gay and lesbian activists have at least partially succeeded in redefining homosexuality from a pathological condition to a normal human variation. More broadly, in recent years, books, magazines, television shows, and popular organizations devoted to teaching people to care for their own health rather than relying on medical care have proliferated. For example, in the early 1970s, the Boston Women's Health Book Collective published a 35-cent mimeographed booklet titled *Our Bodies, Ourselves*. From this, they have built a virtual publishing empire, which has sold millions of books on childhood, adolescence, aging, and women's health to consumers around the world.

Social Control and the Human Genome Project

The potential for medicine to act as a form of social control may soon grow through the work of the internationally funded Human Genome Project. The Project's goal is to map the locations of all human genes and to determine the role each gene plays in health and illness.

Genes affect health in two ways: by causing "true" genetic diseases and by increasing one's predisposition to develop a given disease. True genetic diseases, such as hemophilia, are caused directly by specific genes. Such diseases are relatively uncommon and typically become apparent at birth or early in life. Some can be treated, but currently none can be cured. As researchers learn which genes cause these diseases and develop tests to determine the presence of those genes, they can offer individuals the opportunity to learn whether they, their children, or (for pregnant women) their fetuses carry the gene. Individuals who learn they have a genetic defect may choose to avoid becoming pregnant; to abort any fetuses that also carry the defect; or to continue a pregnancy to term, knowing that the fetus carries the defect and hoping that this foreknowledge will better prepare them for the birth of an ill or disabled child. Finally, individuals who know they have a genetic defect but who want to have a child that is biologically theirs can have fetuses created through *in vitro* fertilization (in which eggs removed from the woman's body are mixed with the man's sperm in the laboratory). They can then have their doctors test the resulting fetuses for genetic defects and implant any nondefective fetuses in the woman's uterus. This strategy is very rare, as the physical costs to the woman and the financial and psychological costs to the couple are extremely high, while the odds of success are low.

In other cases, genes do not directly cause disease but can increase the likelihood of disease developing. For example, no single gene causes Alzheimer's disease, breast cancer, heart disease, or diabetes. These diseases occur more often in some families than others, however, which suggests that the diseases occur only in those who have some genetic predisposition. In these cases, if doctors can learn which genes correlate with the disease and develop ways of identifying which individuals have those genes, doctors might find it easier to convince at-risk individuals to take potentially health-preserving actions. For example, women who learn that they have the BRCA-1 gene, which correlates with an increased risk of breast cancer, might choose to adopt a low-fat diet or to have their breasts removed before any cancer appears.

The Human Genome Project brings with it tremendous potential for both good and harm. Not only can those who learn they are at increased risk adopt healthier behaviors, but also those who learn that they are *not* at risk can gain peace of mind, as might those who carry a genetic defect but learn that their fetuses or children do not. Testing could even benefit those who learn that they have or will develop a genetic disorder, for some will prefer certainty to the anxieties of uncertainty.

Yet the potential harm this knowledge can cause is also great. First, although some might cope well with the knowledge that they or their children will develop an unpreventable genetic disease later in life, others will be overwhelmed by this knowledge. It is hard, for example, to imagine how it can help individuals to learn at age 21 that by their forties they will develop Huntington's disease, a devastating neurological disorder that invariably causes progressive insanity, disability, and death.

Second, as the knowledge and technologies developed by the Human Genome Project increase and become part of everyday medicine, the use of genetic testing will undoubtedly spread rapidly. Genetic *counseling,* on the other hand, will probably spread more slowly, because it is considerably more expensive to provide. As a result, in the future more people, especially those who are poor or live far from medical centers, will receive complicated, confusing, and potentially devastating information from genetic tests without receiving the counseling necessary to help them understand and cope with this information.

Third, individuals identified through genetic testing as having an illness or being at high risk for illness might experience discrimination and stigma as a result. Researchers already have documented instances in which individuals have been refused jobs, health insurance, or life insurance because they are carriers of a genetic disease, have a genetic defect although they are still asymptomatic, or are suspected of having or carrying a genetic disease (Billings et al., 1992; Hubbard and Henifin, 1985; Natowicz et al., 1992). Such discrimination will only grow as genetic testing spreads, although whether it would be legal remains open to question. The **Americans with Disabilities Act** (described in the next chapter) outlaws discrimination against ill or disabled persons, and the federal Equal Employment Opportunity Commission has interpreted it as outlawing employment discrimination based on genetic makeup. It remains unclear, however, whether it applies to discrimination in other areas of life (Gostin et al., 1999). At any rate, the law only helps those who know about it and have the time and resources needed to pursue legal redress.

Fourth, genetic tests can tell whether an individual carries the gene for a disease, but not how soon or how severely he or she will be affected. For example, although doctors can tell if a fetus has the chromosomal defect that causes Down Syndrome, they cannot tell if the fetus will become a child who could be self-supporting or a child who could neither walk nor talk. Increasingly, too, tests are identifying genetic anomalies whose effects, if any, are unknown. As a result, couples often must decide whether to abort a genetically abnormal fetus with little idea what their child's life might be like.

Fifth, except for true genetic diseases, genetic tests only can suggest the *probability* that a fetus, child, or adult will develop an illness, not whether it will or will not happen. For example, prospective parents might learn that their fetus has a 60 percent chance of developing breast cancer as an adult. No one can offer any logical rules for making decisions based on such probabilities. As a result, parents in these circumstances will face a far more complex decision than will parents who know their child would have a genetic disease. Moreover, genetic testing cannot tell the former group of parents any more than the latter regarding when or how severely the illness will affect their children.

Finally, the Human Genome Project raises the potential for genetic controls far beyond anything now available. Relatively few persons oppose programs to prevent the birth of children with Tay-Sachs disease, which causes initially healthy children to deteriorate totally—both mentally and physically—and to die between the ages of 3 and 5. Yet many geneticists hope in the future to expand vastly the number of conditions for which genetic tests are run. Already many fetuses are aborted simply because they are female, as described in Chapter 4 (Jeffery et al., 1984; Wertz and Fletcher, 1998). Will the world really be a better place if we can abort fetuses because they will have below average intelligence or a genetic predisposition toward fatness?

The potential impact of the Human Genome Project is magnified by the treatment it has received in the news media. Like illness, news is a social construction, for news media first decide which stories are newsworthy and then decide how those stories will be told. Research conducted by sociologist Peter Conrad (1997) suggests that the media consistently overplay the impact of genes in presenting news stories. Conrad looked at all coverage of genetics in five major newspapers (including the *Los Angeles Times* and the *Wall Street Journal*) and three news magazines (*Time, Newsweek,* and *U.S. News and World Report*) between 1965 and 1995 and found that the media routinely gave prominent coverage to the discovery of a supposed link between a gene and a condition or illness, but either ignored later disconfirmations of the link or relegated them to back pages. For example, all eight news outlets gave prominent and optimist coverage to a 1990 article published in the *Journal of the American Medical Association* that reported a link between a specific gene and alcoholism. Yet none of the magazines and only a few of the newspapers covered an article published eight months later in the same journal that refuted the findings of the first article. Moreover, all news stories on the second article were relegated to the back of the newspapers, and all suggested that new evidence of genetic links would surely be found soon.

These findings led Conrad to conclude that the news media has adopted a **genetic paradigm,** a way of looking at the world that emphasizes genetic causes. This paradigm

> has considerable appeal. It promises primary causes, located on a basic level of biological reality. Genes are often depicted as an essence, what one is really made of. ... We now can be tempted by the lure of specificity, associating specific genes and particular problems. Identifying specific genes seems so much neater than complex, messy, epidemiological and social analyses. This specificity feeds hopes for genetic "magic bullets" to alleviate human problems (Conrad, 1997:142).

Social Control and the Sick Role

Until now, we have looked at how medicine functions as an institution of social control by defining individuals either as sick or as biologically defective. Medicine also can work as an institution of social control by pressuring individuals to *abandon* sickness, a process first recognized by Talcott Parsons (1951).

Parsons was one of the first and most influential sociologists to recognize that illness is deviance. From Parsons' perspective, when people are ill, they cannot perform the social tasks normally expected of them. Workers stay home, housewives tell their children to make their own meals, students ask to be excused from exams. Because of this, either consciously or unconsciously, people can use illness to evade their social responsibilities. To Parsons, therefore, illness threatened social stability.

Parsons also recognized, however, that allowing some illness can *increase* social stability. Imagine a world in which no one could ever "call in sick." Over time, production levels would fall as individuals, denied needed recuperation time, succumbed to physical ailments. Morale, too, would fall while resentment would rise among those forced to perform their social duties day after day without relief. Illness, then, acts as a kind of pressure valve for society—something we recognize when we speak of taking time off work for "mental health days."

From Parsons' perspective, then, the important question was how did society control illness so that it would increase rather than decrease social stability? Parsons' emphasis on social stability reflected his belief in the broad social perspective known as **functionalism.** Underlying functionalism is an image of society as a smoothly working, integrated whole, much like the biological concept of the human body as a homeostatic environment. In this model, social order is maintained because individuals learn to accept society's norms and because society's needs and individuals' needs match closely, making rebellion unnecessary. Within this model, deviance, including illness, is usually considered **dysfunctional** because it threatens to undermine social stability.

Defining the Sick Role

Parsons' interest in how society manages to allow illness while minimizing its impact led him to develop the concept of the **sick role.** The sick role refers to social expectations regarding how society should view sick people and how sick people should behave. According to Parsons, the sick role as it currently exists in Western society has four parts. First, the sick person is considered to have a legitimate reason for not fulfilling his or her normal social role. For this reason, we allow people to take time off from work when sick rather than firing them for malingering. Second, sickness is considered beyond individual control, something for which the individual is not held responsible. This is why, according to Parsons, we bring chicken soup to people who have colds rather than jailing them for stupidly exposing themselves to germs. Third, the sick person must recognize that sickness is undesirable and work to get well. So, for example, we sympathize with people who obviously hate being ill and strive to get well and question the motives of those who seem to revel in the attention their illness brings. Finally, the sick person should seek and follow medical advice. Typically, we expect sick people to follow their doctors' recommendations regarding drugs and surgery, and we question the wisdom of those who do not.

Parsons' analysis of the sick role moved the study of illness forward by highlighting the social dimensions of illness, including identifying illness as deviance and doctors as agents of social control. It remains important partly because it was the first truly sociological theory of illness. Parsons' research also has proved important because it stimulated later research on interactions between ill people and others. In turn, however, that research has illuminated the analytical weaknesses of the sick role model.

Critiquing the Sick Role Model

Many recent sociological writings on illness—including this textbook—have adopted a **conflict perspective** rather than a functionalist perspective. Whereas functionalists envision society as a harmonious whole held together largely by socialization, mutual consent, and mutual interests, those who hold a conflict perspective argue that society is held together largely by power and coercion, as dominant groups impose their will on others. Consequently, whereas functionalists view deviance as a dysfunctional element to be controlled, conflict theorists view deviance as a necessary force for social change and as the conscious or unconscious expression of individuals who refuse to conform to an oppressive society. As a result, conflict theorists have stressed the need to study social control agents as well as, if not more than, the need to study deviants.

The conflict perspective has helped sociologists to identify weaknesses in each of the four elements of the sick role model (Table 5.1). That model declares that sick persons are not held responsible for their illnesses. Yet, as we saw earlier in this chapter, and as Eliot Freidson (1970b), Parsons's most influential critic, has noted, society often does hold individuals responsible for their illnesses. In addition, ill persons are not necessarily considered to have a legitimate reason for abstaining from their normal social tasks. Certainly no one expects persons with end-stage cancer to continue working, but what about people with arthritis or those labeled malingerers or hypochondriacs because they cannot obtain a diagnosis after months of pain, increasing disability, and visits to doctors (Ziporyn, 1992)? Parsons' model also fails to recognize that the social legitimacy of adopting the sick role depends on the socially perceived seriousness of the illness, which in turn depends not only on biological factors but also on the social setting; a nonunionized factory worker, for example, is less likely than a salaried worker with good health benefits to take time off when sick.

Table 5.1	Diseases for Which the Sick Role Model Does and Does Not Fit	
ELEMENTS OF THE SICK ROLE	DISEASES THAT FIT THE MODEL	DISEASES THAT FIT THE MODEL POORLY OR NOT AT ALL
Individual considered to have legitimate reason for not fulfilling obligations	Appendicitis, cancer	Undiagnosed chronic fatigue
Individual not held responsible for condition	Measles, hemophilia	AIDS, lung cancer
Individual expected to try to get well	Tuberculosis, broken leg	Diabetes, epilepsy
Individual expected to seek medical help	Strep throat, syphilis	Alzheimer's, cold

The other aspects of the sick role model are equally problematic. The assumption that individuals will attempt to get well fails to recognize that much illness is **chronic** and by definition not likely to improve. Similarly, the assumption that sick people will seek and follow medical advice ignores the many people who lack access to medical care. In addition, it ignores the many persons, especially those with chronic rather than **acute** conditions, who have found mainstream health care of limited benefit and who therefore rely mostly on their own experience and knowledge and that of other nonmedical people. Finally, the concept of a sick role ignores how gender, ethnicity, age, and social class affect the response to illness and to ill people. For example, women are both more likely than men to seek medical care when they feel ill and less likely to have their symptoms taken seriously by doctors (Bernstein and Kane, 1981; Corea, 1985; Council on Ethical and Judicial Affairs, 1991; Schappert, 1993:2; Steingart, 1991; Zola, 1991).

In sum, the sick role model is based on a series of assumptions about both the nature of society and the nature of illness. In addition, the sick role model confuses the experience of *patienthood* with the experience of *illness* (Conrad, 1987). The sick role model focuses on the interaction between the ill person and the mainstream health care system. Yet interactions with the medical world form only a small part of the experience of living with illness or disability, as the next chapter will show. For these other reasons, research on the sick role has declined precipitously; whereas *Sociological Abstracts* listed 71 articles on the sick role between 1970 and 1979, it listed only 7 articles between 1990 and 1999, despite the fact that far more academic articles overall were published during the 1990s than during the 1970s.

CONCLUSIONS

The language of illness and disease permeates our everyday lives. We routinely talk about living in a "sick" society or about the "disease" of violence infecting our world, offhandedly labeling anyone who behaves in a way we don't understand or don't condone as "sick."

This metaphoric use of language reveals the true nature of illness: behaviors, conditions, or situations that powerful groups find disturbing and believe stem from internal biological or psychological roots. In other times or places, the same behaviors, conditions, or situations might have been ignored, condemned as sin, or labeled crime. In other words, illness is both a social construction and a moral status.

In many instances, using the language of medicine and placing control in the hands of doctors offers a more humanistic option than the alternatives. Yet, as this chapter has demonstrated, increasing medical social control also carries a price. The same surgical skills and technology for cesarean sections that have saved the lives of so many women and children now endanger the lives of those who have cesarean sections unnecessarily. At the same time, forcing cesarean sections on women potentially threatens women's legal and social status. Similarly, the development of tools for genetic testing has saved many individuals from the anguish of rearing children doomed to die young and painfully, but has cost others their jobs or health insurance.

In the same way, then, that automobiles have increased our personal mobility in exchange for higher rates of accidental death and disability, adopting the language of illness and increasing medical social control bring both benefits and costs. These benefits and costs will need to be weighed carefully as medicine's technological abilities grow.

SUGGESTED READINGS

Conrad, Peter, and Joseph W. Schneider. 1992. *Deviance and Medicalization: From Badness to Sickness*. Philadelphia: Temple University Press. Presents a theoretical framework for understanding medicalization, as well as several case studies of this process.

Kirp, David L. 1989. *Learning by Heart: AIDS and Schoolchildren in America's Communities*. New Brunswick, NJ: Rutgers University Press. An engrossing account of how different American communities responded to the presence of children with AIDS in their schools.

Nelkin, Dorothy, and M. Susan Lindee. 1996. *The DNA Mystique: The Gene as a Cultural Icon.* New York: W. H. Freeman. A fascinating exploration of the sources and consequences of the genetic paradigm.

GETTING INVOLVED

Council for Responsible Genetics. 5 Upland Road, Suite 3, Cambridge, MA 02140. (617) 868-0870. www.gene-watch.org. Works to educate the public about the social implications of genetic technologies and to advocate socially responsible use and development of those technologies.

ACT UP. 332 Bleecker St., suite G5, New York, NY 10014. (212) 966-4873. www.actupny.org. Individuals "united in anger and committed to direct action to end the AIDS crisis." Seeks to increase public awareness and government involvement in the fight against AIDS through rallies and demonstrations.

REVIEW QUESTIONS

What does it mean to say that illness is a social construction and a moral status?

How have explanations for illness changed over time, and how have explanations for illness blamed ill people for their illnesses?

What is the medical model of illness and what are some of the problems with that model?

What is medicalization, why does it occur, and what are some of its consequences?

How might the Human Genome Project act as social control?

What is the sick role model, and what are some of the problems with that model?

INTERNET EXERCISES

Although medical sociologists, health psychologists, and doctors are all interested in issues related to illness, their specific interests vary greatly. Using your library or the Web, obtain access to the major online indexes in these three fields: *Medline, Sociofile* or *Sociological Abstracts,* and *PsycInfo.* Search each database for information on *susto* and on medicalization. How does coverage of these issues differ across fields? To what extent does coverage overlap? What does this tell you about researchers who work in these three fields?

Using your library or the Web, obtain access to *Periodical Abstracts, the Readers Guide to Periodical Literature,* or another index of popular magazine articles. Look for articles on premenstrual syndrome (PMS) published in the last five years. Copy the results of your search onto a diskette, or download it to your hard drive. Based on the titles of the articles, sort the articles into those that seem to assume PMS is an objectively defined illness, those that question the meaning or existence of PMS, and those whose position is unclear. What does this tell you about the medicalization of PMS?

CHAPTER 6

The Experience of Chronic Illness and Disability

—

Nancy Mairs is a writer, teacher, social activist, mother, and wife who has multiple sclerosis (MS). She writes,

> *I am a cripple. I choose this word to name me. . . . People—crippled or not—wince at the word "crippled," as they do not at "handicapped" or "disabled." Perhaps I want them to wince. I want them to see me as a tough customer, one to whom the fates/gods/viruses have not been kind, but who can face the brutal truth of her existence squarely. As a cripple, I swagger. . . .*
>
> *I haven't always been crippled. . . . When I was 28 I started to trip and drop things. What at first seemed my natural clumsiness soon became too pronounced to shrug off. I consulted a neurologist, who told me that I had a brain tumor. A battery of tests, increasingly disagreeable, revealed no tumor. About a year and a half later I developed a blurred spot in one eye. I had, at last, the [symptoms] . . . requisite for a diagnosis: multiple sclerosis. I have never been sorry for the doctor's initial misdiagnosis, however. For almost a week, until the negative results of the tests were in, I thought that I was going to die right away. Every day for the past nearly ten years, then, has been a kind of gift. I accept all gifts.*
>
> *Multiple sclerosis is a chronic degenerative disease of the central nervous system. . . . During its course, which is unpredictable and uncontrollable, one may lose vision, hearing, speech, the ability to walk, control of bladder and/or bowels, strength in any or all extremities, sensitivity to touch, vibration, and/or pain, potency, coordination of movements—the list of possibilities is lengthy and, yes, horrifying. One may also lose one's sense of humor. That's the easiest to lose and the hardest to survive without. . . .*
>
> *I don't like having MS. I hate it. My life holds realities—harsh ones, some of them—that no right-minded human being ought to accept*

without grumbling. One of them is fatigue. I know of no one with MS who does not complain of bone-weariness. . . . As a result, I spend a lot of time in extremis and, impatient with limitation, I tend to ignore my fatigue until my body breaks down in some way and forces rest. Then I miss picnics, dinner parties, poetry readings, the brief visits of old friends from out of town. . . . [As a result,] my life often seems a series of small failures to do as I ought. . . .

[Over time], I [have] learned that one never finishes adjusting to MS. I don't know now why I thought one would. One does not, after all, finish adjusting to life, and MS is simply a fact of my life—not my favorite fact, of course—but as ordinary as my nose and my tropical fish and my yellow Mazda station wagon. It may at any time get worse, but no amount of worry or anticipation can prepare me for a new loss. My life is a lesson in losses. I learn one at a time (1986:9–12, 19).

Nancy Mairs' story illustrates some of the central issues faced by those who live with **chronic illness** or disability—how to obtain an accurate diagnosis, come to terms with a body that does not meet social expectations for behavior or appearance, nurture social relationships despite a contrary body, and construct a viable and life-sustaining sense of self. In this chapter, I look at these and other issues in the lives of people who have chronic illnesses or disabilities. I also describe the social context in which these individuals live and show how that context can structure and limit individuals' lives at least as much as the bodily changes Mairs describes.

I begin this chapter by exploring the meaning and history of disability. I then discuss the extent and social distribution of disability in the United States. Finally, I describe the experience of living with chronic illness and disability.

UNDERSTANDING DISABILITY

Defining Disability

As the previous chapter explained, the meaning of the term *illness* is far from obvious. The same is true for the term *disability.*

Competing definitions of disability reflect competing stances in an essentially political struggle. The **World Health Organization**'s definition is probably the most widely used. WHO defines disability in terms of "impairments": any "disturbances in body structures or processes which are present at birth or result from later injury or disease. . . . An impairment is any loss or abnormality of psychological, physiological, or anatomical structure or function" (1980:47). WHO defines disability as "any restriction or lack (resulting from an impairment) of ability to perform an activity in the manner or within the range considered normal for a human being." The concept of disability, then, includes some but not all persons who have chronic illnesses (the majority of those with disabilities) as well as, for example, persons who are born deaf or who become paralyzed in an auto accident.

As many disability activists have noted (including some who are social scientists, such as Harlan Hahn and the late Irving Zola), this definition reflects a **medical model,** which locates impairments, and, consequently, disabilities, solely within the individual mind or body.

At first glance, such a definition seems perfectly reasonable. After all, isn't a disability something that an individual has, a defect in his or her body? According to many people with disabilities, however, the answer is no. Instead, they argue, the disabilities that they experience stem primarily not from their physical differences but from the way others respond to those differences and from the choices others have made in constructing the social and physical environment. For example, a person whose energy waxes and wanes unpredictably during the day might be able to work forty hours per week on a flexible schedule but be unable to do so within a rigid 9 to 5 schedule. Similarly, someone who uses a wheelchair might find it impossible to work in an office where furniture is constructed and arranged to fit persons who walk and are of average height but might have no problems in an office with more adaptable furniture. Disability activists argue that making an office accessible to wheelchair users does not mean providing special benefits for the disabled, but rather compensating for the unacknowledged benefits that existing arrangements offer those who walk, such as chairs to sit in, stools for reaching high shelves, and carpeted floors that make walking easier but wheeling more difficult.

This approach reflects a *sociological* **model of disability** in its emphasis on social forces and public issues rather than on individual physical variations and troubles. In the rest of this chapter, I will use the term *disability* to refer to restrictions or lack of ability to perform activities resulting largely or solely either from social responses to bodies that fail to meet social expectations or from assumptions about the body reflected in the social or physical environment.

These two models of disability—the medical model and the more sociological model used by disability activists—have strikingly different implications. As Paul Higgins (1992:31) notes, "To individualize disability [as the medical model does] is to preserve our present practices and policies that produce disability. If disability is an internal flaw to be borne by those 'afflicted,' then we do not question much the world we make for ourselves. Our actions that produce disability go unchallenged because they are not even noticed." Individualizing disability, therefore, exemplifies the broader process of **blaming the victim,** through which individuals (in this case, people with disabilities) are blamed for causing the problems from which they suffer (Ryan, 1976); an example is the common belief that women would not be battered if they did not provoke their husbands in some way (Dobash and Dobash, 1998). In contrast, the sociological model of disability challenges us to look at the problem of disability from a very different perspective. If we conclude that the problem resides primarily in social attitudes and in the social and built environment, then we can solve the problem most efficiently by changing attitudes and environments, rather than by "rehabilitating" people with disabilities.

People with Disabilities as a Minority Group

Once we start thinking of disability as primarily a matter of social attitudes and built environments rather than as individual deficiencies, strong parallels emerge between people with disabilities and members of minority groups (Hahn, 1985). A **minority group** is defined as any group that, because of its cultural or physical characteristics, is considered inferior and subjected to differential and unequal treatment and that therefore develops a sense of itself as the object of collective discrimination (Wirth, 1985). Few would argue with the assertion that we differentiate disabled persons from others on the basis of physical characteristics. But can we also argue, as the definition of a minority group requires, that people with disabilities are considered inferior and subject to differential and unequal treatment?

Unfortunately, yes. Even a cursory look at the lives of people with disabilities reveals widespread prejudice and discrimination. **Prejudice** refers to unwarranted suspicion, dislike of, or disdain toward individuals because they belong to a particular group, whether defined by ethnicity, religion, or some other characteristic. Prejudice toward disabled persons is obvious in the fact that, throughout history, most societies have defined those who are disabled as somehow physically or even morally inferior and have considered disabilities a sign that either the individual or his or her parents behaved sinfully or foolishly (Albrecht, 1992).

Prejudice typically expresses itself through **stereotypes,** or oversimplistic ideas about members of a given group. Nondisabled people typically stereotype those who are disabled as either menacing and untrustworthy or as childlike—asexual, dependent, mentally incompetent, the passive "victims" of their fate, and suitable objects for pity (Zola, 1985). These attitudes permeate the health care world as well as the general public. In one study, for example, researchers divided a large sample of health care students and practitioners into two groups and showed each group a videotape of a job interview. Both videotapes used the same actors and scripts, but in one the actor playing the job applicant walked and in the other he used a wheelchair. Those who saw the videotape with the "disabled" applicant rated the applicant significantly more cruel, selfish, incompetent, weak, dependent, and mentally unstable than did those who saw the same actor portraying a nondisabled applicant (Gething, 1992).

Stereotypes about people with disabilities are so strongly held that obvious evidence regarding the falsity of those stereotypes scarcely affects social attitudes. For example, attorney Marylou Breslin, the executive director of the Berkeley-based Disability Rights Education and Defense Fund and a wheelchair user, tells of waiting at the airport for a flight in her dressed-for-success businesswoman's outfit, sipping from a cup of coffee. "A woman walked by, also wearing a business suit, and plunked a quarter into the plastic cup Breslin held in her hand. The coin sent the coffee flying, staining Breslin's blouse, and the well-meaning woman, embarrassed, hurried on" (Shapiro, 1993:19).

Stereotypes about people with disabilities are reflected and perhaps reinforced in the popular media, which often portray disabled individuals as pitiful, maladjusted, or evil (Higgins, 1992:80–97; Safran, 1998). In book and film characters from Captain Hook in *Peter Pan* to Freddie Krueger in the *Nightmare on Elm Street* and the selfish and nasty Lizzy Quinn in *Waking Ned Devine,* the media have equated physical deformity with moral deformity. Moreover, when the media do not portray persons with disabilities as horrifying, they usually portray them as pitiful, whether Tiny Tim in Charles Dickens's classic novel, *A Christmas Carol,* or the Elephant Man (for whom death was preferable to life). Although contemporary media sometimes do present more positive images, such as stories about people with disabilities who have "heroically" compensated for their physical differences or have chosen to live "saintly" lives, these stories too have ignored the social nature of disabilities, instead depicting persons with disabilities as inherently different from others and focusing on how particular individuals have overcome individual deficiencies through personal strength. Exceptions to these rules—films such as *Waterdance, My Left Foot,* and *Children of a Lesser God*—remain rare, although they have become far more common in the last twenty years.

All too often, these prejudices against persons with disabilities result in **discrimination,** or unequal treatment grounded in prejudice. As recently as the first decades of the twentieth century, American laws forbade those with epilepsy, leprosy, Down syndrome, and other conditions from marrying and mandated their institutionalization or sterilization (Schneider and Conrad, 1983:32–33; Shapiro, 1993:197). During the 1930s and 1940s, doctors working for the government of Nazi Germany murdered about 100,000 disabled children and adults as *"Lebensunwertes Leben"*—life unworthy of life (Lifton, 1986). Currently, almost half (49 percent) of working-age disabled Americans are unemployed, a higher proportion than among any minority ethnic group and virtually unchanged since 1983 (Trupin et al., 1997). Yet 79 percent of unemployed, disabled Americans say they can and would work if given the opportunity (*Harris Poll*, 1994). Moreover, and controlling for other variables such as education and experience, the income gap between disabled and nondisabled workers has widened over time, with part- and full-time disabled workers now earning only 64 percent and 85 percent respectively of what nondisabled workers earn (LaPlante et al., 1996). In addition, those whose disabilities carry little social disgrace, like amputees, earn higher average wages than do those whose disabilities carry greater social disgrace, like persons with disfiguring burns (Johnson and Lambrinos, 1987).

According to the definition given earlier, however, minority groups not only experience prejudice and discrimination but also consider themselves objects of collective discrimination. This is the weakest link in defining persons with disabilities as a minority group (Higgins, 1992:39–44). Unlike members of most other minority groups, disabled individuals typically have nondisabled parents. As a result, they might have little contact with, let alone sense of connection to, other people with disabilities. Moreover, fewer than 15 percent of people with disabilities are born disabled (Shapiro, 1993:7). Consequently, most establish their sense of individual and group identity before they become disabled and not all will change their sense of identity following disability. In addition, those who develop a sense of community with others who share their disability do not necessarily feel a connection to persons with other disabilities; deaf people, for example, might identify with others who are deaf, but not with those who have arthritis. As with women during the early days of the feminist movement, however, the sense of belonging to a broader group appears to be growing. As a result, disabled Americans increasingly have come to believe that they deserve not charity—as exemplified by the Muscular Dystrophy telethon, with its implications of inferiority and pity—but the same rights as other citizens to live, work, study, and play in the community.

These sentiments are most obvious in the emergence of the **Disability Rights Movement.** This movement emerged from the social turmoil of the 1960s (Scotch, 1989). The rise of the civil rights movement, followed closely by that of the feminist movement, opened public debate on prejudice, discrimination, and rights. This debate helped Americans with disabilities to see similarities between themselves and other socially marginal groups and to adopt the language of rights.

One of the earliest expressions of the Disability Rights Movement was the growth, beginning in the 1970s, of the **Independent Living Movement** (Shapiro, 1993:41–73). The goal of this movement—organized by and for persons with disabilities—is to provide the services people with disabilities need so they can live and function in the community rather than confined to institutions or private homes. As of 1999, almost 500 independent living centers exist around the country, an increase from about 300 in

1995 (personal communication, Institute for Rehabilitation and Research, Baylor College of Medicine, June 1999).

The concept of disability rights took a major step forward with passage of Section 504 of the federal Rehabilitation Act of 1973. This landmark legislation declared discrimination on the basis of disabilities illegal in programs that receive federal funding, including schools, businesses, and social organizations. Two years later, Congress passed the Education for All Handicapped Children Act. This Act requires school districts to educate all children regardless of disability in the least restrictive environment feasible. With passage of this Act, school systems could no longer declare some children uneducable or relegate them to segregated, often-inferior, residential schools.

More recently, in 1990, Congress passed the **Americans with Disabilities Act (ADA),** which took effect in 1992. The ADA's potential effects are far-reaching because it applies to all private as well as public enterprises; all persons who have disabilities, including mental disabilities; and all who are regarded, whether accurately or inaccurately, as having disabilities (including those with **HIV disease**). The Act outlaws discrimination in employment, public services (including public transit), and public accommodations (including restaurants, hotels, and stores). It also requires that existing public transit systems and public accommodations be made accessible, along with all new public buildings and major renovations of existing buildings, and requires employers to make "reasonable accommodations" in job assignments or work environments to enable qualified employees to perform effectively. Box 6.1 describes the work of Disability Rights Advocates, an organization that fights to enforce the legal rights of disabled persons.

To explore how the ADA has affected the work environment, sociologists Sharon Harlan and Pamela Robert (1998) interviewed a nonrandom but diverse sample of disabled, nonmanagerial civil service workers in one state. One-third of their subjects (32 percent) had never requested an accommodation, either because they were not familiar with the procedures or because they assumed their employers would be able to control the process, and so concluded that doing so would be more likely to call attention to their disabilities and threaten their job than to result in meaningful accommodations. Those who did not request accommodations instead tried to compensate for their disabilities through such actions as working longer hours, working even when sick, or refusing promotions that would bring them more problematic work conditions.

Of those workers who had requested accommodations, 69 percent had been granted or were still awaiting resolution. Requests were most often granted for men, for whites, and for persons in higher status jobs. Employers were more likely to grant requests for changes in the physical environment, such as providing adaptable furniture or disabled parking, than for changes in the social environment, such as offering flexible work schedules

---

Box 6.1   *Making a Difference: Disability Rights Advocates*

Disability Rights Advocates (DRA) is a non-profit law firm that uses individual lawsuits, class-action lawsuits, and the threat of lawsuits to fight for the rights of persons with disabilities (www.dralegal.org). Its staff consists of a few paid lawyers (some of whom have disabilities) and numerous volunteer lawyers and law students. In recognition of its excellent work, it has received ongoing funding from various foundations and associations, including the Kaiser Family Foundation, the San Francisco Foundation, and various local Bar Associations.

Although initially DRA worked solely on California cases, it grew rapidly into a national organization. In addition, since 1995, when DRA received a grant from the private, non-profit Soros Foundation and matching funds from the U.S. State Department, it has run an advocacy program for disabled persons in Hungary, as well as leadership training programs open to disabled persons from across Eastern Europe.

Some of DRA's recent successes include increasing access to California public schools by educating disabled students and their parents about their rights and helping school administrators remove barriers; settling a statewide lawsuit against Denny's restaurants, as a result of which all California Denny's are being made accessible; and using legal pressures to con-vince one California city with a particularly poor record to hire more building inspectors, train its inspectors in disability access laws, and address the backlog of complaints from disabled persons against the city. In addition, in the last few years legal pressure was used to convince Greyhound Bus Lines to provide accessible rest stops and assist riders with disabilities, to convince a major national car rental company to make vehicles with hand controls reasonably available, to convince a hospital to provide sign language interpreters for patients and their relatives who are deaf, to convince several hotels to increase the number of accessible guest rooms and to remove physical barriers, and to convince a major supermarket chain to begin providing assistance to disabled shoppers.

Finally, to support those who pro-actively work to advance the rights of disabled Americans, each year DRA gives out its ADA Eagle Awards. In 1997, Awards were given to Nordstrom, Inc. for its commitment to making its stores accessible to persons with disabilities and its use of models with disabilities, to Marriott International and Noah's Bagels for their efforts to hire and accommodate employees with disabilities, and to NBC's "Dateline" show for its coverage of housing and employment discrimination against people with disabilities.

---

or personal assistance. These findings led the authors to conclude that employers will offer accommodations only if those accommodations do not threaten the authority structure of the work place by suggesting that workers should be granted more flexibility or autonomy.

When employers refuse requests for accommodation, workers have the option of filing complaints with the federal Equal Employment Opportunity Commission or filing law suits. Of the 107,000 persons whose complaints have been resolved since 1992, only 11.4 percent both won their cases and received benefits as a result (Equal Employment Opportunity Commission, 1999). Similarly, of cases that have gone to trial, employees have won only 8 percent (American Bar Association, 1998). In most cases, workers have lost their suits because the courts have used very narrow definitions of who is disabled and thus qualified for protection under the ADA (Gostin et al., 1999). For example, employers have argued that individuals whose diabetes is controlled by insulin, whose spinal cord injuries keep them from working rigid hours but not from performing their jobs satisfactorily, or whose back injuries limit them to nonstrenuous work are not disabled and therefore not eligible for ADA protection *even if they experience discrimination due to their physical condition.* In 95 percent of cases tried during 1995 and 1996 in which the definition of disability was an issue, the courts concluded that the individuals were not disabled (Gostin et al., 1999). Similarly, in 1999 the U.S. Supreme Court concluded that persons whose medical problems are corrected by medication or medical devices (such as blood pressure medication or eye-glasses) are not disabled and therefore can be fired or not hired because of those medical problems.

The Social Distribution of Disability

To understand the importance of disability, we need to understand how widespread it is. This is not quite the same as asking how widespread chronic illnesses or other physical problems are, because illness and physical problems do not necessarily translate into disabilities. The same condition can cause serious problems for one person but only minor annoyances for another; one person with diabetes, for example, might need to eat balanced meals but otherwise live a normal life, whereas another might become blind and lose a leg because of diabetic complications. Similarly, arthritis has little impact on some workers but can end the career of those whose work depends on manual dexterity or whose employers refuse to make any accommodations. To understand the impact on individual lives, therefore, we need to look not so much at the distribution of chronic illness or health problems as at the distribution of disability.

A substantial portion of the U.S. population lives with disabilities. According to U.S. government researchers, approximately 14.7 percent of noninstitutionalized persons living in the United States have a disability, defined by these researchers as a chronic health condition that makes it difficult to perform one or more activities generally considered appropriate for persons of a given age—play or study for children, work for adults, or basic activities needed to maintain an independent life (shopping, dressing, bathing, and so on) for the elderly (National Center for Health Statistics, 1998c:101). (Of course, these same chronic health conditions might not cause disabilities in a different physical, social, or financial environment.) Approximately 4.6 percent of American noninstitutionalized adults of working age cannot obtain work, and another 5.5 percent are limited in the amount or kind of work they can obtain because of disabilities. The most common disabilities are orthopedic impairments, arthritis, and heart disease, which together account for about 40 percent of all persons with disabilities (National Center for Health Statistics, 1998c:101).

The proportion of the population living with disabilities has grown significantly over time (Kaye et al., 1996). Only a few decades ago, most paraplegics, babies born with serious birth defects, and persons with serious head or spinal injuries died quickly. Now most live, although often with serious disabilities. In addition, average survival times for various common chronic conditions, such as hypertension and cardiovascular disease, have increased. Finally, as the proportion of the population over age 65 has increased—and in the absence of meaningful attempts to remove the social and physical barriers that can prevent individuals from living independent lives—so has the proportion living with disabilities. As Table 6.1 shows, within each income category, the percentage with activity limitations (i.e., unable to perform some basic life activity such as shopping or dressing oneself) increases as age increases, for longer lives translate into more years in which to have accidents or develop degenerative diseases. (The apparent decrease in disability in the oldest age group reflects the narrow definition of disability used for this age group.) Even among persons above age 70, however, more than half report no disabilities.

As Table 6.1 also shows, within each age group, more lower- than upper-income persons report activity limitations (National Center for Health Statistics, 1998c:102). Ethnicity, too, affects rates of disabilities, largely because of its relationship to income. By their early 30s, about 12 percent of Native American men have been unable to work or limited in the work they can do because of illness or injury for at least six months (Hayward and Heron, 1999). In contrast, disability does not become equally common among African American men until their late 30s. Even more startling, white and Hispanic men do not reach this rate of disability until their early fifties and Asian-American men not until their early 60s. Gender, on the other hand, has almost no independent effect on rates of disability. However, because women live longer than men, the majority of disabled persons are women.

These statistics help us understand the extent and distribution of disability in the United States, but they tell us little about how chronic illness and disability affect individual lives. The next section looks at the experience of living with chronic illness and disability, focusing on the central issues of uncertainty, stigma, and the self.

LIVING WITH CHRONIC ILLNESS AND DISABILITY

Living with chronic illness and disability is a long–term process, which includes responding to initial symptoms, injuries, or diagnoses; making sense

Table 6.1 **Percentage of Persons with Chronic Activity Limitation by Family Income and Age, 1995**

FAMILY INCOME AND AGE	NO ACTIVITY LIMITATION	LIMITATION IN MAJOR ACTIVITY*	LIMITATION IN ANY NONMAJOR ACTIVITY**
UNDER $10,000:			
All ages	71.8%	21.4%	6.8%
Under 18 years	90.8	6.8	2.4
18–44 years	79.6	15.7	4.8
45–64 years	38.3	58.3	8.5
65–69 years	40.7	50.3	9.0
70 years and over	52.7	28.5	18.9
$10,000 – $19,999:			
All ages	77.9	15.8	6.3
Under 18 years	91.0	7.0	1.9
18–44 years	85.0	11.6	3.4
45–64 years	59.2	34.6	6.2
65–69 years	56.1	34.3	9.5
70 years and over	59.6	18.3	22.1
$20,000–$34,999:			
All ages	85.1	10.0	4.9
Under 18 years	93.5	4.8	1.7
18–44 years	90.3	6.5	3.1
45–64 years	75.3	19.1	5.6
65–69 years	62.4	28.3	9.2
70 years and over	66.1	14.6	19.2
$35,000 OR MORE:			
All ages	90.8	5.7	3.5
Under 18 years	95.3	3.0	1.7
18–44 years	92.9	4.4	2.7
45–64 years	86.3	8.9	4.8
65–69 years	77.1	16.8	6.2
70 years and over	67.7	13.8	18.5

*Major activities are defined as play for children under age 5, study for children ages 5 to 17, work or house-keeping for adults under age 70, and activities needed to live independently (shopping, dressing, bathing, and so on) for those over age 70.
** Nonmajor activities are any other activities that the individual regards as important.

Source: National Center for Health Statistics (1998c:101).

of one's situation; and continually reconceptualizing one's future. In this section, I describe this process and also explore how illness and disability affect individuals' lives, relationships with others, and sense of self.

Initial Symptoms and Diagnosis

Becoming a chronically ill or disabled person begins with recognizing that something about the body is troubling. This recognition does not always come easily. Health problems often build gradually, allowing individuals and their families slowly and almost unconsciously to adapt to them and to minimize their importance (Bury, 1982; Charmaz, 1991:24–28; Schneider and Conrad, 1983; Stewart and Sullivan, 1982). In addition, the signs of illness and disability often do not differ greatly from normal bodily variations. A child who doesn't walk by 12 months might have a disability or might simply be a slow developer. Similarly, children with epilepsy, for example, can for many years experience "strange feelings," "headaches," "spaciness," "blackouts," and "dizzy spells" before they or their families recognize these as signs of epilepsy. As one man recalled,

---

Box 6.2   *Ethical Debate: Advertising Drugs Direct to Consumers*

Since 1985, the federal Food and Drug Administration (FDA) has allowed the advertising of legal drugs under their generic (chemical rather than brand) names directly to consumers. Such advertisements, however, were not common because most pharmaceutical companies believed they could sell their products most effectively by advertising to doctors. By the early 1990s, however, doctors no longer had the freedom to prescribe whatever drug they thought best but, rather, faced increasing pressure from insurance companies to prescribe only those drugs the companies considered most cost-effective. As a result, pharmaceutical companies reversed their earlier policies and dramatically increased their advertisements directly to consumers to encourage consumers to request specific drugs from their doctors ·(*Consumer Reports,* 1996). Advertisements directed at consumers grew even more common following the 1997 FDA decision to allow advertising to consumers by

brand name. Pharmaceutical manufacturers budgeted $1.3 billion dollars for consumer advertisements during 1999 and $2 billion for the year 2000 (Deam, 1999). These strategies have paid off for pharmaceutical manufacturers, as sales of advertised drugs have skyrocketed. Sales also have been brisk at the more than a dozen Internet Web sites that, as of 1999, not only advertised but also prescribed and sold drugs direct to consumers, based on brief questionnaires consumers filled out describing their health problems (Stolberg, 1999a).

Pharmaceutical companies that advertise drugs directly to consumers argue that such advertising is simply an extension of normal business practices, no different from any other form of advertising. In addition, these companies argue, such advertising is a public service because it encourages consumers to seek medical care for problems they otherwise might have ignored at their peril. Moreover, the companies

I'd always had the tendency to roll my eyes back in my head . . . to kind of fade out for a while. But I thought that was nothing, but . . . I guess they call them petit mal [epileptic seizures]? I'd lose consciousness for a while. I wasn't really conscious of it and [the only] time anybody would notice it was when the family was all together at the dinner table and I, I'd be like daydreaming for a while and then I'd roll my eyes back and they'd go, "Stop that!" and I'd go "Stop what?," y'know, I didn't know what I was doin' (Schneider and Conrad, 1983:57–58).

Social scientists refer to this process of defining, interpreting, and otherwise responding to symptoms and deciding what actions to take as **illness behavior** (Mechanic, 1995). A review article by anthropologists Vuckovic and Nichter (1997), summarizing twenty years of research studies, concluded that U.S. residents treat between 70 and 95 percent of all illness episodes without a doctor's assistance. Individuals typically begin by medicating themselves or those under their care with nonprescription medications recommended by friends, families, store clerks, or pharmacists or, more rarely, with prescription medicines left over from previous illnesses. (Box 6.2 analyzes the ethics of advertising drugs directly to consumers.)

argue, these advertisements pose no health risks because consumers still must get a prescription before they can purchase a drug, leaving the final decisions in doctors' hands. Thus, any attempts to control such advertisements represent an unfair intrusion on manufacturers' right to earn a living and a paternalistic infringement on consumers' right to make independent choices about their health care.

Those who oppose such advertisements, on the other hand, argue that consumers lack the necessary medical expertise needed to evaluate the information contained in these advertisements. Nor do the advertisements generally contain all the necessary information: a review of 28 drug advertisements, conducted by the nonprofit Consumers Union, found that only about 40 percent fairly described the benefits and risks in the main text (*Consumer Reports,* 1996). The rest either left out this information or buried it in tiny, jargon-laden sidebars or footnotes.

Moreover, since the purpose of direct-to-consumer advertisements is to encourage consumers to pressure their doctors to prescribe certain drugs to them, it is disingenuous of advertisers to argue that doctors will protect consumers from making poor drug choices, especially because the increasing pressures on doctors to see more patients in less time makes it increasingly likely that doctors will acquiesce to patient requests rather than taking the time to explain why those requests might be unwise. Finally, when companies not only advertise drugs but also arrange for their prescription via the Internet, the risk to consumers is far greater because in these circumstances consumers receive drugs based only on a limited list of close-ended questions, without a physical examination, and without any follow-up care. Thus, opponents conclude that direct-to-consumer advertising is necessarily a harmful and, hence, unethical business practice.

Research results are mixed regarding whether gender or ethnicity affects use of self-medication, but age clearly has an impact: persons over age 65 are considerably more likely than others to self-medicate, with the majority using one or more nonprescription drugs regularly (Vuckovic and Nichter, 1997). Social class does not affect the *use* of self-medication but does affect the *reasons* for doing so: affluent persons are more likely to self-medicate to save time, whereas poorer persons are more likely to do so to save money. For all Americans, however,

> cultural demands to be productive and practical contingencies related to job/household responsibilities make time off for illness a luxury few Americans can afford. As popular commercials for cold and flu remedies remind mothers, construction workers, and teachers, there is simply no time to be ill. Pressures of the clock inherent in modern life often prohibit taking time for the extra sleep necessary to care for a cold or for the relaxation required to relieve a "stress" headache. In the past, individuals who were ill might "tough it out," waiting for symptoms to subside. Today, Americans can avoid delays by taking products "strong enough to tackle even the toughest cold." Medicines obviate the need to devote time and energy to healing activities, or to the "down time" necessitated by ill health (Vuckovic and Nichter, 1997:1289).

Similarly, parents sometimes use medications to control a child's symptoms so that the child can go to school or day care and the parents won't have to take time off from work. In these cases, parents treat children's symptoms most aggressively when they occur midweek, hoping to ward off the illness until the weekend—a pattern referred to by one working mother as the "Oh my God, not on Wednesday" syndrome (Vuckovic and Nichter, 1997:1290).

When and whether individuals seek formal diagnosis for acute or chronic medical problems depends on a variety of factors. According to the illness behavior model developed by David Mechanic (1995) and summarized in Table 6.2, the likelihood of seeking formal diagnosis depends, first, on the presence of alternative explanations for symptoms and the frequency, visibility, and severity of those symptoms (including most importantly how much they interfere with usual daily activities). In turn, how individuals interpret these factors depends on the social context; symptoms that seem serious to a middle-class professional who generally enjoys good health might seem quite minor to a homeless or elderly person who expects a certain amount of bodily discomfort. Social networks of friends and relatives also play a large role in determining how individuals will interpret and respond to symptoms because those networks can reinforce either a medical or a nonmedical interpretation of the problem and of how to treat it (Pescosolido, 1992). Finally, access to care and attitude toward health care providers also affect how quickly individuals seek care;

Table 6.2	*Some Factors Predicting Illness Behavior*
ILLNESS BEHAVIOR LIKELY WHEN:	**ILLNESS BEHAVIOR UNLIKELY WHEN:**
Symptoms appear frequently or persistently (e.g., coughing blood once per day for a week)	Symptoms appear infrequently (e.g., coughing blood every few months)
Symptoms are very visible (e.g., rash on face)	Symptoms are not very visible (e.g., rash on lower back)
Symptoms are severe enough to disrupt normal activities (e.g., epileptic convulsions)	Symptoms are mild (e.g., annoying but tolerable headaches)
Illness is only likely explanation for physical problems	Alternative explanations available (e.g., recent stresses could explain headaches)
Ready access to health care (e.g., good health insurance)	Poor access to health care (e.g., no health insurance)
Positive attitude to health care providers (e.g., trusts doctors' abilities and motives)	Negative attitude to health care providers (e.g., distrusts doctors' abilities and motives)

those who can afford care only from public clinics and whose experience of clinics has taught them to expect long waits and rude treatment often put off seeking care for some time.

Eventually, however, if symptoms persist and, especially, if they progress, individuals and their families are likely to reach a point where they cannot avoid recognizing that something is seriously wrong. As their previous interpretations of their symptoms crumble, individuals find themselves in an intolerable situation, torn by uncertainty regarding the changes in their bodies and their lives. Once they reach this point, the incentive grows to seek diagnosis and treatment because any diagnosis can become preferable to uncertainty.

Seeking a diagnosis, however, does not necessarily mean receiving one. Although some problems are relatively easy to diagnose—a 45-year-old white man who complains to his doctor of pains in the left side of his chest will probably quickly find himself getting tested for a heart attack—others are far less obvious. Persons with multiple sclerosis, for example, often find that doctors initially dismiss their symptoms as psychosomatic or trivial (Register, 1987; Stewart and Sullivan, 1982). In addition, the same symptoms may more rapidly produce a diagnosis for some than for others. For example, and as mentioned earlier, doctors more often ascribe women's than men's complaints to emotional problems rather than to physical illness (Bernstein and Kane, 1981; Corea, 1985; Council on Ethical and Judicial Affairs, 1991; Steingart, 1991; Zola, 1991;).

Initially, both women and men can find these alternative diagnoses comforting and welcome—after all, it is far easier to hear that one is suffering from stress than that one has a serious illness. When symptoms persist, however, individuals find themselves torn by ambiguity and uncertainty, suffering anxiety about their failing health but receiving little sympathy or help from relatives and colleagues (Bury, 1982; Schneider and Conrad, 1983; Stewart and Sullivan, 1982; Waddell, 1982).

These problems can motivate individuals to seek more accurate diagnoses. Some go from doctor to doctor, seeking a more believable diagnosis; others research their symptoms, diagnose themselves, and then press their doctors to confirm their self-diagnoses through testing. The story of one such individual, interviewed by writer Cheri Register, illustrates this process:

> It started about ten years ago, and my first symptom was a lot of dizziness. I spent about three months just crawling on the floor with my children. I went through the Mayo Clinic and was told that I had an inner ear disorder that I would just have to live with. Time went on and it got better and I forgot about it. Then, as time went by, other symptoms started appearing: numbness in my legs, inability to walk at times, double vision, a lot of bladder and kidney problems, a lot of bowel problems. I was told that if I just got busy and away from my children, all of these things would clear up by themselves. The doctors called it "housewife's syndrome."
>
> So I became a supermom. I was the kind of mother who made her own noodles. . . . I also went to work about that time and worked for several years at a job I really enjoyed, until one day I literally fell in a heap and was carted off to the emergency room.
>
> This all happened over a period of five years [during which time I was hospitalized several times for observation, testing, and surgery]. Through this time it caused a great deal of struggle within our family, because I knew there was something wrong with me, and yet I felt I had to keep up. . . .
>
> Finally, when I had my Big Heap, I was introduced to my doctor, my neurologist, who specializes in multiple sclerosis. He put me in the hospital and diagnosed me as having M.S. about two days later. And with that came this wonderful sense of relief. I giggled and laughed. I was joyous. My husband was the same way. We were just like two kids running through a park. We had a name to something. We could deal with it. I was not a neurotic lady (Register, 1987:3–4).

As this story suggests, even diagnosis with a serious illness can prove easier to live with than uncertainty.

Responding to Illness or Injury

Once newly diagnosed or newly disabled individuals learn the nature of their conditions, responses vary widely. Some individuals with HIV disease, for example, find it easiest to cope by immediately considering their diagnosis a "death sentence," eliminating any uncertainty from their minds (Weitz, 1991). Others initially assume they can "beat" their illness, refusing to take seriously any dire predictions about their future. Still others cope by accepting their diagnoses intellectually but denying them emotionally. For example, one young man told how, two months after learning he had AIDS, he thought that he had picked up someone else's medical file when he noticed that his file read, "Caution: Patient has AIDS" (Weitz, 1991). Similarly, following traumatic injuries, some individuals refuse to participate in rehabilitation because they consider their situation hopeless, whereas others refuse because they consider their injuries temporary.

Faced by the uncertainties and loss of control that accompany chronic illness and disability, individuals must reconstruct their image of their futures. Two basic strategies are available to these individuals, as to all who confront uncertainty—**avoidance** and **vigilance** (Janis and Mann, 1977; Weitz, 1989). Some cope by avoiding knowledge about their conditions so they can maintain earlier images of their futures and ward off depression. Others cope by seeking knowledge vigilantly so that they can feel prepared to respond appropriately to any changes in their bodies. Both strategies reduce uncertainty and give individuals ways of understanding and, thus, responding to their health problems.

Although learning the nature of one's condition answers some questions, it raises new questions about why this has happened. Consequently, those who experience serious illness or injury must not only reconceptualize their futures but also their pasts. Only by doing so can individuals make their situation comprehensible and, consequently, tolerable.

This search for explanations is often a painful one, set as it is in the context of a culture that continues at least partially to believe that individuals deserve their illnesses and disabilities. Nevertheless, some individuals do manage to avoid allocating blame to themselves. For example, one gay man with HIV disease stated in an interview: "Nobody deserves it [HIV disease]. I have friends that say 'Well, hey, if we weren't gay, we wouldn't get this disease.' That's bullshit. I mean, I don't want to hear that from anybody. Because no germ has mercy on anybody, no matter who they are—gay, straight, babies, adults" (Weitz, 1991:68).

Other individuals, however, readily conclude—whether accurately or not—that they caused their own health problems by acting in ways that either contravened "divine laws" or put them at risk (such as smoking tobacco, having multiple sexual partners, or driving fast). As another man with HIV disease stated, "I should have helped people more, or not have yelled at somebody, or been better to my dad even though we have never gotten along. . . . Maybe if I had tried to get along better with him, maybe this wouldn't be happening" (Weitz, 1991:68). Increasingly, too, individuals conclude that they caused their health problems through their psychological conflicts. As described in the previous chapter, this theory has gained considerable public exposure through the writings of Bernie Siegel and others, who have theorized that individuals become ill because they "need" their illnesses.

Interruptions, Intrusions, and Immersions

To understand what it is like to have a chronic illness, sociologist Kathy Charmaz interviewed more than one hundred persons during the course of a decade. These interviews led her to conclude that illness can be experienced as an interruption, an intrusion, or something in which one is immersed. Although her research addressed only chronic illness, similar patterns undoubtedly apply to at least some individuals with disabilities, especially those that worsen over time.

When illness or disability is an **interruption,** it remains only a small and temporary part of one's life (Charmaz, 1991:11–40). Viewing it as an interruption means regarding it essentially as an acute problem—something one has to deal with at the moment, but not something that will have a significant long-term impact. This strategy can work as long as episodes of illness are minor or rare or the disability is a mild one. For example, because of unexpected physical problems, someone with multiple sclerosis may need to change plans for a given day but not necessarily for the next week.

If the illness or disability progresses, however, it can become an **intrusion,** demanding time, accommodation, and attention and requiring that one "live day to day" (Charmaz, 1991:41–72). For example:

> I just take each day as it comes and I don't worry about tomorrow. I know that when I'm feeling good I should try to do as much as I can without overdoing, because sometimes I won't be able to do that (Register, 1987:190).

Similarly:

> My life feels like such a day-to-day event. That's a real difference between my husband and me and a source of conflict between us. He's always making long-range plans. My God, he talks about retirement! It's like each day is a new event for me. The days run together for him. A week just races by. A week is seven days to me (Register, 1987:190).

If the illness or disability progresses still further, individuals can find themselves immersed in their bodily problems (Charmaz, 1991:73–104). Once this stage of **immersion** is reached, individuals must structure their lives around the demands of their bodies rather than structuring the demands of their bodies around their lives. Social relationships often wither, and individuals often withdraw into their selves. Dealing with the body and illness can take most of one's day and require the assistance of others. One woman, for example, told Charmaz that her kidney dialysis "just about takes up the day. . . . I'm supposed to be on at 12:30, but sometimes don't get on until 1:00, then I'm dialyzed for four and a half hours and then it takes approximately half an hour to be taken off the machine and to have it clot. So quite often it's 6:00 or 6:30 before I ever leave there. So the day is shot" (1991:83).

Managing Health Care and Treatment Regimens

Use of Conventional Health Care

Living with chronic illness or disability often means living a life bound by health care regimens. However, in the same manner that, following injury or diagnosis with a chronic illness, some individuals seek and some avoid knowledge, some will strictly follow prescribed regimens of diet, exercise, or medication and others will not. Researchers traditionally have framed this issue as a matter of compliance—whether individuals do as instructed by health care workers.

The most commonly used framework for studying compliance is the **health belief model.** Although, as Chapter 2 described, the model was originally designed to explain why individuals adopt preventive health behavior, it also can be used to understand why people who have acute or chronic health problems comply with medical advice regarding treatment. The model suggests that individuals will be most likely to comply if they believe they are susceptible to a health problem that could have serious consequences, believe compliance will help, and perceive no significant barriers to compliance. For example, individuals who have diabetes will be most likely to comply with their prescribed diet if they believe that they face substantial risks of blindness as a result of diabetes-induced glaucoma, that blindness would substantially decrease their quality of life, that the prescribed diet would substantially reduce their risk of blindness, and that the diet is neither too costly nor too inconvenient.

The health belief model is a useful but limited one for understanding compliance with medical treatment because it largely reflects the medical model of illness and disability. First, the health belief model assumes that noncompliance with medical recommendations stems primarily from psychological processes internal to the patient. Although this is sometimes true, in other cases patients do not comply because health care workers did not sufficiently explain either the mechanics of the treatment regimen or the benefits of following it (Conrad, 1985). Patients also might not comply because they lack the money, time, or other resources needed to do so.

Second, the health belief model implicitly assumes that compliance is always good (that is, that health care workers always know better than patients what patients should do). Yet, although health care workers often can help their patients considerably, this is not always the case, especially with chronic conditions (and is one of the reasons the **sick-role** model does not fit chronic illnesses well). Bodies rarely respond precisely as medical textbooks predict. Nor can those textbooks determine whether an individual will consider a given treatment worth the impact it has on his or her quality of life. For example, persons with bipolar disorder (manic-depression) often resist taking medications because the medications leave them feeling sedated and deprive them of the sometimes–pleasurable highs of mania. Moreover, for numerous chronic conditions, the only available treatments are disruptive to normal routines, experimental, only marginally effective, unpleasant, or potentially dangerous. As a result, many individuals who at first diligently follow prescribed regimens eventually abandon them and lose some of their faith in mainstream health care (Conrad, 1985). Meanwhile, health care providers who do not understand why their patients did not respond to treatment as expected often blame the problem on patient noncompliance, further eroding relationships between patients and providers and leading to future noncompliance.

As individuals' faith in mainstream medicine declines, some begin experimenting with their treatment regimens, learning through trial and error what works best for them, not only physically, but also socially, psychologically, and economically (Conrad, 1985). Others begin using **alternative or "complementary" therapies** (defined broadly as treatments not widely integrated into medical training or practice in the United States).

Use of Alternative Therapies

Interest in alternative therapies has grown rapidly in the United States, among both healthy persons interested in avoiding illness and among those with chronic or acute illnesses. The most widely cited data on use of alternative therapies comes from two national, **random** surveys of English-speaking U.S. residents, conducted by the same Harvard-based research team in 1990 and again in 1997 (Eisenberg et al., 1998). The researchers looked at use of 16 alternative therapies, including chiropractic, acupuncture, megavitamins, "folk" remedies, and biofeedback. Forty-two percent of respondents reported using at least one alternative therapy in 1997, compared with 34 percent in 1990, and 46 percent of respondents reported visiting an alternative health care provider in 1997, compared with 36 percent in 1990. Respondents made more visits in 1997 to alternative health care providers than to general and family practitioners combined.

Users of alternative therapies are disproportionately likely to be female, upper-income, middle-aged, college-educated, not African-American, and suffering from chronic health problems (Astin, 1998; Eisenberg et al., 1998; Fairfield et al., 1998; Krauss et al., 1998). The most commonly used therapies are relaxation techniques, followed by herbal medicine, massage, and chiropractic. Respondents typically use alternative therapies to complement rather than to replace mainstream medicine: whereas 32 percent of those who visited a medical doctor because of injury or illness also used an alternative therapy, virtually all—96 percent—of those who visited an alternative therapist also visited a medical doctor (Eisenberg et al., 1998).

The popularity of alternative therapies rests on belief—or at least hope—in the efficacy of these treatments, combined with a belief that "natural" treatments are unlikely to do harm. These beliefs are supported both by personal experience and by recommendations from friends and acquaintances who believe alternative therapies have helped them. In some of these cases, the therapies no doubt did help, either because of the biological effects of the therapies or because consumers' belief in the therapy enabled the body to heal itself, as happens in about 30 percent of all persons treated with **placebos** (drugs known to have no biological effect). In other cases, individuals attribute cures to therapies when in fact the cures were accomplished by the body's natural healing abilities. Finally, individuals sometimes convince themselves that the therapies helped them even though their health did not actually improve.

To convince people to try alternative therapies and to believe in their efficacy, manufacturers and retailers now spend millions yearly on promotion. For example, GNC, which sells nutritional supplements and other alternative and natural products, recently contracted with the Rite Aid drug store chain to open outlets in 1500 Rite Aid stores and to jointly run an Internet Web site where consumers can learn about and purchase their products. The two companies agreed to spend $30 million during the first year to market the stores and Web site (Janoff, 1999).

Other, less obvious, means are also now used to promote alternative therapies. Mainstream supermarkets routinely devote large sections in prime locations to "wellness products" and alternative therapies, and newsstands are filled with magazines devoted to informing consumers of the reputed health benefits of various alternative therapies and laced with advertisements for those products. Mainstream media, too, regularly run articles and advertisements promoting alternative therapies; a review for this textbook of 68 articles on chiropractors indexed in the *Reader's Guide to Periodical Literature* during 1998 found only four that gave at least equal space to the risks as well as to the potential benefits of chiropractic treatment.

The huge amounts spent on promoting alternative therapies are justified by the even larger amounts of money spent by consumers on such services and products. Eisenberg and his colleagues (1998) conservatively estimate that Americans spent $21.2 billion in 1997 on fees for alternative practitioners, as well as $8.9 billion for herbal therapies and megavitamins and $7.7 billion on books, classes, and equipment related to alternative therapies.

A fascinating study by Matthew Schneirov and Jonathan David Geczik (1996) suggests that neither marketing campaigns nor the potential health benefits of alternative therapies can fully explain the appeal of these thera-pies. Instead, the authors suggest, alternative healing appeals to individuals as a **new social movement,** a term first coined by German sociologist Jür-gen Habermas (1981). Habermas argued that whereas older social move-ments arose out of discontent with material social conditions such as poverty, the new social movements stem from discontent with modern so-ciety's emphasis on science and rationality and its disvaluing of the "**life-world**" of everyday human interaction, identity, and needs. Because new social movements focus on the lifeworld, they are less concerned with po-litical strategies for social change and more concerned with creating ways of living that reflect their values. Thus new social movements depend less on formal organizations and more on "submerged networks" (Melucci, 1995) in which like-minded individuals can trade resources and obtain so-cial support for adopting nonnormative ways of life. Although more recent writers tend to argue that movements cannot be neatly dichotomized into "new" versus "old," Habermas's insight regarding the importance of the lifeworld to social movement growth is nonetheless an important one.

Using Habermas's model, Schneirov and Geczik argue that the rise of alter-native healing reflects dissatisfaction with the lack of match between doctors' concerns and patients' concerns: whereas doctors typically are concerned with solving the puzzle of diagnosis and identifying a specific body part that re-quires treatment, patients are primarily concerned with the impact of illness on their lives (Mechanic, 1995). This mismatch can leave patients feeling like depersonalized objects and deeply dissatisfied with the care they receive, even if it is technically competent. In contrast, Schneirov and Geczik argue, alterna-tive healing offers patients the opportunity to work as collaborators with health care providers and the promise to look holistically at the sources of their health problems and the consequences of any treatments.

Using interviews, ethnographic observations, and focus groups, Schneirov and Geczik uncovered two slightly overlapping submerged networks linked to alternative healing in the Pittsburgh area: one made up of working-class conservative Christians, the other of college-educated followers of spiritual, Eastern, or New Age philosophies. The researchers conclude that

> at the core of alternative health is a commitment to an ecological conception of the body, in which biochemical processes, emotional states, beliefs, lifestyle prac-tices (especially nutrition), and spiritual phenomena are thought to be intercon-nected. Beyond this emphasis on holism is also a commitment to low-tech care; individualized treatment regimes (treating the person not the symptom), in which the patient's intuitions and perceptions of his or her illness is an impor-tant part of diagnosis and treatment; an emphasis on the self-healing capacities of the body; a commitment to something more than the absence of disease—to "wellness" or some positive conception of health; a desire to narrow the power

imbalances between practitioner and patient; and finally an effort to critically appropriate healing traditions that lie outside of Western allopathic medicine (Schneirov and Geczik, 1996:630–31).

Most members joined these networks when confronted by a chronic illness and dissatisfied with the treatment they received from mainstream health care providers, and most of the rest joined while going through some other sort of life crisis. Network members were united by several beliefs: that modern medicine focuses too much on surgery and medication to treat symptoms rather than on lifestyle changes to prevent illness, that government regulation of health care endangers both personal freedom and health, that individuals should commit themselves to taking responsibility for their own health, and that doing so means adopting stringent behavior regimens, such as restrictive diets and regular use of laxatives. Through these shared beliefs, users of alternative healing constructed not only a philosophy of health care but also a shared sense of identity and community. Thus, Schneirov and Geczik conclude, "the alternative health movement may be seen as part of a larger wave of discontent with the bureaucratic-administrative state, its reliance on expert systems, and the way it coordinates people's health care practices 'behind their backs'—without their knowledge and participation" (1996:642).

A New Information Source: The Internet

Whether individuals rely primarily on mainstream or alternative therapies, many seek information about their conditions on their own, rather than relying solely on information provided by health care professionals. In the last few years, public access to information has exploded because of the exponential growth in use of the Internet.

During 1998, approximately 22.3 million adults in the United States are believed to have sought health care information from the estimated 15,000 to 17,000 Internet health sites (Stolberg, 1999a). The most heavily-used of these web sites are those devoted to specific diseases and specific drugs, suggesting that people use the Internet primarily to get information about their own chronic health problems.

Unfortunately, there are no controls on the quality of materials posted on the Internet, and its vast size makes it impossible to police for fraudulent information. In fact, in only one day searching the web during 1999, Federal Trade Commission investigators identified 400 Internet sites purveying apparently fraudulent information (Stolberg, 1999a). Partly in response to concerns about the potential for health fraud, the U.S. Department of Health and Human Services in 1997 established its own Web site (www.healthfinder.gov), which links consumers to 5,000 other sites it regards as reliable sources of health information. During the first half of 1999, 2.5 million people explored the site (personal communication, Department of Health and Human Services, August, 1999).

Dealing with Service Agencies

For those who experience disabilities, whether or not they are chronically ill, dealing with social service agencies can become a major part of life. Unfortunately, and despite the best intentions of many social service providers, the philosophies and structures of those agencies create systems that sometimes harm more than help those they serve (Albrecht, 1992; Higgins, 1992:151–187).

Typically, social service agencies adopt a medical model of disability, focusing on how individuals can compensate for their individual deficiencies rather than on how social arrangements handicap them (Phillips, 1985). This has several **unintended negative consequences.** First, to accept someone as a client, agencies must define him or her as disabled. As a result, workers spend much of their time certifying individuals as disabled—identifying internal individual problems rather than looking for individual strengths. Through this process, individuals learn to think of themselves as disabled. According to Paul Higgins (1992:132), "When service agencies evaluate, place, categorize, transfer, educate, rehabilitate, and so much more, the agencies are informing people who they are and who they are becoming." At the same time, because agencies receive funding based on how many clients they serve, agencies sometimes unintentionally encourage individuals to remain dependent on their services.

Second, because agencies use a medical model that defines persons with disabilities as inherently flawed, agencies typically define "progress" as making persons with disabilities as much like the nondisabled as possible (Albrecht, 1992; Higgins, 1992). As a result, rehabilitation workers might for example, encourage someone to use a false leg even though the individual could move more quickly and less painfully on crutches or in a wheelchair; Box 6.3 describes how this philosophy has affected the education of deaf persons.

Third, the medical model encourages agencies to adopt a hierarchical pattern of care. This pattern of care is based on the premise that social service providers understand clients' needs, desires, problems, and strengths better than the clients themselves do and that social service providers are thus better equipped than clients to make decisions regarding clients' lives. Like other health care professionals, those who work in service agencies "evaluate, plan, treat, monitor, revise, discharge, and in other ways manage people. Disabled people (and their families) are expected to do what they are told" (Albrecht, 1992:178). Thus, unwittingly, agencies encourage dependency.

Social Security, the major governmental program for persons with disabilities, further encourages dependency by economically penalizing those who obtain paid employment. Persons with disabilities who accept paid employment risk losing their government benefits, including both financial assistance and health care. Yet the costs of living with a disability are high; for example, as of 1999, modifying a van for a wheelchair-using driver costs anywhere from $10,000 to $25,000 depending on the changes needed, and medications can cost thousands more per year. As a result, unless individuals can get well-paid professional jobs with full health benefits, they may find employment unaffordable (Burns et al., 1993).

Illness, Disabilities, and Social Relationships

For better or worse, chronic illness and disability alter relationships not only with health care providers and service agencies but also with friends, relatives, and colleagues. Illness and disability can strengthen social relationships, as families pull together to face health problems, old wounds are healed or put aside, and individuals realize how much they mean to each other. Illness and disability also, however, can strain relationships. Friends and family might help each other willingly during acute illnesses, the first few acute attacks of a chronic illness, or the first few weeks following a traumatic injury, but might become more loath to do so if the need for assistance continues over weeks, months, or years. This is especially true for male children, friends, lovers, and spouses, who less often than females are socialized to be caregivers (Cancian and Oliker, 2000; Fine and Asch, 1988b). Moreover, the growing burden of gratitude can make those who have chronic illnesses or disabilities loath to ask for needed help. Problems are especially acute among elderly persons who have outlived their close relatives and friends, and thus must rely on more distant social connections. For all these reasons, relationships may wither.

Relationships also suffer if individuals no longer can participate in previous activities. How do you maintain a relationship with a tennis partner once you no longer can hold a racket? How do you maintain a relationship with a friend when architectural and transportation barriers keep you from going to movies or restaurants? And how do you maintain a relationship with a spouse or lover when your sexual abilities and interests have changed dramatically—or when your partner no longer finds you sexually attractive?

Declines in financial standing also strain relationships. An individual might, for example, have the physical ability to go to a movie with a friend but lack the admission fee. Women and minorities are especially hard hit because they typically earned lower wages and had more erratic work histories before becoming ill or disabled, and so qualify for lower Social Security benefits, if any (DeJong et al., 1989). At the same time, the stress caused by financial pressures can damage relationships with children, lovers, and spouses.

Managing Stigma

Illness and disability affect not only relationships with friends and family but also less intimate relationships. Most basically—and despite the predictions of the sick-role model—living with illness or disability means living

---

Box 6.3   *American Sign Language and the Education of Deaf Children*

American Sign Language (ASL) is the native language of the U.S. deaf community. (English Sign Language is quite different.) Until recently, nonsigners considered ASL little more than a crude collection of gestures. In fact, however, ASL is a fully functioning language with a coherent and unique grammatical structure that allows people to communicate complex ideas as quickly and fluently as any spoken language (Klima and Bellugi, 1979). The history of ASL and its place in the education of deaf children demonstrates the disabling impact of prejudice (Lane, 1992; Neisser, 1983; Shapiro, 1993).

Before the nineteenth century, no national American sign language existed, although deaf individuals, scattered around the country, typically developed their own "home signs." Although European schools had begun teaching deaf children during the eighteenth century, American educators still considered them ineducable. In 1813, Thomas Gallaudet, a Congregationalist minister, distressed by the isolation of a neighbor's deaf child, decided to travel to Europe to observe deaf education there.

In France, Gallaudet for the first time saw sign language used to teach deaf children. He became convinced that deaf children could learn if taught in a language they could understand. Gallaudet returned to the United States accompanied by Laurent Clerc, a deaf teacher who communicated via French Sign Language.

Once back in the United States, Gallaudet and Clerc opened a school in Hartford, Connecticut. Most of the teachers were deaf, and all could sign fluently in the new language—American Sign Language—that developed naturally out of the combination of French Sign Language and American home signs. The school boasted impressive results, as deaf children, taught to communicate in ASL, learned a wide variety of other academic skills, including reading and writing English.

This "golden age" of ASL was brief, however. In 1867, the Clarke School for the Deaf was established to promote "oralism," the philosophy that deaf children would learn to lipread and speak English only if forbidden to use ASL. In 1880, the International Congress of Educators of the Deaf—a Congress that included only one deaf educator—voted to make

---

with **stigma.** Stigma refers to the social disgrace of having a deeply discrediting attribute, whether a criminal record, a gay lifestyle, or a socially unacceptable illness. The term *stigma* does not imply that a condition *is* immoral or bad, only that it is commonly viewed that way.

Some illnesses, especially acute illnesses such as influenza or streptococcal infections, produce relatively little stigma, but others, such as leprosy or HIV disease, are so stigmatized that they can affect even relationships with health care providers. Individuals whose illnesses carry a heavy burden of stigma can manage that stigma in various ways. First, individuals can attempt to **pass,** or to hide their illnesses or disabilities from others (Charmaz, 1991:68–70, 110–119; Goffman, 1963:73–91; Schneider and Conrad, 1983; Weitz, 1991:128–132). For example, an elderly man who

oralism the sole method for teaching deaf children. This decision remained in force for more than a century.

The decision to adopt oralism reflected the times (Neisser, 1983). With immigration rising in the United States, many Americans feared (as some do now, more than one hundred years later), that "inferior" languages would soon replace English. The movement to eliminate ASL from the classroom paralleled the movement to ban these other languages. ASL seemed especially foreign and even sinister because its reliance on gestures made it seem less like English and more like the stigmatized languages of low-status Jewish and Italian immigrants. In addition, ASL seemed heretical to many because it seemed to refute the popular belief that God had separated humans from animals through speech.

Following the adoption of oralism, schools removed deaf teachers from the classrooms and in some cases began punishing children caught using ASL. Yet, except for the very small proportion of deaf children and adolescents who lose their hearing after learning English—and even for many of those—communicating in English usually remained an empty promise.

Students would now spend hours each day practicing lip-reading and forming sounds they could not hear. Despite these hours of practice, by the time they graduated, the vast majority could lip-read only a small fraction of spoken English and spoke English so poorly even their teachers could not understand them (Lane, 1992:129). Moreover, the hours devoted to studying oral skills left little time for scholarly subjects, which, at any rate, were taught in incomprehensible spoken English. It was as if U.S. public schools taught children mathematics in Japanese! As a result, most deaf people remained functionally illiterate (Lane, 1992:130).

Since the 1970s, the ban on manual communication in the classroom has eased. In its place, though, most educators have adopted not ASL but artificial systems that substitute signs for words within grammatically English sentences. Whether deaf students are best taught in English-based systems or in ASL remains a highly contentious subject among educators and the deaf community, and the average reading level of 18- to 19-year-old deaf students remains no better than that of 9- to 10-year-old hearing students (Paul, 1998:23).

bumps into furniture because of failing eyesight might try to convince others that he is merely clumsy, and one who sometimes does not respond to questions because of hearing problems might try to convince others that he is merely absentminded. Similarly, those who have chronic illnesses can choose to go out only on days when their symptoms are least noticeable.

Although passing offers some protection against rejection, it carries a high price. Fear of disclosure means constant anxiety. Relationships with friends and families suffer when disabled or ill individuals lie about their conditions. In addition, those individuals forfeit the emotional or practical support they might receive if others understood their situations. Individuals also risk losing jobs or flunking courses when they cannot explain their reduced productivity and increased absences.

Those who cannot tolerate the stresses of passing can instead adopt a strategy of **covering**—no longer hiding their condition but instead trying to deflect attention from it (Goffman, 1963:102–104). A woman with a visible leg brace can wear eye-catching jewelry, and persons with mobility limitations can arrive early to social gatherings to accustom themselves to the setting, identify potential physical hazards, and find accessible seats. Similarly, elderly persons who no longer see well enough to drive at night can schedule their social activities during daylight hours.

Conversely, those who have invisible disabilities sometimes find advantages in **disclosing** their disability to elicit sympathy or aid (Charmaz, 1991:119–133). For example, a woman might choose to wear a leg brace or tell co-workers about her arthritis to avoid being labeled lazy when she cannot do certain tasks.

Other individuals deal with the potential for stigma through a process of **deviance disavowal** (that is, convincing others that they are the same as "normal" people) (Davis, 1961). These individuals do not try to pass or cover their deviance, but instead try to prove that their illnesses or disabilities make them no different from others. Such "supercrips"—in the slang of disabled activists—often appear in the pages of popular magazines: the quadriplegic who paints holding a brush between her teeth, the blind man who is a champion skier, the participants in Special Olympics, and so on.

Each of these strategies can ease individuals' lives in an intolerant society. None, however, challenges the basis of that intolerance. Those who pass or cover in no way threaten the prejudices of those who would reject them. Even those who attempt to disavow their deviance do not challenge social prejudices regarding disabilities as much as proclaim they are not like others who have disabilities.

In contrast, other individuals take the more radical step of rejecting their rejecters and **challenging** the stigma of illness and disability. These individuals reject the social norms that denigrate them and refuse to adopt the accommodative strategies of passing, covering, or disavowing deviance. Instead, they argue that their deviations from bodily norms should not limit their civil rights or social status. Rather than accepting the stigma of illness and disability, these individuals attempt instead to label those who discriminate against them as foolish or immoral (Weitz, 1991:132–133). They disclose their illness or disability not to elicit sympathy or aid but to affirm their dignity and pride in the lives they have made for themselves despite—or perhaps because of—the ways their bodies differ from social expectations. For example, a woman born without a hand who, after a year of wearing a hot, uncomfortable, and functionally useless artificial hand decided to switch to a metal hook told an interviewer of her habit of looking at herself when passing store windows:

I never failed to get a reaction from people, so I always looked too. What the hell are they looking at? I looked and I saw a woman with a *surprisingly* short arm! But when I got the [cosmetic] hand, I looked and I thought, oh my God, that's what I would have looked like [if I had been born with a hand]! And I saw this person that I would have been. But maybe I would have been an asshole just like all the rest of them [the nondisabled]. . . . And [now] when I see the hook, I say, boy, what a *bad* broad. And that's the look I like the best (Phillips, 1990:855).

This quote illustrates how individuals can construct an alternative view of both themselves and "normals"—in this case, redefining the self as feisty, independent, and rebellious and "normals" as voyeuristic "assholes."

Similar sentiments underlie the student rebellion at Gallaudet University (Lane, 1992:186–191). Although Gallaudet is the only American college or university devoted to serving deaf and hearing-impaired students, all its presidents before 1988 had been hearing. That year, when the college's Board of Trustees (80 percent of whom were hearing) once again chose as president a hearing person who could not communicate in sign language, the students, along with many faculty, staff, and others, rose in protest. To these students, there is nothing wrong with deaf people, only with those who consider them inferior; many refer to themselves as "Deaf" rather than "deaf" to signify that they are linked by a minority culture rather than by a physical deficit. The protesters' anger grew after the Board's chairperson was reported to have told a group of students that the university needed a hearing president because "deaf people are incapable of functioning in a hearing world" (Lane, 1992:188); the chairperson's later disclaimer that her remark had been mistranslated into sign language only highlighted the incongruity of allowing hearing people who knew no sign language to run Gallaudet. The students' protests and the groundswell of support they received from alumni, staff, faculty, and the general public led to the resignation of the chairperson, the appointment of a new board with a majority of deaf and hard-of-hearing people, and the appointment of a new president who is deaf and knows sign language.

The Body and the Self

All disabilities and chronic illnesses challenge the self, especially when, as is usually the case, they occur after adult identity is established (Brooks and Matson, 1987; Bury, 1982; Charmaz, 1991; Corbin and Strauss, 1987; Fine and Asch, 1988a, 10–11; Schneider and Conrad, 1983). Those whose bodies differ in some critical way from the norm must develop a self-concept in the context of a culture that interprets bodily differences as signs of moral as well as physical inferiority. The resulting stigma leads individuals to feel set apart from others (Conrad, 1987; Kutner, 1987; Weitz, 1991).

Illness and disability threaten self-concept in a variety of ways. Individuals who become physically deformed or less attractive often find it difficult to maintain their self-images, as do those who lose their financial standing or their social roles as worker, student, spouse, or parent (Brooks and Matson, 1987; Weitz, 1991:97). In addition, the need to rely on others for assistance can shake individuals' images of themselves as competent adults.

Disability and illness create different problems for women compared with men. American society expects men to be emotionally, physically, and financially independent, and the threat to self-esteem when men cannot meet these expectations can be great. Conversely, American society expects women (with the exception of African-American women) to be dependent, so disability typically does not threaten women's self-esteem as much as it threatens men's self-esteem. For African-American women, however, and for all other American women who cherish their independence, illness or disabilities can hamper the struggle to obtain that independence, as prejudice and discrimination based on illness and disability compound prejudice and discrimination based on gender.

Similarly, the sexual changes that accompany disability and illness affect women and men differently. Social norms for both persons with and without disabilities expect men to be sexually active while regarding women's sexual desires with suspicion. Following disability, men can lose esteem in both their eyes and those of their partners if they no longer can perform as before. Women, on the other hand, often find that others assume they have no sexual feelings at all once they no longer meet social norms of sexual attractiveness. This denial of women's sexuality narrows women's lives and diminishes their self-images (Lonsdale, 1990).

To cope with these threats to the self, individuals sometimes attempt intellectually to separate their essential selves from their recalcitrant bodies. Cheri Register (1987:33), a writer who has a rare, chronic disease, describes "a need that many of us feel to visualize the illness as smaller than ourselves. Rather than letting the illness overtake our identities, we try to find some confined space within ourselves or our lives to contain it, and then draw boundaries around it: '*Here* is the illness. I will only let it make *this* much difference.'" This strategy succeeds best when symptoms follow a predictable course and the problem affects only one part of the body.

Yet the impact of disability and illness on the self is not solely negative. Indeed, disability and illness can bring improved self-esteem and quality of life. Over time, individuals may learn to devalue physical appearances, derive self-esteem from other sources, and focus on the present rather than on an intangible future (Weitz, 1991:136–140). They may learn to set priorities in their lives so that, often to a greater extent than before, they accomplish their most important goals rather than wasting precious energy on trivial concerns (Charmaz, 1991:134–166). Finally, they may come to define their condition simply as part of who they are, with good points and bad points, and to recognize that much of their personalities and accomplishments exist not *despite* their physical condition, but *because* of it (Higgins, 1992:141). As Barbara Rosenblum, a sociologist and artist who died of breast cancer at age 44, wrote,

> I am a very different person now: more open, much more honest, and more self-knowing. . . . I turned it [cancer] into a possibility of opening up to myself, for discovering, and for exploring new areas.
>
> I've realized that I want to list the ways in which cancer can do that. You can get courage to take larger risks than you ever have before. I mean, you're already sick, so what can happen to you? You can have much more courage in saying things and in living than you ever had before. . . .
>
> And you can do things you've always wanted to do. Cancer, by giving you the sense of your own mortality, can entice you into doing those things you have been postponing. . . .
>
> You have this sense of urgency. And you can turn this urgency—you can harness this energy that propels you—so that you go ahead and do these things and discover new parts of yourself. All the things you ever wanted to do, all the dreams you had. And the dreams that you couldn't even dream, because you didn't allow yourself. . . .
>
> Cancer has put me in touch with that. And then also, it has taught me to enjoy the tenderness and the preciousness of every moment. Moments are very important because there may not be any after that—or you may throw up. Cancer exquisitely places you in the moment.
>
> I have become very human to myself in a way that I would never have imagined. I've become a bigger person, a fuller person. This to me is one of the greatest lessons: just being human. Having cancer doesn't mean that you lose yourself at all. For me it meant that I discovered myself (Butler and Rosenblum, 1991:160–61).

CONCLUSIONS

Given the progressive aging of the American population and the increasing ability of medical technology to keep alive ill and disabled individuals, many more of us can expect eventually to live with illness and disability, whether our own, our parents', or our children's. Consequently, understanding what it means to live with illness and disability has never been more important.

As both social constructions and social statuses, illness and disability affect all aspects of life. In addition to forcing those who are ill or disabled to interact with health care providers and to manage health care regimens, illness and disability affect relationships with family and friends, work and educational performance and opportunities, and, perhaps most important, one's sense of self and relationship with one's own body. Living with illness and disability also requires individuals to come to terms—or to refuse to come to terms—with uncomfortable questions and harsh realities regarding one's past, present, and future.

Illness and disability can confer social disadvantages similar to those experienced by members of traditionally recognized minority groups. Yet the impact of illness and disability is not always negative, for illness and disability at times can provide individuals with the basis for increased self-esteem and enjoyment of life. Moreover, like other minorities, those who live with illness and disability have in recent years moved from pleas for tolerance to demands for rights. Those demands have produced significant changes in American architecture, education, transportation, and so on, and have laid the groundwork for the changes still needed.

SUGGESTED READINGS

Butler, Sandra, and Barbara Rosenblum. 1991. *Cancer in Two Voices.* San Francisco: Spinsters. A moving, first-person account, written jointly by a sociologist who had breast cancer and her lover.

Hockenberry, John. 1995. *Moving Violations: War Zones, Wheelchairs, and Declarations of Independence.* New York: Hyperion. A vivid and honest memoir by radio and television correspondent John Hockenberry, who has been a paraplegic since age 19.

Register, Cheri. 1987. *Living with Chronic Illness: Days of Patience and Passion.* New York: Free Press. A wonderfully written analysis of the experience of chronic illness, based on interviews and first-hand experience.

Schneider, Joseph P. 1993. *No Pity: People with Disabilities Forging a New Civil Rights Movement.* New York: Random House. A history of the disability rights movement.

GETTING INVOLVED

Disability Rights Education and Defense Fund. 2212 6th Street, Berkeley, CA 94710. 415-644-2555. www.dredf.org. Activist group promoting independent living and civil rights for persons with disabilities.

REVIEW QUESTIONS

How do the medical and sociological models of disability differ?

Are disabled people a minority group?

What is the Americans with Disabilities Act?

How common is disability and which social groups have the highest rates of disability?

What difficulties do individuals face in responding to initial symptoms of illness or disability, obtaining diagnoses, and coming to terms with their diagnoses?

What is illness behavior?

How can illness serve as an interruption, an intrusion, or something in which one is immersed?

Why do individuals sometimes ignore medical advice?

Why do individuals use alternative health care?

How can illness or disability affect social relationships and self-image?

How can individuals manage the stigma of illness or disability?

INTERNET EXERCISES

Find three sites devoted to disability rights. (Hint: each site will probably have links to other sites.) Browse the sites. In what ways are the problems identified by these sites similar to or different from the problems identified in this chapter?

Find a Web site that sells Human Growth Hormone (HGH) direct to the public. (Hint: search for "purchase HGH.") Critique the Web site: what kinds of information is the Web site highlighting? What kinds of necessary information about the drug is either not available on the Web site, hard to find, or hard to read? What techniques is the Web site using to convince the viewer to purchase the drug (for example, suggesting that the drug is more "natural" than other available drugs, or recommended by medical "experts")?

A homeless, mentally ill person, warming himself over a New York City subway grate.

CHAPTER 7

The Sociology of Mental Illness

—

At the age of 18, Susan Kaysen was committed to a private mental hospital, where she spent the next two years. In the book *Girl, Interrupted*, she describes her experience making the transition from mental hospital to the outside world:

> The hospital had an address, 115 Mill Street. This was to provide some cover if one of us were well enough to apply for a job while still incarcerated. It gave about as much protection as 1600 Pennsylvania Avenue would have.
>
> "Let's see, nineteen years old, living at 1600 Pennsylvania Avenue— Hey! That's the White House!"
>
> This was the sort of look we got from prospective employers, except not pleased.
>
> In Massachusetts, 115 Mill Street is a famous address. Applying for a job, leasing an apartment, getting a driver's license: all problematic. The driver's license application even asked, Have you ever been hospitalized for mental illness? Oh, no, I just loved Belmont so much I decided to move to 115 Mill Street.
>
> "You're living at One fifteen Mill Street?" asked a small basement-colored person who ran a sewing-notions shop in Harvard Square, where I was trying to get a job.
>
> "Uh-hunh."
>
> "And how long have you been living there?"
>
> "Oh, a while." I gestured at the past with one hand.
>
> "And I guess you haven't been working for a while?" He leaned back, enjoying himself.
>
> "No," I said. "I've been thinking things over."
>
> I didn't get the job.
>
> As I left the shop my glance met his, and he gave me a look of such terrible intimacy that I cringed. I know what you are, said his look (Kaysen, 1993:123–124).

As Susan Kaysen's story suggests, mental illness is a social as well as a psychiatric condition, and mental hospitalization has social as well as psychiatric consequences. I begin this chapter by describing the extent and distribution of mental illness. I then contrast the medical model of mental illness, which views mental illness as an objective reality (if subjectively experienced), with the sociological model, which views mental illness as largely a social construction. Finally, I look at the history of treatment and the experience of mental illness.

THE EPIDEMIOLOGY OF MENTAL ILLNESS

The importance of understanding mental illness becomes clearer once we realize how many people are affected. In the following section, we look at research on the extent, distribution, and causes of mental illness.

The Extent of Mental Illness

Since the 1920s, social scientists have tried to ascertain the extent of mental illness. These researchers essentially have adopted medical definitions of mental illness (which, as we will see later in this chapter, are somewhat problematic). However, whereas doctors and other clinicians have focused on how biological or psychological factors can foster mental illness, social scientists have focused on how *social factors* can do so.

Over the years, researchers using a variety of methods have reached two consistent conclusions regarding the extent of mental illness. First, all societies, from simple to complex, include some individuals who behave in ways considered unacceptable and incomprehensible (Horwitz, 1982:85–103). Second, mental disorder is fairly common. According to the National Comorbidity Survey, the largest and latest national **random** survey on the topic (Kessler et al., 1994), 30 percent of working age adults had experienced a diagnosable mental illness in the previous year (or 18 percent, if substance abuse was not counted). Almost half (48 percent) had experienced a diagnosable mental illness at some point in their lives (21 percent if substance abuse was not counted). The most common illnesses were major depression and problems with alcohol use, reported by 10 percent and 7 percent respectively. Severe mental illness was far more rare: only 17 percent reported ever having a major depressive episode, 2 percent reported ever having a manic episode, and 0.7 percent reported ever having a psychotic illness.

The Distribution of Mental Illness

Mental illness is not distributed randomly among the population, but rather falls more on some social groups than on others. In this section, I look at the effects of ethnicity, gender, and social class.

The Impact of Ethnicity: Social Class or Discrimination?

Researchers have uncovered few significant ethnic differences in rates of mental illness except with regard to affective disorders (depression or manic episodes). These disorders strike African Americans *more* often than non-Hispanic whites and strike Hispanics *less* often than non-Hispanic whites, regardless of individuals' education or income (Kessler et al., 1994). In addition, researchers have found that African Americans have significantly higher rates than whites of psychological distress, which overlaps with but is not the same as diagnosable mental illness. These ethnic differences in levels of distress occur at all income levels, although they taper off as income rises. Researchers theorize that psychological distress among African Americans results from the daily stresses of living with racism and tapers off at upper income levels because those with higher incomes can better shield themselves from at least some of the effects of racism (Kessler and Neighbors, 1986).

Little recent research is available on the causes of psychological distress among other U.S. minority groups. Studies typically have found, however, that Mexican-Americans report less distress than whites or African Americans (Rogler et al., 1991). These results have led some to hypothesize that the emphasis within Mexican culture on strong extended families provides some protection against psychological distress in two ways: by offering social support to single and childless persons who might otherwise be isolated and by offering sources of self-esteem to those who lack higher education or prestigious careers (Mirowsky and Ross, 1980). This hypothesis, however, has been challenged by research suggesting that acculturation into Anglo culture does not harm or might even improve Mexican-Americans' mental health. This debate remains unresolved: a review of 30 studies on the impact of acculturation on mental health found the studies evenly divided between those suggesting that acculturation improved mental health and those suggesting it harmed mental health (Rogler et al., 1991). At this point, then, why Mexican-Americans have lower rates of psychological distress remains unclear.

The Impact of Gender: Sex-Role Socialization

The impact of gender on mental illness is at least as complex as the impact of ethnicity. Gender has no consistent effect on the rate of schizophrenia or other severe mental disorders that leave individuals out of touch with reality. However, men consistently display higher rates of substance abuse problems and personality disorders (conditions characterized by **chronic,** maladaptive personality traits, such as compulsive gambling or antisocial tendencies), whereas women consistently display higher rates of anxiety disorders and of major and minor depression (Kessler et al., 1994).

These differences in mental illness parallel differences in gender roles. Consistently, men display higher rates of disorders linked to violence, such as paranoid schizophrenia and antisocial personality disorder. As a result, some researchers hypothesize that these forms of mental illness occur when men become "oversocialized" to their gender role. The symptoms of antisocial personality disorder (listed in Box 7.1), for example, essentially parallel expectations within many lower-class communities for male behavior. Within these communities, men who meet these expectations are typically considered dangerous but not mentally ill, because their behavior is comprehensible. Although they might be labeled criminal, they are unlikely to be labeled mentally ill unless they somehow come to the attention of doctors from outside their communities.

Similarly, researchers have hypothesized that depression results from oversocialization to a traditional female role (Radloff, 1975). According to **learned helplessness theory,** depression develops when individuals learn to believe they cannot control their lives (Seligman, 1975). Researchers hypothesize that women more often than men become depressed because women (with the possible exception of African Americans) are more often socialized to be helpless—to depend on men for emotional strength, physical assistance, and financial support. In addition, women are taught that

---

Box 7.1   *Diagnostic Criteria for Antisocial Personality Disorder*

A. Current age at least 18

B. Before age 15, . . . a history of three or more of the following:

   (1) was often truant

   (2) ran away from home overnight at least twice . . .

   (3) often initiated physical fights

   (4) used a weapon in more than one fight

   (5) forced someone into sexual activity with him or her

   (6) was physically cruel to animals

   (7) was physically cruel to other people

   (8) deliberately destroyed others' property . . .

   (9) deliberately engaged in fire-setting

   (10) often lied

   (11) has stolen without confrontation of a victim on more than one occasion

   (12) has stolen with confrontation of a victim

C. Since the age of 15, . . . at least four of the following:

   (1) is unable to sustain consistent work behavior . . .

   (2) fails to conform to social norms with respect to lawful behavior . . .

   (3) is irritable and aggressive, as indicated by repeated physical fights or assaults . . .

   (4) repeatedly fails to honor financial obligations . . .

   (5) fails to plan ahead, or is impulsive . . .

   (6) has no regard for the truth . . .

   (7) is reckless regarding his or her own or others' personal safety, as indicated by driving while intoxicated or recurrent speeding

   (8) if a parent or guardian, lacks ability to function as a responsible parent . . .

   (9) has never sustained a totally monogamous relationship for more than one year

   (10) lacks remorse

their failures and problems derive from their personal deficiencies rather than from external problems such as sex discrimination. Women therefore learn that they cannot and, indeed, should not, control their fates, while learning to blame themselves for their lack of control. Support for this theory comes from studies suggesting that women who have the most control over their lives—working women and single women without children—or who *feel* most in control of their lives have considerably lower rates of depression (Gore and Mangione, 1983; Radloff, 1975; Rosenfield, 1989). The few studies that have looked at both African-American and white women have found few if any differences in rates of depression, perhaps because the combined effects of racism and sexism leave African-American women feeling equally helpless (Gore and Mangione, 1983; Rosenfield, 1989; Somervell et al., 1989).

The Impact of Social Class: Social Stress or Social Drift?

Of all the demographic variables researchers have investigated, social class shows the most consistent impact on mental illness. Researchers agree that as social class increases, the rate of both diagnosable mental illness and psychological distress decreases (Eaton, 1980; Kessler et al., 1994; Link and Dohrenwend, 1989). However, researchers are sharply divided on whether lower social class status causes mental illness or mental illness causes lower social class. In other words, do the social stresses associated with lower-class life lead to greater mental disorder, or do those who suffer from mental disorder drift downward into the lower social classes? These two theories are referred to, respectively, as **social stress** versus **social drift.**

Researchers interested in social class have focused primarily on schizophrenia, the disease that shows the most consistent relationship to social class; studies have found that schizophrenia and related disorders occur two to five times more often among those who have not graduated college compared with those who have (Link and Dohrenwend, 1989). Those who favor the social drift argument have shown that, at first admission to a mental hospital, schizophrenic patients hold jobs lower in social class than one would expect given their family backgrounds. This suggests that mental problems caused these individuals to drift downward in social class (Eaton, 1980).

Those who favor the social stress argument, on the other hand, argue that instead of looking at the jobs schizophrenic patients held at *first admission* to a mental hospital, we should instead look at schizophrenic patients' *first jobs.* When this is done, researchers find no difference in educational attainment or in prestige levels of first jobs between schizophrenic patients and comparable others in their communities (Link et al., 1986). Therefore, these researchers argue, whatever causes downward social drift occurs *after* completing one's education and obtaining a first job but *before* first admission to a mental hospital. They further note that compared with the general public, a higher proportion of schizophrenic patients have worked in unusually noisy, hazardous, hot, cold, smoky, or humid environments, leading researchers to conclude that the social stress of these working conditions precipitated mental disorder in vulnerable individuals. Similarly, other researchers have found that mental health problems increase among workers laid off because of plant closings, again suggesting that the stresses of unemployment and lower class status lead to mental disorder, rather than mental disorder leading to lower class status (Kessler et al., 1987).

The Impact of Social Stress and Life Events

This research on the relationship between social class and mental illness is part of a broader literature on the relationship among life events, stress, and mental illness. In the same way some researchers have investigated whether the stresses of factory work or unemployment can lead to mental illness, others have looked at whether social stress in general can lead to mental illness.

Much of the research on stress has focused on how stressful **life events,** such as divorce, losing a job, or a death in the family, might affect psychological distress. Although initially research in this area focused on the sheer number of life events individuals experience, most current research instead focuses on how the *meaning* life events have for people and the *resources* individuals have for dealing with those life events affects psychological distress. For example, an unplanned pregnancy means something quite different to a poor woman who already has four children than to a financially secure woman with one child. Similarly, some individuals have resources that can reduce the stresses of life events (such as money, social support networks, and psychological coping skills), whereas others lack such resources (Ensel and Lin, 1991; Pearlin and Aneshensel, 1986). For example, someone whose marriage fails but who earns enough to maintain his or her current lifestyle, has close friends who provide companionship and social support, and has learned adaptive ways of dealing with stress will probably experience less distress than will someone whose economic standing following divorce plummets, who has few friends, and who responds to stress by withdrawing emotionally or drinking.

Although life-events theory has produced a great outpouring of increasingly sophisticated research, it remains problematic. First, those who use this theory assume that because life events and psychological distress typically appear together, life events must *cause* distress. Yet one could just as easily conclude that mental illness causes life events. For example, psychological distress easily can lead individuals to become divorced, lose a job, go to jail, and so on.

Second, considerable research suggests that the impact of life events on individuals diminishes fairly rapidly (e.g., Avison and Turner, 1988; Norris and Murrell, 1987). As a result, researchers increasingly have de-emphasized discrete life events and shifted to investigating how long-term stresses, especially **role strain,** might affect psychological functioning (Pearlin, 1989). Role strain refers to problems individuals experience within their major social roles (Pearlin, 1989). These problems include having to adapt rapidly to changing role responsibilities, having to balance conflicting demands from multiple roles (such as being a worker as well as a parent), having to meet role demands that exceed one's resources and coping abilities, and having to adopt unwanted roles.

Like research on life events, research on role strain has generated criticism. Most important, some researchers have questioned whether role strain leads to psychological distress or whether psychological distress leads to problems in role functioning (which is related to, but not quite the same as, role strain). For example, if a woman who must balance the demands of her husband, child, and job experiences psychological distress, did the role stress lead to the psychological distress or did psychological distress make it difficult for her to function in her social roles?

Unfortunately, because the scales used to measure role strain overlap significantly with those used to measure role functioning, researchers looking at very similar data can reach opposite conclusions: those researchers who use role strain scales conclude that role strain causes psychological distress, whereas those who use role functioning scales conclude that psychological distress causes problems in role functioning (Link et al., 1990). Consequently, researchers will need to develop better measures before we will be able to understand the nature and direction of the relationship between role stress and psychological functioning.

DEFINING MENTAL ILLNESS

As with **disability** and physical illness, doctors and sociologists typically have very different ways of thinking about mental illness. In this section, I contrast the medical model of mental illness with the sociological model. Neither of these models is absolute, however, for both sociologists and doctors often blend elements from each in their work. Nevertheless, the contrast between these two "ideal types" provides a useful framework for understanding the broad differences between the two fields.

The Medical Model of Mental Illness

To doctors and most other clinicians in the field, mental illness is an illness essentially like any other. To understand what this means, it helps to understand the history of medical treatment for syphilis, the disease that first demonstrated the potential power of medicine to control conditions labeled mental illness.

Since the fifteenth century, doctors had recognized syphilis as a discrete disease. Because of its mild initial symptoms, however, only in the late nineteenth century did doctors realize the full damage syphilis can inflict on the nervous system, including blindness, deformity, insanity, and death. Unfortunately, doctors could do little to help those with syphilis. The best available treatment consisted, essentially, of poisoning patients with arsenic and other heavy metals in the hopes that these poisons would kill whatever had caused the disease before the poisons killed the patients.

In 1905, scientists first identified the bacterium *Treponema pallidum* as the cause of syphilis. Five years later, Paul Ehrlich discovered the drug salvarsan as a cure for syphilis. Salvarsan, an arsenic derivative, was the first drug that successfully targeted a specific microorganism. As such, it opened the modern era of medical therapeutics. Doctors now could cure completely those who sought early treatment for syphilis, whereas those who put off treatment risked irreversible neurological damage and a horrible death.

The history of salvarsan and syphilis provided ideological support for a **medical model of mental illness.** This medical model has four components. First, the medical model assumes that, in the same way that the presence of a specific bacterium defines syphilis, objectively measurable conditions define mental illness. Second, the medical model assumes that mental illness stems largely or solely from something within individual psychology or biology, even if researchers (like those who studied syphilis before 1905) have not yet identified its sources. Third, the medical model assumes that mental illness, like syphilis, will worsen if left untreated, but, depending on the diagnosis, can diminish or disappear if treated promptly by a medical authority. Fourth, the medical model assumes that treating mental illness, like treating syphilis with salvarsan, rarely if ever harms patients. It therefore further assumes that it is safer to treat someone who might really be healthy than to refrain from treating someone who might really be ill (Scheff, 1984).

The Sociological Model of Mental Illness

The sociological model of mental illness questions each of these assumptions (Table 7.1). Perhaps most important, those who use this model argue that definitions of mental illness, like the definitions of physical illness and disability discussed in Chapters 5 and 6, reflect subjective social judgments more than objective scientific measurements of biological problems.

What do we mean when we say someone is mentally ill? Why do we diagnose as mentally ill people as disparate as a teenager who uses drugs, a woman who hears voices, and a man who tries to kill himself? According to sociologist Allan Horwitz (1982), behavior becomes labeled mental illness when persons in positions of power consider that behavior both unacceptable and inherently incomprehensible. In contrast, we tend to define behavior as crime when we consider it unacceptable but comprehensible; we do not approve of theft, but we understand greed as a motive. (Judging some-

| Table 7.1 | *Premises of the Medical and Sociological Models of Mental Illness* |

THE MEDICAL MODEL	THE SOCIOLOGICAL MODEL
Mental illness is defined by objectively measurable conditions.	Mental illness is defined through subjective social judgments.
Mental illness stems largely or solely from something within individual psychology or biology.	Mental illness reflects a particular social setting as well as individual behavior or biology.
Mental illness will worsen if left untreated but could improve or disappear if treated promptly by a medical authority.	Persons labeled mentally ill might experience improvement regardless of treatment, and treatment might not help.
Medical treatment of mental illness can help but never harm.	Medical treatment for mental illness sometimes can harm patients.

one not guilty by reason of insanity falls on the border between crime and mental illness.) Similarly, we might not understand why physicists do what they do, but we assume that those with appropriate training find their behavior comprehensible.

According to Peggy Thoits (1985), behavior leads to the label of mental illness when it contravenes **cognitive norms, performance norms,** or **feeling norms.** Someone who thinks he is Napoleon, for example, breaks cognitive norms (that is, norms regarding how one should think), whereas someone who can't hold a job breaks norms regarding proper role performance. Thoits argues that the last category—breaking feeling norms—accounts for most behavior labeled mental illness. Feeling norms refer to socially defined expectations regarding the "range, intensity, and duration of feelings that are appropriate to given situations" and regarding how one should express those feelings (Thoits, 1985:224). For example, laughing is highly inappropriate at a Methodist funeral but perfectly acceptable at an Irish wake, and feeling sad that one's pet cat died is considered reasonable for a few days but unreasonable if it lasts for a year.

Different social groups consider different behaviors comprehensible and acceptable. The friends of a drug-using teenager, for example, might consider drug use a reasonable way to reduce stress or have fun. Their views, however, have little impact on public definitions of drug use. Similarly, members of one church might consider a woman who reports talking to Jesus a saint, whereas members of another church consider her mentally ill. The woman's fate will depend on how much power these opposing groups have over her life. The definition of mental illness, then, reflects not only socially accepted ideas regarding behavior but also the relative power of those who hold opposing ideas.

Researchers who use this sociological definition of mental illness do not mean to imply that emotional distress does not exist or that people do not feel real pain when they cannot meet social expectations regarding thought, behavior, or emotions. Nor do these researchers mean to imply that biology has no effect on behavior or thought. They do, however, question the purpose and consequences of using medical language to describe such problems and question why we label certain behaviors and individuals but not others.

Not all sociologists raise these questions, however. Many, especially those working in health care settings and in **epidemiology,** employ a **sociology *in* medicine** approach and use essentially medical definitions of mental illness in their research and writing. Nevertheless, sociologists are united in assuming that mental illness, like physical illness and disability, stems at least in part from social life rather than solely from individual psychology or biology. For example, during the last thirty years the number of young women diagnosed with eating disorders such as anorexia and bulimia has skyrocketed (Brumberg, 1997). Those who use a medical model trace these disorders to biological defects such as endocrine or biochemical imbalances or to psychological factors such as poor adjustment to normal life changes, a need for personal perfection, poor relationships with parents, and adolescent identity crises (see, for example, Costello and Costello, 1992:151–152). In contrast, those who use a sociological model of mental illness argue that eating disorders have mushroomed partly because of the increased cultural pressures on women to be slim (Brumberg, 1997). Thus, sociologists shift the focus from individual biology and psychology to the social context.

The Problem of Diagnosis

The sociological model of mental illness gains credibility when we look at research on the problems with psychiatric diagnosis. These problems became a political embarrassment for psychiatrists (medical doctors who specialize in treating mental illness) following a well-publicized experiment by psychologist David Rosenhan (1973). Rosenhan and seven of his assistants had presented themselves to twelve mental hospitals and complained of hearing voices, but otherwise had acted normally. The hospitals diagnosed all eight "pseudopatients" as mentally ill and admitted them for treatment. Once admitted, all behaved normally, leading 30 percent of the other patients to identify them as frauds. None of the staff, however, noticed anything unusual about these pseudopatients. It took an average of nineteen days for them to win their release, with their symptoms declared "in remission."

When these results were published, psychiatrists objected vociferously that the results were some sort of fluke. In response, Rosenhan agreed to send pseudopatients to another hospital and challenged the staff at that hospital to identify the pseudopatients. During the three months of the experiment, the staff identified 42 percent of their new patients as pseudopatients, even though Rosenhan really had not sent any!

These two experiments vividly demonstrate the subjective nature of psychiatric diagnosis and its susceptibility to social expectations. Within the context of a mental hospital, staff members quite reasonably assume patients are ill and interpret everything patients do accordingly. When, for example, one bored pseudopatient began taking notes, a worker officially recorded this "note-taking behavior" as a symptom. Conversely, when staff members expected to find pseudopatients, they interpreted similar behaviors as signs of mental health.

The problems with diagnosis are particularly acute when therapist and patient do not share the same culture. With the rise in immigration to the United States during the last two decades, doctors increasingly must diagnose and treat patients whose symptoms do not appear in western textbooks (Goleman 1995). For example, whereas Americans sometimes fear that their bodies will embarrass *them,* Japanese people sometimes experience disabling fears (known as "taijin kyofusho") that their bodies will embarrass *others.* Meanwhile, Malaysian men may be stricken by "koro," the sudden and intense fear that their penises and testicles will recede into their bodies and kill them, and Latin Americans can succumb to "boufee delirante"—sudden outbursts of excited, confused, violent or agitated behavior. In response to growing concerns about cross-cultural misunderstandings, the American Psychiatric Association (APA) in 1995 adopted new guidelines that explicitly recommend considering cultural and ethnic factors when diagnosing or treating patients. In addition, APA adopted new guidelines requiring psychiatric training programs to include training in cross-cultural issues.

The Politics of Diagnosis

To reduce the problems with diagnosis, psychiatrists over the years have attempted to refine the definitions of illnesses in the **Diagnostic and Statistical Manual of Mental Disorders (DSM).** Since it was first published by the APA in 1952, virtually all psychiatrists have relied on this manual for assigning diagnoses to patients. So, too, do most other clinicians, because insurers usually require a DSM diagnosis before they will reimburse clinicians for treating a patient. DSM and the subsequent DSM-II, published in 1968, instructed clinicians to reach diagnoses based on the clinicians' inferences about such intrapsychic processes as defenses, repression, and transference. Because clinicians cannot measure these processes, the same behavior often elicited quite different diagnoses from different clinicians (Helzer et al., 1977).

Partly because of these problems, the APA in 1974 announced its decision to revise DSM-II (Spitzer et al., 1980). Ironically, although the resulting DSM-III, published in 1980, was designed to quiet questions about the ambiguities of psychiatric diagnosis, it instead illuminated those ambiguities because its writing became an overtly political battle, involving active lobbying by both professional and lay groups (Kirk, 1992). This battle revealed wide differences among clinicians regarding what behaviors signified mental illness, what caused those behaviors, who should treat them, and how they should be treated.

The extent of these differences already had surfaced during earlier and openly contentious battles regarding homosexuality (Conrad and Schneider, 1992). DSM-I and -II had listed homosexual behavior and desires as conclusive evidence of mental illness. By the early 1970s, however, gay rights activists had begun to challenge this definition of the situation, arguing instead that homosexuality was a natural human variation. Active lobbying by gay activists and sympathetic professionals led the APA to hold a referendum in 1974, in which 58 percent of APA members voted to drop homosexuality from DSM-II. This decision was based as much on political and moral considerations as on new scientific evidence.

The battle over the meaning of homosexuality began again with the writing of DSM-III. In the end, a compromise was reached declaring only "ego-dystonic" homosexuality a mental illness. Ego-dystonic homosexuality referred to individuals whose homosexuality caused them emotional pain and who had proved unsuccessful in changing their sexual orientations. (DSM-IV, published in 1994, differs little from DSM-III, but includes the symptoms of ego-dystonic homosexuality under "sexual disorder not otherwise specified" rather than as a separate disease.) This compromise did not, however, end differences over treatment, for those who considered homosexuality a viable alternative sexual orientation treated ego-dystonic homosexuality by helping individuals become comfortable with their sexuality, whereas those who considered homosexuality pathological treated it by trying to change individuals' sexual orientation.

Debate over other diagnoses revealed equally divergent views on causation and treatment (see, for example, Scott, 1990). Clinicians trained in Freudian psychiatry (described later in this chapter) traced the roots of mental illness to unresolved childhood sexual conflicts and favored treating it with intensive psychoanalysis. Other clinicians traced mental illness to problematic interpersonal relationships, inappropriate social learning, or biological defects and favored treating it with, respectively, psychotherapy, behavioral conditioning, or drug therapies.

To encourage support for DSM-III and to avoid open political battles among psychiatrists, its authors decided to stress symptomatology and avoid discussing either causation or treatment (Kirk, 1992). In addition, to increase the odds that clinicians would use DSM-III, the authors described the various diagnoses based not on available research but, rather, on the consensus among practicing psychiatrists. These two strategies, they hoped, would produce a widely used and highly reliable document. **Reliability** refers to the likelihood that different people who use the same measure will reach the same conclusions—in this case, that different clinicians, seeing the same patient, would reach the same diagnosis. Yet even this modest goal was not achieved, for studies continue to find high rates of disagreement over diagnosis (Kirk, 1992; Mirowsky and Ross, 1989). Moreover, reliability in the absence of validity is not particularly useful. **Validity** refers to the likelihood that a given measure accurately reflects what those who use the measure believe it reflects—in this case, that persons identified by DSM-III as having a certain illness actually have that illness. As Phil Brown (1990:393) notes, "anyone can achieve interrater reliability by teaching all people the 'wrong' material, and getting them to all agree on it. . . . The witch trials [of earlier centuries] showed a much higher degree of interrater reliability than any DSM category, yet we would not impute any validity to those social diagnoses."

Finally, even if the diagnostic categories used by clinicians are reliable and valid, clinicians will not necessarily apply them in an objective fashion. Research on DSM-III suggests that ethnicity and gender of both patient and clinician affect diagnosis. For example, Marti Loring and Brian Powell (1988) asked 290 randomly selected psychiatrists to diagnose two cases based on a brief description. Both cases had experienced hallucinations and extreme anxiety, had symptoms severe enough to damage their family life, and had proved unable to keep a job. Both also met the DSM-III definition of undifferentiated schizophrenic disorder with a dependent personality disorder, a serious psychiatric illness with roots in childhood or adolescence.

The case descriptions the psychiatrists received were identical except for the descriptions of the cases' sex and ethnicity. When sex and ethnicity either were not given or matched those of the psychiatrist, the psychiatrists' diagnoses matched those of the researchers. In the other situations, however, bias seemed to affect the diagnoses. Male psychiatrists proved more likely to diagnose the female cases as having either depression or histrionic personality disorder, a diagnosis given to individuals with a long-standing tendency to express emotions intensely, act charmingly and seductively, feel helpless and therefore act dependent, and engage in romantic fantasies. Both depression and histrionic personality disorder fit stereotypical notions of female psychology and are diagnosed more often in women. In addition, white clinicians and, to a lesser extent, African-American clinicians, more often diagnosed African Americans as paranoid schizophrenics. Paranoid schizophrenia is characterized by violence and is considered extremely difficult to treat, and so is considerably more serious than the researchers' diagnosis.

Only nineteen psychiatrists could not reach a diagnosis based on the information they had received. Of these, almost two-thirds (63 percent) had not received information about sex or ethnicity, further suggesting that psychiatrists base their diagnoses at least in part on social stereotypes of gender and ethnicity rather than on symptoms.

A HISTORY OF TREATMENT

The history of treatment for mental illness further reveals the role social values play in medical responses to problematic behavior. In this section, I trace the treatment of mental illness from the pre-scientific era to the present.

Before the Scientific Era

Although the concept of mental illness is relatively new, all societies throughout history have had individuals whose behavior set them apart as unacceptably and incomprehensibly different. However, pre-modern societies more often could find informal ways of coping with such individuals (Horwitz, 1982). First, pre-modern societies could offer acceptable, low-level roles to those whose thought patterns and behaviors differed from the norm. Second, because work roles rarely required individuals to function in highly structured and regimented ways, many troubled individuals could perform at marginally acceptable levels. Third, in pre-modern societies, work occurred within the context of the family, whether at home or in fields or forests. As a result, families could watch over those whose emotional or cognitive problems interfered with their ability to care for themselves. These three factors enabled families to **normalize** mental illness by explaining away problematic behavior as mere eccentricity. As a result, unless individuals behaved violently or caused problems for civil authorities, their families and communities could deal with them informally.

In some cases, however, individuals behaved too disturbingly, unacceptably, or incomprehensibly for their communities to normalize. In these cases, and as is true with all illnesses (as described in Chapter 5), communities needed to find explanations to help them understand why such problems struck some people and not others. Such explanations helped to make the world seem more predictable and safe by convincing the community that such bad things would never happen to "good people" like themselves.

Until the modern scientific age, societies typically looked to the supernatural for explanations of disturbed and disturbing behavior. In general, societies viewed such behavior as a punishment for sin or for violating a taboo; a sign that the afflicted individual was a witch; or a result of evildoing by devils, spirits, or witches. These views logically led societies to assign treatment to religious authorities, whether shamans, witch doctors, or priests, who relied on prayer, exorcism, spells, and sometimes physical interventions such as bloodletting or trepanning (drilling a hole in the skull to let "bad spirits" out).

Religious control of socially disturbing behavior reached a spectacular climax with the witchcraft trials of the fifteenth to seventeenth centuries, during which religious authorities brutally killed at least 100,000 people (Barstow, 1994). Many of these individuals were singled out because of behavior we would now label mental illness. (However, considerably more were targeted because they worked as midwives or lay healers, whose skills threatened to make priests and prayer unnecessary, or because, as unmarried women, they threatened social ideas about acceptable female behavior [Barstow, 1994].)

As a capitalist economy began to develop, both religious control and informal **social control** began to decline (Horwitz, 1982; Scull, 1977). Under capitalism, work moved from home and farm to workshops and factories. As a result, families found it increasingly difficult to care informally for their more problematic members. In addition, a capitalist economy could less readily absorb those whose productivity could not be scheduled and regimented. At the same time, widespread migration from the countryside to cities weakened families and other social support systems, as did migration from Europe to the United States in subsequent centuries. Meanwhile, other changes in society weakened religious systems of social control.

These changes fostered a need for new, formal institutions to address mental illness. By the end of the eighteenth century, however, only a few hospitals devoted to treating the mentally ill existed, along with a few private "madhouses" run by doctors for profit. Instead, most of those we would now label mentally ill were housed with the poor, the disabled, and the criminal in the newly opened network of public **almshouses,** or poorhouses.

Conditions in both almshouses and madhouses were generally miserable, but were especially bad for those considered mentally ill. Both doctors and the public typically considered persons with mental illness incurable and essentially animals. As a result, institutions felt free to treat them as animals—chaining them for years to basement walls or cells, often without clothing or proper food, and beating them if they caused problems.

The Rise of Moral Treatment

By the late eighteenth century, however, attitudes toward persons with mental illness began to moderate (Scull, 1989:96–117). The two individuals most identified with this change were the English reformer, William Tuke, and the French doctor, Philippe Pinel. Pinel was appointed head of the insane asylum at Bicêtre in 1793. He immediately ordered chains removed

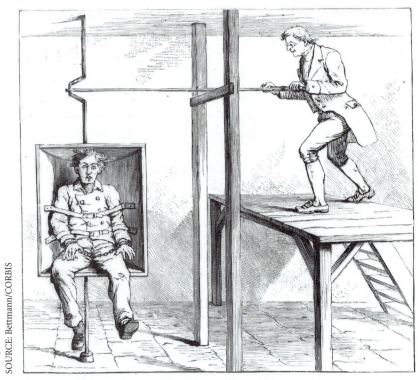

SOURCE: Bettmann/CORBIS

Nineteenth-century doctors developed many odd "treatments" for mental illness, like this "circulating swing."

from the inmates, abolished physical punishment, and improved living conditions. In place of punishment and warehousing, Pinel proposed **"moral treatment"**: teaching individuals to live in society by showing them kindness and giving them opportunities to work and play. Tuke mandated similar changes in England at the York Retreat for Quakers, but differed from Pinel in viewing mental illness more as a moral than a medical issue. At both Bicêtre and York, these changes produced stunning results, as inmates who previously had been chained, mute, and essentially animalistic recovered sufficiently in some cases to leave the hospital altogether. These successes convinced the public that mental illness was curable and stimulated interest in moral treatment.

In the United States, the first formal provision for the treatment of the mentally ill had come in 1751, when the country's first hospital, the Pennsylvania Hospital, built a mental ward in its basement. Two decades later, in 1773, Virginia established the first institution devoted to treating mental illness. The first American institution designed to provide moral treatment, the Friends' (or Quakers') Asylum, was founded in 1817; not surprisingly, given its religious origins, it drew heavily on Tuke's model.

The Decline of Moral Treatment

Despite this strong beginning, moral treatment in the end could not compete with medical models of mental illness (Scull, 1989:137–161). Because those who promoted moral treatment continued to use the language of medicine, talking of illnesses and cures, medical doctors could argue successfully that only they should control this field. In addition, because moral treatment required only kindness and sensitivity, which theoretically any professionals could offer, no professional group could claim greater expertise than doctors. As a result, by 1840, doctors largely had gained control over the field of mental illness both in the United States and Europe.

As care gradually shifted from laypersons to doctors, custodial care began to replace moral treatment. Several factors encouraged this shift. First, the primary motivation of those who funded mental hospitals was not treatment but social control—protecting society from those whom it feared or rejected. This was especially true in the United States, where fear of the mentally ill combined with fear of foreign immigrants and African Americans, two groups that the public considered especially susceptible to mental illness and that increasingly filled public mental hospitals. In addition, although communities might have been willing to pay for moral treatment for the middle-class persons who had predominated in private asylums, they were far less willing to do so for the poor persons who increasingly filled public mental hospitals. Second, and given these concerns, most communities considered it an extravagance to provide the decent living conditions and low staff-to-patient ratios that moral treatment required. Third, late nineteenth-century science emphasized the genetic basis of illness and encouraged the public to view mental illness as genetically determined and untreatable. Finally, once the first burst of enthusiasm for reform had passed and conditions in asylums began regressing, cures became rarer and the justification for moral treatment faded. For all these reasons, then, custodial care largely replaced moral treatment by the 1870s.

The abandonment of moral treatment, however, did not mean the abandonment of institutional care. Rather, the number of mental hospitals grew exponentially during the nineteenth century (Rothman, 1971). Whereas before 1810, the United States had only one public mental hospital and a few private ones, by 1860, twenty-eight of the thirty-three states had public mental hospitals. From the 1830s on, the numbers of hospitalized mental patients grew far faster than did the population in general. Historians refer to this change, and the similar but earlier developments in Europe, as "The **Great Confinement.**"

The rise of institutions reflected the need to respond to public **deviance.** The Great Confinement drew energy from the well-meaning efforts of reformers—most notably, Dorothea Dix—to close down the brutal and anarchic almshouses and to provide facilities specifically designed to care for the mentally ill, instead of warehousing them with criminals, disabled persons, and the poor (Sutton, 1991). Because no agreed-upon definitions of mental illness existed, however, families and communities found it relatively easy to move troublesome relatives into the newly established mental hospitals. Indeed, a substantial proportion of those found in these new hospitals suffered primarily from old age and poverty coupled with a lack of relatives who could or would care for them (Sutton, 1991). As a result, with the exception of those wealthy enough to obtain care in small, private mental hospitals, most of those labeled mentally ill continued to find themselves housed with others whom society had rejected. The only difference was that instead of residing in institutions filled with a varied group of deviants, they now lived in large institutions officially devoted to the "care" of the mentally ill.

Freud and Psychoanalysis

By the beginning of the twentieth century, then, doctors controlled the mental illness field. Yet medicine was torn by internal divisions. From the nineteenth century to the present, although doctors overwhelmingly traced mental illness to sources internal to individuals, some emphasized the emotional roots of mental illness while others emphasized organic, or physical, causes of mental illness.

This split grew wider with the rise of Freudian psychiatry. According to Sigmund Freud, a Viennese doctor, to become a mentally healthy adult one had to respond successfully to a series of early childhood developmental issues. Each issue occurred at a specific stage, with each stage linked to biological changes in the body and invested with sexual meanings. For example, during the oral stage, infants and toddlers derived their greatest satisfaction from sucking a breast or bottle. Those who did not learn how to signal and fulfill those needs, Freud concluded, would later develop traits such as dependency and narcissism.

The phallic stage (between about ages 3 and 6) plays an especially important role in Freud's model because that is when the **superego**—that portion of the personality that has internalized social ideas about right and wrong—is hypothesized to develop. During the phallic stage, according to Freud, children start noticing and responding to their genitalia. As a result, they begin experiencing sexual attraction toward the opposite-sex parent and viewing their same-sex parent as a rival. When boys first learn that girls do not have penises, however, they immediately and naturally (according to Freud) conclude that girls have been castrated by their fathers as punishment for some wrongdoing. Fearing the same fate, boys abandon their attraction to their similarly castrated mothers and identify with their fathers, whose love they try to obtain by adopting their fathers' values. Through this process, boys develop a strong superego.

But what of girls, who lack penises? According to Freud, once they realize they lack penises, girls immediately recognize their inferiority (1925 [1971:241–260]. As a result, girls descend into jealousy and narcissism, which they can relieve only partially and only by marrying and having baby boys who vicariously give them penises of their own. Thus, girls can never develop strong superegos because they lack the fear necessary for their development.

Freud based this theory on his interpretations of the lives and dreams of his upper-middle-class patients; no scientific data underpins this theory. Looking back at this theory from the present, it is hard to comprehend how anyone could have believed in such notions as three-year-olds lusting after their parents or girls naturally feeling jealous of boy's penises (as opposed to feeling jealous of the social power maleness confers). Yet Freudianism's long-standing popularity should not surprise us, given its congruence with gender **norms.** Freudianism both reflected and supported contemporary cultural notions that held that men's anatomy, intellect, and moral capabilities naturally and obviously surpassed women's, that women lacked the necessary maturity and selflessness to hold positions of authority in society, and that women were destined to become wives and mothers. These notions have not been totally abandoned; although no longer widely used in its pure form and rarely used by modern psychiatrists, Freud's conception of human nature and of mental illness continues to permeate American culture and to affect ideas about both normal and abnormal psychology. For example, a widely read book by psychiatrist Nancy Andreasen (1984) argues that psychiatry entered the modern era when it abandoned intrapsychic explanations of mental illness and adopted biological explanations. Yet the book nevertheless praises Freud for discovering "many interesting patterns of human thinking and behavior—the meanings hidden in the unconscious life of dreams, the sexual attraction that young children feel toward parents of the opposite sex, the jealousy they feel toward parents of the same sex, [and] the traumatic effects that early life events may have" (p. 13).

For those who accepted Freud's theory, the only way to cure mental illness was to help patients come to a more satisfactory resolution of their developmental crises. To do so, Freud and his followers relied on psychoanalysis, a time-consuming and expensive form of psychotherapy geared to patients without major mental illnesses. In psychoanalysis, patients recounted their dreams and told a largely silent therapist whatever came to mind in order to recover hidden early memories and understand their unconscious motivations.

Because psychoanalysis was so costly, most mental patients during the first half of the 1900s instead received far-cheaper physical interventions (Valenstein, 1986). Insulin therapy became immediately popular from its inception in 1933, followed by electroconvulsive (shock) therapy in 1938. These therapies caused comas or seizures, which psychiatrists believed improved mental functioning. Neither therapy had received scientific testing before becoming popular, nor did later studies find evidence of their effectiveness. Similarly, lobotomies—operations that permanently destroy part of the brain—became popular during the 1940s and 1950s. An estimated 50,000 Americans received lobotomies, and its originator, Dr. Egas Moniz, received the Nobel Prize in Medicine in 1949. Yet the only proven effects of lobotomies are diminished memory, intelligence, creativity, and emotional capacity (Valenstein, 1986). At any rate, therapy of any sort occupied only a minuscule proportion of patients' time in mental hospitals. Instead, patients spent their days locked in crowded wards with little other than radio or, later, television to ease their boredom.

The Antipsychiatry Critique

By the middle of the twentieth century, mental hospitals had become a huge and largely unsuccessful system (Mechanic, 1989). Patients with mental illnesses occupied half of all hospital beds in the United States. Virtually all (98 percent) of these beds were located in public mental hospitals; insurance rarely covered mental health care, so private hospitals had no interest in the field. At their peak in 1955, public mental hospitals held 558,000 patients, the majority of them involuntarily confined, for an average of eight years.

Beginning in the 1960s, many voices would challenge this system. The Civil Rights, antiwar, and feminist movements all brought issues of individual rights to the forefront and stimulated a broader questioning of authority and social arrangements. These ideas contributed to a growing critique of mental health treatment by sociologists, psychologists, and even some psychiatrists such as R. D. Laing (1967) and Thomas Szasz (1970, 1974).

One of the most powerful critiques of large mental institutions appeared in a classic study by sociologist Erving Goffman (1961). Goffman's work fell within the tradition of **symbolic interactionism** theory. According to this theory, individual identity develops through an ongoing process in which individuals see themselves through the eyes of others and learn through social interactions to adopt the values of their community and to measure themselves against those values. In this way, a **self-fulfilling prophecy** is created, through which individuals become what they are already believed to be. So, for example, children who constantly hear that they are too stupid to succeed in school might conclude that it is senseless to attend classes or study. Consequently, they might fail in school, fulfilling the prophecies about them.

Goffman used symbolic interactionism theory to analyze mental hospitals and the experiences of mental patients. He pointed out that mental hospitals, like the military, prisons, and monasteries, were **total institutions**—institutions where a large number of individuals lead highly regimented lives segregated from the outside world. Goffman argued that these institutions necessarily produced **mortification** of the self. Mortification refers to a process through which one's self-image is damaged and is replaced by a personality adapted to institutional life.

Several aspects of institutional life foster mortification. Persons confined to mental hospitals lose the supports that usually give people a sense of self. Cut off from work and family, the only available role is that of patient. That role, meanwhile, is a **master status**—a status considered so central that it overwhelms all other aspects of individual identity. Within the mental hospital, a patient is viewed solely as a patient—not as a mother or father, husband or wife, worker or student, radical or conservative. According to Goffman's observations and as in Rosenhan's (1973) experiment, all behavior becomes interpreted through the lens of illness. In addition, because each staff member must manage many patients, staff members necessarily deal with patients *en masse*. In these circumstances, patients typically lose the right to choose what to wear, when to awaken or sleep, when and what to eat, and so on. Moreover, all these activities occur in the company of many others. Individuals thus not only experience a sense of powerlessness but also can lose a sense of their identity—their desires, needs, personalities—in the mass of others. As a result, patients experience **depersonalization**—a feeling that they no longer are or are considered fully human.

At the same time, the hierarchical nature of mental hospitals reinforces the distinctions between inmate and staff and constantly reminds both of the gulf between them. Consequently, patients can avoid punishment and eventually win release only by stifling their individuality and accepting the institution's beliefs and rules. These forces producing mortification are so strong that even Rosenhan's pseudopatients—knowing themselves sane and hospitalized only briefly—experienced depersonalization.

Implicit in Goffman's work is the idea that mental hospitals may be one of the worst environments for treating mental problems. More recent research supports this conclusion. A review of ten **controlled** studies on alternatives to hospitalization, including halfway houses, day care, and supervised group apartment living, found that all could boast equal or better results than traditional hospitalization, as measured by subsequent employment, reintegration into the community, life satisfaction, and extent of symptomatology (Kiesler and Sibulkin, 1987).

Deinstitutionalization

By the time the antipsychiatry critique appeared, the Great Confinement already had begun to wane. Beginning in 1955, the number of mental hospital inmates declined steadily, even though demand for care, as measured by rates

Table 7.2 **Average Daily Inpatient Census of Mental Patients, by Type of Organization, 1969–1988**

TYPE OF ORGANIZATION	1969	1975	1979	1983	1986	1988
State and country mental hospitals	367,629	193,380	138,600	116,236	107,056	99,869
Veterans Administration medical centers	47,140	32,123	28,693	20,342	21,242	19,602
Private psychiatric hospitals	11,608	12,058	13,901	16,467	23,475	29,698
Nonfederal general hospitals with psychiatric services	17,808	22,874	23,110	34,328	34,437	35,902
Residential centers for children	12,406	16,164	18,054	15,826	22,650	23,092
All other organizations (includes community mental health centers)*	6,240	10,989	11,026	20,970	19,670	19,673
Total	468,831	287,588	233,384	224,169	228,530	227,836

* Note: Large growth in 1983 occurred because the federal government began including nonhospital partial care facilities in this category.

of both first and total admissions, continued to rise (Scull, 1985). The median length of stay would decrease from forty-one days in 1970 to twenty-three days in 1980 and fifteen days in 1992 (Manderscheid and Sonnenschein, 1992:288). Similarly, the number of beds per 100,000 persons would fall from 207.4 in 1970 to 56.1 in 1984 (Mechanic and Rochefort, 1990). Meanwhile, **outpatient** care would become vastly more common; whereas in 1955 the number of persons treated as **inpatients** (i.e., formally admitted to hospitals) was almost three and a half times the number treated as outpatients, by 1983 that ratio had almost reversed (Mechanic and Rochefort, 1990). The first two lines of Table 7.2 show the tremendous decline in large public institutions.

The process of moving mental health care away from large institutions, known as **deinstitutionalization,** received support from an influential report, *Action for Mental Health,* published in 1961 by the U.S. Joint Commission on Mental Health and Illness. The report, drawing on social science research, called for replacing institutional, custodial care with outpatient care in the community. Two years later, Congress passed the Community Mental Health Centers Act, which provided seed money for opening a network of such centers. (The last line in Table 7.2 shows the growth in these and other outpatient facilities.) During the 1970s, deinstitutionalization gained further support in the courts, as mental patients successfully fought against involuntary treatment, against hospitals that provided custodial care rather than therapy, and for the right to treatment in the "least restrictive setting" appropriate for their care. (However, over time these court victories have had only modest impact on actual policies [Appelbaum, 1994].)

Explaining Deinstitutionalization

Those who adopt a medical model of illness typically assume that deinstitutionalization resulted from the introduction, beginning in 1954, of drugs known as phenothiazines (see, for example, Brill and Patton, 1962). These drugs, such as chlorpromazine (Thorazine), significantly reduce severe symptoms such as hallucinations in many patients. To these drugs would later be added antidepressants and anti-anxiety drugs such as diazepam (Valium). Yet during the first ten years after the introduction of these drugs, the number of patients in public mental hospitals decreased by less than 2 percent per year. In contrast, a massive exodus from mental hospitals occurred in later years (Gronfein, 1985). Something other than drugs, then, must primarily account for deinstitutionalization.

Historical evidence suggests that although phenothiazines facilitated deinstitutionalization by making mental patients compliant enough for communities to tolerate their release, financial changes more fully explain this shift (Gronfein, 1985; Mechanic and Rochefort, 1990). Increasingly during the 1960s and 1970s, private insurers covered the costs of mental health care, making the treatment of mental illness profitable for private hospitals. As a result, these hospitals began aggressively developing psychiatric facilities and, gradually, care of patients who had private insurance began shifting from large public institutions to private psychiatric hospitals (Mechanic, 1999; Mechanic and Rochefort, 1990). Table 7.2 shows the growth in private facilities for treating mental illness. General hospitals also sought psychiatric patients as a means of filling beds emptied first by the overbuilding of general hospitals during the 1950s and 1960s and then, beginning in the 1980s, by pressures from insurers to control costs by releasing patients with physical illnesses from hospitals as soon as feasible (Brown, 1985:116–117; Gray, 1991).

Changes in public health insurance played an even more important role in fostering deinstitutionalization. With the establishment in 1965 of **Medicare** and **Medicaid**—the two major federal programs for providing health care to the poor, disabled, and elderly (described in the next chapter)—funds became available to pay for the care of some individuals with chronic mental problems in **nursing homes.** Partly as a result, the number of nursing home beds mushroomed. States, meanwhile, were happy to move patients from mental hospitals funded through state tax dollars to nursing homes largely paid for by the federal government via Medicare, thereby shedding responsibility for what was the single most expensive item in many states' budgets.

During the 1960s, the federal government also expanded financial benefits available under Social Security for persons with physical or mental disabilities. These benefits allowed individuals some independence by providing them with funds to buy food, clothing, and shelter. As a result, mental hospitals could now release patients who previously would have been unable to support themselves. Three-quarters of the reduction in the number of mental patients occurred after these financial changes occurred, suggesting that these changes were the most important factor behind deinstitutionalization.

Finally, as sociologists Allan V. Horwitz and Jeffrey S. Mullis argue, at a broader level deinstitutionalization also stemmed from the rise of **individualism**—a set of "sociocultural beliefs and practices that encourage and legitimate the autonomy, equality and dignity of individuals" (1998:122). In past generations, individuals' identities depended on their places within family or community. Because families and communities were far more important than individuals as social units, laws typically upheld the right of these groups over any rights of the individual. Thus, for example, until about 1900, parents had near-absolute rights to discipline their children without interference from the law. Similarly, most psychiatric inpatients were committed by their families, and most requests by families to commit individuals were honored (Horwitz and Mullis, 1998).

During the last few decades, however, this "moral sovereignty" of the family has weakened, as families are no longer assumed to know what is best for their members and as family ties of all sorts have weakened. In its stead, individualism has become dominant. Although families still are the most common source of requests for commitment, they now must demonstrate that commitment is in the best interest of the individual. Similarly, mental hospitals now must demonstrate that an individual needs continued treatment rather than the individual having to demonstrate that he or she does not.

Perhaps more important, as family ties have weakened, increasingly families simply abandon their more problematic members, rather than either care for them or arrange for them to be cared for elsewhere. At the same time, now that laws increasingly protect the right of individuals to dress and behave in unusual ways, communities no longer police unusual public behavior so closely. For both these reasons, the rise of individualism has resulted in fewer commitments to mental hospitals.

The Consequences of Deinstitutionalization

Following deinstitutionalization, persons with mental illness no longer found themselves locked for years in the often-brutal conditions of large mental institutions. Yet the promise that deinstitutionalization would herald a new era in which individuals would receive appropriate therapy in the community, avoiding the **stigma,** degradation, and mortification of mental hospitalization, has been met only partially.

Currently, most of those receiving inpatient care at private psychiatric hospitals or general hospitals suffer relatively mild symptoms and would not have received treatment before insurance coverage for mental illness became common. Not surprisingly, given the combination of mild presenting problems and limited insurance coverage, inpatient treatment is usually short, averaging less than two weeks. Patients are disproportionately female, under age 18, and middle- or upper-class, and typically pay their bills through private insurance (Brown, 1985:118–119; Mechanic, 1999:131).

The same population also largely accounts for the tremendous growth in outpatient care for mental illness, most of it received at community mental health centers. According to the National Comorbidity Survey, 13.3 percent of working-age Americans—nearly half of whom had no mental disorder diagnosable by the survey instrument—had sought outpatient treatment during the preceding twelve months (Kessler et al., 1999). Use of outpatient care rose substantially during the 1980s and early 1990s, especially in self-help groups such as Overeaters Anonymous.

But what of those who have severe, chronic mental illnesses and who previously would have been sent to large state hospitals? Unfortunately, for many, deinstitutionalization meant being sent home to relatives with no hope of treatment or cure. Many others exchanged life on locked wards for life in nursing homes or **board and care homes** (residential facilities that provide assistance in daily living but not nursing or medical care), which sometimes proved as degrading as large mental hospitals. (Both nursing homes and board and care homes are described more fully in Chapter 10.) Others were left with no source of care and became homeless.

How did deinstitutionalization lead to these results? One answer lies in the vested economic interests of those who benefited from the old system. Many mental hospitals were located in isolated rural communities whose economies depended heavily on the hospitals. When deinstitutionalization began, both hospital employees and local governments lobbied successfully to retain funding for public hospitals (Mechanic, 1989). As a result, only enough funds were available to build about 25 percent of the initially proposed network of **community mental health centers (CMHCs),** which, at any rate, from the start focused on care of mild, **acute** mental illness (Scull, 1985; Brown, 1985:47–73).

This situation worsened significantly during the 1980s, when the federal government cut funding for social services and low-income housing, while tightening eligibility for Medicaid, Medicare, and Social Security Disability benefits. As a result, many chronically mentally ill persons could pay for neither treatment nor housing. Consequently, many persons with chronic mental illness began to cycle between homelessness, brief jail stays when they prove too troublesome for local authorities who lack other alternatives, and acute episodes in public mental hospitals, which continue to provide the bulk of chronic inpatient care (Brown, 1985:118–119).

More recently, the federal government has continued underfunding social and mental health services while dramatically increasing funding for prisons (Butterfield, 1999). As a result, public mental hospitals now find that the best way to fill beds and pay their bills is to accept for treatment persons sent to them by the criminal justice system: mentally ill prison inmates, people found innocent by reason of insanity, and violent sex offenders who under new "sex predator" laws can be involuntarily confined for treatment after finishing their prison sentences. At California's Napa State Hospital, for example, almost 75 percent of patients during 1999 came from the criminal justice system (Kligman, 1999). Conversely, the same shift in funding from mental health to criminal justice has created a burgeoning population of mentally ill persons in the nation's prison system; a report released by the U.S. Department of Justice in 1999 estimated that 16 percent of jail and prison inmates have a mental illness (Butterfield, 1999).

Despite the severe gaps in the mental health system, however, observers generally agree that the bulk of seriously mentally ill persons are more satisfied with life and better off since deinstitutionalization than they were before (Grob, 1997; Allan Horwitz, personal communication, 1999). This is especially true for those who have participated in so-called "**model programs**" such as Training in Community Living in Madison, Wisconsin, and Fountain House in New York City. These programs aim to reduce the stigma and dehumanization previously associated with psychiatric care by fostering a sense of partnership between staff members and clients and deemphasizing hierarchical relationships. In addition, the programs seek to avoid mortification by empowering patients to take charge of their lives. Thus programs combine psychiatric care with a range of services (including financial support and housing assistance) designed to improve patient's quality of life and ability to function in the community. When compared with similar patients who receive care through traditional programs, model program patients on average have fewer rehospitalizations, more social relationships, greater economic self-sufficiency, and greater satisfaction with life (Rosenfield, 1997).

The Remedicalization of Mental Illness

Despite the success of model programs that combine social and psychological interventions, in general, treatment during the last twenty years has shifted toward the **remedicalization** of mental illness (Brown, 1990). Psychiatrists have developed new techniques for diagnosis and treatment and new theories of illness **etiology** that link mental illness to individual abnormalities in biochemistry, neuroendocrine functioning, brain structure, or genetic structure and downplay the effects of social factors.

The data for this "biological revolution" consist primarily of simple correlations between biological abnormalities and certain serious mental disorders (Brown, 1990); no studies have uncovered significant biological differences between those who have minor mental disorders and the rest of the population. None of this research adequately sorts out other factors that might account for these correlations (such as differences in nutrition or in the use of various drugs) or determines whether either the mental disorders or treatment for them might have caused, rather than resulted from, biological abnormalities.

Despite these weaknesses in the biological model of mental illness, most psychiatrists have adopted it. As a result, psychiatrists now present a more united front in their struggles for control against other mental health occupations such as psychology and social work. In addition, they have increased their political power relative to these other occupations because, having declared mental illness a biological problem, they now can argue that only persons trained in medicine can properly diagnose and treat it (Brown, 1990).

Reflecting this medical model, doctors now rely primarily on psychoactive drugs not only to treat mental illness but also to diagnose it. In a process first brought to public attention by psychiatrist Peter Kramer (1993) in his popular book, *Listening to Prozac*, doctors now "listen to drugs," assuming that the reaction to a drug tells us something basic about an individual's mental state. So if Prozac (fluoxetine hydrochloride) or another selective serotonin reuptake inhibitor (SSRI), which increases levels of the neurotransmitter serotonin in the brain, somehow makes an individual feel less depressed, then physicians typically conclude that lack of serotonin must have caused the depression. Yet as Kramer points out, pneumonia is not caused by a lack of antibiotics nor headaches by a lack of aspirin, but both drugs make ill people feel less ill. Similarly, doctors increasingly decide whether a patient is clinically depressed based not on whether that patient meets standard criteria for that diagnosis but on whether the patient responds favorably to SSRIs. Yet most people feel better when they take a mood-enhancing drug, whether it is Prozac or cocaine. As a result, by 1997 28 million Americans were taking SSRIs (*Wall Street Journal*, 1997).

Most of the drugs now used to treat mental illness fall into one of three main categories: antipsychotics, mood-stabilizers, and antidepressants. Psychiatrists use antipsychotic drugs, such as Clozaril and Risperdal, to help control severe symptoms in persons with major mental illnesses such as schizophrenia. These drugs are considerably less likely than are older drugs such as Haldol and Thorazine to produce loss of alertness and a condition known as "tardive dyskinesia" (uncontrollable, severe, and sometimes permanent muscular spasms). To control anxiety, obsessions, compulsions, and the severe mood swings of bipolar disorder, doctors commonly use mood-stabilizers such as Tegretol and Depakote. Finally, psychiatrists use antidepressants to alleviate major depressions. Unlike previous generations of antidepressants, SSRIs and other new drugs have fewer side effects and cannot be used to commit suicide. However, effectiveness can lessen over time, and their long-term consequences remain unknown (Kramer, 1993).

The Rise of Managed Care

During the 1990s, insurance coverage for mental illness became considerably more common, although most insurers offer less coverage for mental illness (especially chronic illness) than for physical illness (Frank and McGuire 1998; Mechanic, 1999:128–132). Increasingly, too, that coverage is offered through **managed care** organizations, which, as of 1998, provided mental health coverage to nearly 142 million Americans (Durham, 1998). Managed care is described more fully in the next chapter, but essentially refers to any system that controls health care spending by closely monitoring where patients receive health care, what sorts of providers patients use, what treatments they receive, and with what consequences.

It is too soon to fully assess the impact of managed care on either the cost or quality of care. However, early research suggests that managed care may be able to reduce the costs of mental health treatment, at least for less severe illnesses, by encouraging shorter rather than longer inpatient stays, outpatient rather than inpatient care, conservative rather than aggressive interventions, use of lower level clinicians (such as social workers) rather than psychologists or psychiatrists, and use of individual therapy rather than group therapy (Mechanic, 1995; Mechanic, 1999:160–162). Similarly, according to David Mechanic, probably the most influential sociologist in the area of mental health care,

> managed care has potential to improve mental health practice. . . . Most important, . . . by reducing inpatient admissions and length of stay, managed care programs potentially make available considerable resources for substitute services and other types of care. Managed care provides incentives to seek closer integration between inpatient and outpatient and primary and specialized services to achieve cost-effective substitutions.
>
> Managed care also offers the potential to bring . . . science-based mental health care into the mental health system more quickly than traditional programs. . . . Many individual practitioners resist practice guidelines and scientific findings, preferring their own clinical experience, but managed care can put systems in place to measure performance and to enforce adherence to established standards.

[Yet] the risks of managed care should also be recognized. Most obvious is the risk of withholding services or failing to maintain them over an appropriate interval. . . . Patients may be diverted to clinicians and service programs that lack the expertise and experience for patients with complex problems. Managed care may also resist use of new and more effective drugs [if these drugs] are more expensive or fail to maintain therapy [long enough] as a cost-saving measure (1997:45–46).

At this point, it remains unclear whether the benefits of managed care will outweigh the disadvantages.

THE EXPERIENCE OF MENTAL ILLNESS

The previous sections described the nature, causes, distribution, and history of mental illness. In this section, we will move from macro-level analysis to micro-level analysis and look at the experience of mental illness, from the process of labeling individuals mentally ill to the long-term effects of hospitalization.

Becoming a Mental Patient

As already noted, in any given year about 30 percent of working-age adults experience a diagnosable mental illness (Kessler et al., 1994). However, the majority of these receive no treatment. Ironically, as the stigma among the middle class against seeking counseling for minor problems has diminished and insurance has increased, levels of treatment have increased among basically well-functioning individuals who experience situational stress, sadness, or lowered self-esteem (Kessler et al., 1999). Nearly half of those who receive outpatient treatment have no mental disorder that can be identified through surveys, although some of these might have disorders that could be identified by clinicians (Kessler et al., 1999). What explains this discrepancy between experiencing symptoms and receiving treatment?

According to Allan Horwitz, "Symptoms of mental disorder are usually vague, ambiguous, and open to a number of varying interpretations. . . . Labels of 'mental illness,' 'madness,' or 'psychological disturbance' are applied only after alternative interpretations have failed to make sense of the behavior" (1982:31). The key question, then, is how do individuals come to be defined by themselves or others as mentally ill?

Self-Labeling

Regardless of how others define their situation, at least initially individuals usually define themselves as mentally healthy, using a process Whitt and Meile (1985) refer to as **aligning actions,** or actions taken to align one's behavior with social expectations. If individuals' problems increase, however, these aligning actions become less convincing. In a process Whitt and Meile refer to as **snowballing,** each additional problem becomes more difficult to deal with than the previous one, so a person with four problems experiences more than twice the difficulty of a person with two problems. As this snowballing occurs, individuals become more likely to define themselves as mentally ill and to seek care.

Peggy Thoits (1985) has provided a more detailed model of how self-labeling works among those—the majority—who experience only acute or mild problems. Her model, like that of Erving Goffman, draws on the theory of symbolic interactionism. Thoits applies this to mental illness by hypothesizing that well-socialized individuals sometimes label themselves as mentally ill when their behavior departs from social expectations, even if others do not consider their behavior disturbed or disturbing.

Because individuals recognize the stigma attached to mental illness, however, they work to avoid this label. According to Thoits, and as described earlier, most of the behavior that can lead to the label of mental illness involves inappropriate feelings or expressions of feelings. To avoid the label of mental illness, therefore, individuals can attempt to make their emotions match social expectations, through what Arlie Hochschild (1983) refers to as "**feeling work.**"

Feeling work can take four forms. First, individuals can change or reinterpret the situation that is causing them to have feelings others consider inappropriate. For example, a working woman distracted from her work by worries about how to care for an ill parent and distracted while with her parent by worries about her work can quit her job. Second, individuals can change their emotions physiologically, through drugs, meditation, biofeedback, or other methods. The woman with the ill parent, for example, could drink alcohol or take Prozac to control anxiety. Third, individuals can change their behavior, acting as if they feel more appropriate emotions than they really do. Fourth, individuals can reinterpret their feelings, telling themselves, for example, that they only feel tired rather than anxious.

When feeling work succeeds, individuals can avoid labeling themselves mentally ill. This is most likely to happen when the situations causing the emotions are temporary and brief and when supportive others legitimize their emotions. If, for example, the woman with the ill parent has similarly situated friends who describe similar emotions, she might conclude that her emotions are understandable and acceptable. If, on the other hand, her colleagues do not sympathize with her concerns and continually tell her to put her work first, her attempts at feeling work could fail, and she might conclude that she has a mental problem.

Ironically, some individuals label themselves mentally ill or are labeled by others because they succeed too well at feeling work. For example, those who rely too heavily on drugs to manage their feelings can lose control of their lives, and those who consistently reinterpret their emotions—telling themselves that they are not angry, for example, even while punching a wall or a spouse—can find that others label them crazy when their emotions and behavior don't match. In addition, those who consistently engage in feeling work can lose the ability to interpret their feelings accurately and experience them fully. The resulting sense of numbness and alienation eventually can lead individuals to define themselves as mentally ill.

Labeling by Family, Friends, and the Public

Like individuals, families only reluctantly label their members mentally ill (Horwitz, 1982). Instead, families can deny that a problem exists by convincing themselves that their relative's behavior does not depart greatly from the norm. If they do recognize that a problem exists, they can convince themselves that their relative is lazy, a drunkard, "nervous," responding normally to stress, or experiencing physical problems rather than mental illness. Finally, families might recognize that their relative is experiencing mental problems but define those problems as temporary or unimportant.

Two factors explain how and why families can ignore behavior others would label mental illness for so long. First, those who share cultural values, close personal relationships, and similar behavior patterns have a context for interpreting unusual behavior and therefore can interpret behavior as meaningful more easily than outsiders could. Second, families often hesitate to label one of their own for fear others can reject or devalue both the individual and the family. As a result, families have a strong motive to develop alternative and less stigmatizing explanations for problematic behavior.

Surprisingly, strangers as well as intimates tend to avoid interpreting behavior as mental illness. In one study, for example, researchers had subjects read vignettes describing individuals who met the criteria for various psychiatric diagnoses (D'Arcy and Brockman, 1976). The researchers found that the proportion of subjects who defined the described individuals as mentally ill declined from 70 percent for the vignettes of paranoid schizophrenics to 34 percent for the vignettes of simple schizophrenics, 25 percent for the vignettes of alcoholics, and less than 10 percent for the vignettes of neurotics (that is, persons who experience psychological distress but are in touch with reality and able to function). This evidence suggests that the public applies the label of mental illness only when disordered behavior is public, violent, dramatic, or otherwise unignorable.

Moreover, even when relatives and other intimates define an individual as mentally ill, they do not necessarily bring the individual to treatment. Instead, they can continue to protect the individual against social sanctions through a process Lynch (1983) refers to as **accommodation.** Accommodation refers to "interactional techniques that people use to manage persons they view as persistent sources of trouble" and to avoid conflict (Lynch, 1983:152).

Based on analyzing essays in which college students described how they handled family members, workmates, fraternity brothers, and others whom they regarded as disturbed, Lynch identified three forms of accommodation. First, students could minimize contact with problematic individuals—avoiding them, ignoring them when they could not be avoided, or restricting interactions to a minimal and superficial level when they could not be ignored. Second, students could limit the trouble individuals could cause through such actions as taking over the individuals' responsibilities or humoring their wishes and beliefs. Third, they could manage the *reactions* to the problematic individual through such actions as providing excuses when the individual did not meet social expectations or hiding the individual from others' view—for example, keeping a "crazy" fraternity brother out of sight when outsiders were present during parties.

Nevertheless, despite these attempts to normalize and accommodate mental illness, families and friends may eventually conclude that an individual needs treatment. At that point, they must either get the individual to agree or coerce the individual into getting treatment despite his or her active resistance. A recent study of all persons seeking care for a serious mental illness for the first time found that 42 percent had actively chosen to seek care and 23 percent had been coerced (Pescosolido et al., 1998). Coercion was most common among those with bipolar disorder, who often enjoyed the "highs" of mania even though others regarded them as seriously disturbed, and among those with large, tight social networks. In another 31 percent of cases, families "muddled through," and either individuals went along with treatment decisions made by others without either accepting or rejecting those decisions or no one in the family seemed to have been in charge of the decision-making process.

Labeling by the Psychiatric Establishment

Once individuals enter treatment, a different set of rules applies, for whereas the public tends to normalize behavior, mental health professionals tend to assume illness. This happens for several reasons. First, as explained earlier, the medical model of mental illness stresses that treatment usually helps and rarely harms and so is the safest option in ambiguous situations. This ideology, coupled with the inherent incentives for any profession to expand its territory and power, encourages mental health workers to define mental illness broadly. Second, because mental health workers see prospective patients outside of any social context, behavior that might seem reasonable in context often seems incomprehensible. This is especially likely when mental health workers and prospective patients come from different social worlds, whether because they differ in gender, ethnicity, social class, or some other factor. Third, mental health workers reasonably can assume that individuals would not have been brought to their attention if they did not need care. Finally, because normalization and accommodation are so common, mental health workers often do not see individuals until the situation has reached a crisis, making it relatively easy to conclude that the individuals are mentally ill.

The Post-Patient Experience

Research on the post-patient experience has focused on the sources, consequences, and extent of stigma experienced by former patients. This is a critical issue, for it challenges the medical model's assumption that psychiatric treatment is benign.

Those who support a medical model of illness point to several studies suggesting that the public stigmatizes only those former patients who continue to engage in problematic behavior (Link et al., 1987). Other researchers, however, have questioned these results, noting that former patients consistently report experiencing substantial stigma and that surveys of the general public find that mental illness evokes as much stigma and social rejection as drug and alcohol addiction, prostitution, and having a criminal record (Albrecht et al., 1982). Additional evidence for the importance of stigma can be gleaned from the fact that 16 percent of claims filed in 1999 alleging discrimination under the **Americans with Disabilities Act** came from people with mental illnesses (Curtis, 1999).

To explain why some studies find high rates of stigma toward former mental patients and others do not, Bruce Link and his colleagues (1987) asked a random sample of survey respondents to fill out questionnaires regarding their attitudes towards persons with mental illness and to respond to a description of a person whose behavior met the definition of mental illness. None of the respondents was told that the person was mentally ill, but half were told that he was a former mental patient. Respondents who believed mentally ill persons are dangerous proved *more* likely to reject a person who was described as a former mental patient, whereas those who believe persons with mental illness are generally harmless proved *less* likely to reject the former patient. The authors conclude that previous studies found no evidence of stigma because they unintentionally had combined these two groups.

In two further studies, Link and his colleagues argued that labeling an individual mentally ill has negative effects not only because of how the general public responds but also because of how the labeled individual responds (Link, 1987; Link et al., 1989). These studies found that both former patients and the general public believe that most people devalue and reject former mental patients. This leads former patients to devalue themselves, which damages their self-esteem and their work performance. In addition, these processes lead former patients to expect rejection and therefore to engage in defensive behaviors such as secrecy and emotional withdrawal, creating additional problems in social relationships.

Box 7.2   *Ethical Debate: Confidentiality and the Duty to Warn*

In the fall of 1969, Prosenjit Poddar entered outpatient psychotherapy at the University of California-Berkeley Student Health Center. During the course of therapy, he told his therapist, Dr. Lawrence Moore, that he planned to kill his girlfriend, fellow-student Tatiana Tarasoff.

Therapists, like medical doctors and clergy, always have regarded their discussions with patients as privileged communication in which, both legally and morally, confidentiality must be safeguarded. In a situation such as this one, however, therapists must weigh the danger to their patients if they breach confidentiality against the danger to others if they do not.

Dr. Moore's first response was to consult with his two supervisors. All three concurred that Poddar needed to be hospitalized for observation. Moore's supervisor then notified the campus police and asked them to bring in Poddar. When the police detained and interviewed him, however, they concluded that he was rational and not dangerous. As a result, Moore's supervisor rescinded the original commitment order.

Not surprisingly, Poddar felt betrayed by his therapist's breach of confidence and broke off therapy. Two months later, when Tarasoff returned from a long trip, Poddar killed her.

After Tarasoff's death, her parents learned that Poddar had told his therapist of his intentions. In Tarasoff v. Regents of the University of California (131 California Reporter 14, July 1, 1976), the parents successfully sued Dr. Moore and the university on the grounds that therapists must abandon confidentiality when another life is endangered and that, specifically, they must inform intended victims as well as legal authorities.

At first reading, the message of the Tarasoff case seems obvious: if a therapist reasonably suspects a client is dangerous, the therapist must warn both the legal authorities and the intended victims. This same reasoning has been applied to clients who tell their therapists

---

These findings, of course, do not necessarily mean that the hazards of stigma outweigh the benefits of treatment. In the last decade, research has found substantial evidence that both psychotherapy and drug treatment can reduce symptoms and prevent relapse, at least in the short term (Link et al., 1997). Other research, however, suggests that the negative effects of stigma coexist with the benefits of treatment, partially canceling each other out (Link et al., 1997; Rosenfield, 1997). These results led Bruce Link and his colleagues to conclude that "stigma has important effects, effects that remain even when people improve while participating in treatment programs. Health care providers are therefore faced with the challenge of how to address stigma in its own right if they want to maximize the quality of life for those they treat and maintain the benefits of treatment beyond the short term" (1997:187). The potential for stigmatizing mental patients and

of suicidal thoughts. More recently and in a somewhat different vein, some have argued that health care workers must breach confidentiality when they learn of **HIV**-infected clients having unprotected sex without informing their sexual partners of their infection. The codes of ethics of both the American Medical Association and the American Psychiatric Association, as well as various legal decisions, declare that doctors must breach confidentiality when the health or welfare of either a client or others in the community is endangered.

A closer look at the Tarasoff case, however, reveals some of the difficulties of reaching any simple conclusion. On the one hand, one could argue that if Tatiana had been informed, she could have protected herself. Yet women are killed daily who know full well that their husbands or lovers want to kill them. Police often can offer little protection to these women, and the women often can do little to protect themselves.

In addition, in the Tarasoff case, the one documented result of informing the police was that Poddar ended therapy. One could argue, therefore, that, far from protecting the intended victim, breaching confidentiality placed her in greater danger by convincing Poddar to end therapy, thus reducing the chances that he would find a nonviolent way of managing his anger.

Finally, the argument that therapists must breach confidentiality regarding dangerous clients assumes that therapists know which clients are dangerous. Yet, as various studies have shown and as the American Psychiatric Association and several other professional organizations argued in briefs filed on behalf of the therapists in the Tarasoff case, this is far from true. Moreover, if psychiatrists wrongly conclude that clients are dangerous and therefore breach confidentiality, they can subject the clients to substantial stigma, sometimes with permanent consequences. Indeed, with the growth of large, all-too-accessible, computerized data banks of medical records and the growth in access to those records by insurers, peer review organizations, and the like, the more serious issue facing therapists in the future may be how to *protect* confidentiality, not when to breach it (Alpert, 1993).

the problems that arise when the interests of mental patients conflict with the interests of others are discussed in Box 7.2.

CONCLUSIONS

This chapter has presented and compared the sociological and medical models of mental illness. As we have seen, and as with the medical models of physical illness and disability discussed in Chapters 5 and 6, the medical model of mental illness asserts that mental illness is a scientifically measurable, objective reality, requiring prompt treatment by scientifically trained personnel. As such, this model downplays the role of social and moral values in the definition and treatment of mental illness and the effect of mortification and stigma on those who receive treatment.

Entering the twenty-first century, we find ourselves facing a situation uncomfortably similar to that of past centuries. As in the years before the Great Confinement, thousands of persons who have mental illnesses now live on the streets and support themselves at least in part by begging. Many more are confined in nursing homes, board and care homes, or prisons, along with other social rejects, in the same way that earlier societies confined persons with mental illness in almshouses along with the poor, the disabled, and those without families. Although drugs largely have replaced shackles, society still allocates far too few resources to provide humanely for those who suffer mental illnesses. We can only hope that, in the future, with a greater understanding of the nature of mental illness and the social response to it, we can develop more compassionate and effective means of coping with mental illness.

SUGGESTED READINGS

Brown, Phil. 1985. *The Transfer of Care: Psychiatric Deinstitutionalization and Its Aftermath.* Boston: Routledge & Kegan Paul. Describes the sources and consequences of the deinstitutionalization of mental patients.

Goffman, Erving. 1961. *Asylums.* Garden City, NY.: Doubleday. The classic text on the nature of mental hospitals and other total institutions. Still fascinating reading.

Kaysen, Susan. 1993. *Girl, Interrupted.* New York: Random House. A memoir of mental illness and its treatment.

GETTING INVOLVED

American Civil Liberties Union. 132 W. 43rd Street, New York, NY 10036. (212) 944-9800. www.aclu.org. Among other things, works for the civil rights of mental patients.

National Alliance for the Mentally Ill. 200 North Glebe Road, Suite 1015, Arlington, VA 22203-3754. (800) 950-6264. www.nami.org. The nation's leading grassroots, self-help and family advocacy organization devoted to improving the lives and treatment of persons with severe mental illnesses. Supports the medicalization of mental illness.

REVIEW QUESTIONS

How and why do ethnicity, gender, and social class affect rates of mental illness?

What is the relationship between life events and mental illness?

What are the differences between the medical and sociological models of mental illness?

What are the problems embedded in psychiatric diagnoses?

What was moral treatment and why did it fail?

What was the antipsychiatry critique?

What were the sources and consequences of deinstitutionalization?

What is the remedicalization of mental illness?

How is managed care affecting the treatment and experience of mental illness?

How do individuals become mental patients?

What are the consequences of labeling an individual mentally ill?

INTERNET EXERCISES

Browse the Web site for the National Alliance for the Mentally Ill (www.nami.org) the major national organization promoting the interests of persons with mental illness and their families. What is its approach to mental illness? How is it similar to or different from the perspective presented in this chapter?

To ascertain the extent to which Freudian ideas now permeate American culture, obtain access through your library or the Internet to *Periodical Abstracts,* the *Readers Guide to Periodical Literature,* or another index of popular magazine articles. Then search for the term "Freudian." In what ways is the term now used, by what sorts of persons and organizations, and for what purposes?

PART THREE

Health Care Systems and Settings

CHAPTER EIGHT
The U.S. Health Care System and the Need for Reform

CHAPTER NINE
Alternative Health Care Systems

CHAPTER TEN
Health Care Settings

In Part Two, we looked at illness primarily from the perspective of the ill individual. In this part, we move to a macrosociological level, looking at health care systems and settings. In Chapter 8, I describe the history and current nature of the U.S. health care system, examining why and how millions of Americans have found themselves uninsured, underinsured, or precariously insured—threatened with the loss of health insurance at any moment. I begin Chapter 9 by presenting a series of measures useful for evaluating any health care system, and then describe five alternative health care systems—those of Canada, Great Britain, Germany, the People's Republic of China, and Mexico. With this as a basis, I conclude the chapter by looking at the prospects for reforming the U.S. health care system. Finally, in Chapter 10, I investigate the major settings in which health care is offered in the United States (other than individual doctors' offices). As I will show, hospitals' strength lies in dealing with acute rather than chronic health problems. Consequently, other institutions, including nursing homes, board and care homes, and hospices, have emerged to respond to chronic illness and dying. Meanwhile, as I will describe, most health care still takes place within the family home.

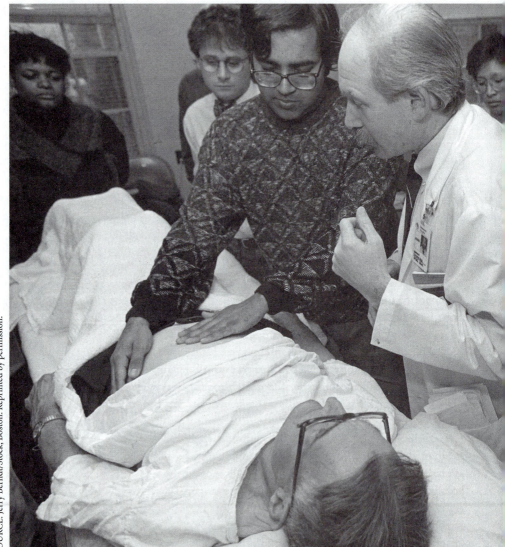

Harvard Medical School students learning how to palpate a patient.

CHAPTER 8

The U.S. Health Care System and the Need for Reform

Health care in the United States is a system in crisis. The lives of Stephen Beidner, Kay Nichols, and John Andrusyshyn suggest some of the problems with this system (*Consumer Reports*, 1990:534–537).* For Stephen Beidner, health insurance became unaffordable as soon as he needed it:

> [When] Stephen Beidner, a part-time worker at a California winery, . . . first took out a policy with a company called Consumers United Insurance in 1985, he paid $912 a year. By 1989, his premium had jumped to nearly $3,600. [Later that year,] after Beidner had arthroscopic surgery for a knee injury, the company hiked his premium a whopping 93 percent to $6,900.

Kay Nichols, on the other hand, still has health insurance but can keep it only by remaining locked into her current job:

> Nichols, a fitness counselor at a Gainesville, Florida, health club, is in the pink of health except for glaucoma, an eye disease that can cause blindness if not treated. Not long ago, her employer wanted to switch insurance carriers to take advantage of lower premiums. When the health club found another insurer, the agent told Nichols that she would not be covered, even though her glaucoma is under control.
>
> Nichols looked into a conversion policy from her present company [so that she could continue her old insurance on her own] but found she would have to pay $6,000 for six months of coverage for her family. She tried Blue Cross [a major insurance company once known for accepting

*all applicants, regardless of health risks], but its policy would have ex-
cluded coverage for glaucoma.*

*When her employer learned of her plight, he decided to keep the cur-
rent policy despite its higher premiums. "If the premiums get phenome-
nally high, they can't keep the policy just for me, and I understand that,"
Nichols says. At the same time, she realizes she has a problem that won't
go away. "Maybe I don't want to stay with this company the rest of my
life," she says. "It makes me worry."*

Nichols is 38.

Finally, consider John Andrusyshyn's story:

*John Andrusyshyn worked in a Nevada casino. Three summers ago, he
noticed a mole growing on his chest, but said nothing about it to his
family. He could not afford to pay another bill, so he put off seeing a
doctor. Andrusyshyn was not eligible for insurance from his employer
until he had been at his job for a year; he couldn't afford his own coverage on the $880-a-month he was bringing home to support his wife,
Karen, and two children, Laura and Nikolai.*

*Several months went by before Karen insisted he go to a doctor. Be-
cause dermatologists in Reno were booked up, three more months passed
before a doctor examined him. By then, the mole had ulcerated, and John
was so desperate for treatment he paid for the visit with a bad check.*

*The diagnosis was a malignant melanoma [cancer] that was already
coursing through his body. By the time he underwent surgery, he was eligible for insurance from the casino. But Karen had to scrape together
$56 a week to pay his share of the premiums, forgoing food and other
necessities. The policy covered the hospital bill, but not the $4,000 sur-
geon's fee. On John's medical records, doctors noted, "Patient has no
money; we'll do the best we can."*

Last fall, at the age of 54, John Andrusyshyn died.

HEALTH INSURANCE IN THE UNITED STATES

Despite this chapter's title, the United States does not have a health care
system. Nor, unlike every other **industrialized nation,** does it have any
mechanism for guaranteeing health care to its citizens. Rather, the United
States has an agglomeration of public and private health care insurers and
providers functioning autonomously in myriad and often competing ways.
This section describes the four main forms of health insurance coverage in

the United States—Blue Cross and Blue Shield, commercial insurance, health maintenance organizations, and federal insurance programs—and the movement of all these forms toward **managed care.**

Blue Cross and Blue Shield

The American health insurance industry experienced its first growth spurt during the Great Depression of the 1930s. During those years, nongovernmental hospitals faced two serious threats: Americans' growing inability to pay their hospital bills and growing belief that the government should provide social services, including health care, for its needy citizens (Rothman, 1997). These two threats led the American Hospital Association to found **Blue Cross,** a nonprofit company that sells insurance to cover individuals' hospital bills. By offering reasonably priced insurance, Blue Cross freed both hospitals and much of the middle class from worries about unpaid hospital bills. By so doing, Blue Cross substantially reduced political pressures to create a national health system among that segment of the U.S. population most likely to vote (Rothman, 1997).

The success of Blue Cross led the American Medical Association to found **Blue Shield** shortly thereafter. Whereas the purpose of Blue Cross was to protect hospitals, the purpose of Blue Shield, which provides coverage for medical bills, was to protect doctors. As of 1995, 66 percent of privately insured Americans have Blue Cross or Blue Shield coverage.

Until the 1980s, both Blue Cross and Blue Shield (collectively known as "the Blues") established their fees based on **community rating.** Under community rating, each individual pays a "group rate" premium (or yearly fee) based on the average risk level of his or her community as a whole. Even if a particular individual is a bad insurance risk because of a preexisting illness, a dangerous job, or a family history of illness, the insurer need not charge that individual a high premium because most members of the community will have much lower risks, keeping the average costs to the insurer low. This explains why those who purchase insurance as part of a large group, such as all employees of IBM, pay far lower premiums than do the approximately 9 percent of Americans under age 65 who purchase insurance as individuals.

Typically, individuals who have Blue Cross/Blue Shield insurance may seek care from the hospitals and doctors of their choice and are charged on a **fee-for-service** basis; that is, they are billed a fee for each office visit, test, or other service they receive. Under this **fee-for-service insurance,** individuals must first pay these bills and then request reimbursement from their insurance providers. However, individuals typically must pay on their own the first $100 to $500 in bills they receive each year (known as the **deductible**), as well as 20 percent of their remaining medical bills, up to 20 percent of their hospital bills, and all costs for preventive medical care. To keep Blue Cross/Blue Shield insurance attractive, many plans now cover higher proportions of patient bills and charge lower deductibles if members use doctors who have agreed to charge lower, pre-set fees to plan members.

Both Blue Cross and Blue Shield usually establish lifetime and sometimes annual maximums. Individuals who exceed their maximums must pay their remaining bills themselves, a serious problem for those with **chronic illnesses.** Individuals who exhaust their savings can only hope the government will pay for their health care.

Commercial Insurance

During their first few years, Blue Cross and Blue Shield had little competition. Beginning in the 1940s, however, **commercial insurance** plans (i.e., insurers that function on a for-profit basis) grew rapidly. Commercial insurers typically offered fee-for-service insurance similar to that offered by the Blues. However, these plans could offer lower premiums than the Blues because they used **actuarial risk rating** rather than community rating. Under actuarial risk rating, insurers maximize their profits by insuring only individuals whose health risks appear low or insuring higher-risk persons only if they will pay higher premiums. For example, commercial insurers typically charge higher premiums to those who have allergies, back strain, kidney stones, or ulcers and deny coverage to those who have ulcerative colitis, diabetes, or severe obesity. In addition, these insurers frequently deny coverage to individuals who work in fields that they believe either entail high risks or attract persons who enjoy risk-taking, including aviation, auto sales, construction, and law. In addition, since the start of the **HIV epidemic,** insurers have grown reluctant to insure beauticians, restaurant workers, and flight attendants, among others, on the assumption that these fields attract disproportionate numbers of gay men. (In most jurisdictions, discrimination based on occupation or sexual orientation is legal.)

The success of commercial insurers in luring low-risk individuals away from the Blues has increased the average costs the Blues must pay to provide care to their remaining customers. To avoid raising their rates to prohibitive levels, many Blue Cross/Blue Shield companies now use actuarial risk rating.

The Rise of HMOs

The 1930s and 1940s also saw the rise of a very different type of health insurance program: **health maintenance organizations (HMOs).** Unlike the Blues and the commercial insurers, the first HMOs to attract national attention—Kaiser Permanente and the Group Health Cooperative of Puget Sound—were organized by individuals whose primary aim was providing affordable, high-quality health care to their communities. Whereas the Blues and the commercial insurers used **retrospective reimbursement,** reimbursing individuals for health care costs after they fell ill, the HMOs used **prospective reimbursement** in an attempt to keep people from falling ill in the first place.

Under prospective reimbursement, HMOs paid doctors a salary, rather than paying them on a fee-for-service basis. Because doctors received the same salary regardless of the number of times they saw their patients or the number of procedures they performed, they could not increase their income by providing unnecessary medical care. Instead, doctors would earn the highest net income by keeping patients healthy so the patients would require less of their time and resources in the long run.

In line with their emphasis on restraining costs by keeping members healthy, HMOs, unlike the Blues and commercial insurers, paid the full cost of preventive care. Patients, meanwhile, paid nothing beyond the cost of their insurance premiums as long as they used only doctors affiliated with their HMO and saw specialists only if referred by their **primary care doctor** (known as a **gatekeeper** in systems of this sort).

The Changing Structure of HMOs

Over time, research on HMOs began to suggest that HMOs could provide a quality of care at least equal to that of other insurance plans but at a lower cost (e.g., Leape, 1992). As a result, interest in using HMOs to generate corporate profits began to grow. By 1998, 25.2 percent of Americans belonged to HMOs (U.S. Department of Health and Human Services, 1998:365), and 62 percent of those individuals belonged to for-profit HMOs.

The interest in HMOs as cost-saving mechanisms has altered the structure of HMOs substantially. To discourage unnecessary visits to doctors, most HMOs now charge **copayments**—fees consumers must pay each time they see a care provider, ranging from $5 to 20 percent of costs. To discourage primary care doctors from unnecessarily referring patients to more expensive specialists, many HMOs now set aside annually a pool of money to pay for referrals to specialists and allow the primary care doctors to divide among themselves any money left over at the end of the year. A California survey of primary care HMO doctors found that 57 percent felt pressured to limit referrals (Bodenheimer, 1999); those who do not limit referrals are less likely than others to have their contracts renewed. On the other hand, to reduce patient dissatisfaction, HMOs recently have reduced use of gatekeepers and most now offer members the option of seeing doctors outside their HMO in exchange for higher copayments (Bodenheimer, 1999).

To further increase doctors' incentive to control the costs of health care, only 3.1 percent of HMOs still pay doctors on salary (Himmelstein et al., 1999). Instead, HMOs contract either with several group medical practices or with individual doctors to provide services to their members. HMOs that contract with several groups are known as **network model HMOs** and those that contract with individual doctors are known as **individual practice associations (IPAs)**; most of the remainder combine these two models. Both network model HMOs and IPAs typically contract only with primary care providers and a few specialists, paying to send members to outside specialists as needed. Network model HMOs typically pay primary care doctors by **capitation,** paying them a set annual fee per person in their practice, plus supplements for newborns, elderly patients, or chronically ill patients. IPAs pay doctors either by capitation or on a fee-for-service basis. Unlike doctors in private fee-for-service practice, however, IPA doctors must abide by a schedule of fees negotiated in advance with the HMO. Those paid by capitation have more of a financial incentive to *undertreat* because their profits depend on spending less than their capitation payments, and those paid fee-for-service have more of a financial incentive to *overtreat.* The latter tendency is somewhat restrained, however, by the knowledge that doctors whose bills are higher than average might not have their HMO contracts renewed.

At 51 percent of all HMOs, IPAs are now the largest and fastest growing form (Committee on Ways and Means, 1997). Patients prefer them because they typically offer care in smaller, less bureaucratic settings than other HMOs do, and doctors prefer them because their fee-for-service structure rewards more work with more income (Bodenheimer, 1999).

Government-Provided Health Care: Medicare and Medicaid

Those who cannot afford any insurance coverage, whether through the Blues, commercial insurers, or HMOs, must turn to the government for care. As Chapter 10 will describe, since the nineteenth century the government has offered free care to indigent persons at hospitals scattered around the country. Many Americans, however, live in areas not served by such hospitals. Moreover, hospitals focus on providing intensive high-technology care, not the primary care individuals more often need.

Concern about lack of basic health care coverage for the poor waxed and waned during the twentieth century. In 1912, Theodore Roosevelt proposed a national health insurance system during his unsuccessful presidential campaign. During the 1930s, President Franklin D. Roosevelt considered implementing some sort of national health insurance along with the Social Security system, but was stymied by political opposition. His successor, Harry Truman, was even more committed to beginning a national health insurance system but was blocked by accusations that national health care was a form of "creeping socialism."

Although political concerns continued to block the development of national health insurance, by the 1960s support had grown considerably for expanding coverage at least to some segments of the population. This sentiment coalesced in 1965 when Congress enacted the Medicare and Medicaid programs.

Medicare

Medicare covers more persons than any other single insurance program in the nation. Virtually all Americans over age 65 receive Medicare, as do some permanently disabled persons (Campbell, 1999). All persons eligible for Medicare receive at no cost coverage for as many as 150 days of hospital care, although these persons must pay substantial deductibles and copayments. In addition, they receive limited coverage for post-hospital nursing services, home health care, and hospice care. Medicare also offers fee-for-service insurance for outpatient medical costs, at a monthly premium of $526 as of 1998. This insurance, too, has substantial deductibles and does not cover many medical costs, such as prescription drugs, long-term nursing home care, and routine eye care. Adding together the costs of copayments, deductibles, premiums, and items not covered by insurance, Medicare recipients over age 65 pay an average of 21 percent of their medical costs out of pocket (Committee on Ways and Means, 1997). As a result, those with incomes below the federal poverty line pay between 35 and 50 percent of their total incomes on medical costs (Iglehart, 1999).

To keep their costs to a minimum, almost all Medicare recipients purchase (or receive from their former employers) additional insurance known as **medigap policies.** (The poorest Medicare recipients may receive additional coverage through Medicaid, the government's program for indigent health care.) Medicare recipients also have the option of receiving their Medicare benefits through HMOs, which offer additional benefits (such as inexpensive prescriptions) in exchange for accepting the restrictions on choice inherent in HMOs. Finally, Medicare recipients can choose to opt out of Medicare and instead set money aside in tax-free **Medical Savings Accounts (MSAs),** which they can use to purchase health care privately. This option gives individuals the most choices, but leaves them vulnerable to catastrophic financial costs should they have a major illness or injury. In addition, this option endangers the system as a whole, as it threatens to draw healthy, wealthy individuals away from Medicare and thus to force up the average costs of providing care to the remaining participants (*Consumer Reports*, 1998).

This economic threat to Medicare comes as the system faces increasing economic pressures from other sides. Medicare is primarily funded through federal Social Security taxes. Essentially, working adults pay taxes into a trust fund that pays the health care bills of the elderly. According to researchers working for the U.S. Health Care Financing Administration (De Lew et al., 1992), "this financing approach is not actuarially sound: Expenses are increasing faster than revenues. . . . As the population ages, there will be fewer workers supporting each beneficiary. There were 5 workers for each beneficiary in 1960, there will be 3 workers per beneficiary in 2000, and 1.9 by the year 2040." As a result of this trend, the 1999 *Social Security and Medicare Annual Report* projects that Medicare will be unable to pay full benefits for hospital care after the year 2015, unless major changes are made. Responding to this problem, Congress in 1997 cut $116 billion from the Medicare budget. As a result, insurance premiums paid by Medicare recipients are scheduled to more than double between 1998 and 2006, while benefits and fees paid to health care providers are being cut, leading many HMOs to cancel or reduce their Medicare programs.

Medicaid

Whereas Medicare provides coverage to individuals based primarily on age, **Medicaid** provides coverage based on income and physical vulnerability. To receive Medicaid, individuals must be both poor and either aged, blind, disabled, pregnant, or the parent (almost always the mother) of a dependent child.

Medicaid is funded through a combination of federal and state taxes. The federal government contributes 50 percent of costs to the wealthier states and as much as 79 percent of costs to the poorer states. Because of this joint funding, states have considerable leeway to determine eligibility and benefits, as long as they provide Medicaid to individuals whose family incomes, as of 1998, are less than $13,650 (for a family of three) and who additionally are either under age 6 or pregnant. Because of these restrictions, many Americans who cannot afford health care are nevertheless ineligible for Medicaid. As of 1998, only 40 percent of poor Americans—and 10.3 percent of Americans overall—receive Medicaid (Campbell, 1999).

The Rise of Managed Care

The most striking change in the U.S. health care system during the last two decades has been the dramatic rise in **managed care.** Managed care refers to any system that controls costs through closely monitoring and controlling the decisions of health care providers. Most commonly, managed care organizations (MCOs) monitor and control costs through **utilization review,** in which doctors must obtain approval from the insurer before they can hospitalize a patient, perform surgery, order an expensive diagnostic test, or refer to a specialist outside the insurance plan. Although the terms *HMO* and *managed care* increasingly are used interchangeably, HMOs represent only one form of managed care, and most fee-for-service insurers now also use managed care. Taken together, about 75 percent of all Americans with private insurance now belong to some form of managed care plan (Iglehart, 1999), as did 40 percent of Medicaid recipients and 10 percent of Medicare recipients as of 1995 (U.S. Department of Health and Human Services, 1998:376–377).

The growth in managed care has raised serious questions regarding whether its emphasis on cost-containment has reduced quality of care. Most of the research currently available suggests that managed care does reduce use of expensive resources. For example, a report produced by the U.S. Congressional Budget Office (1995) found that MCOs use 8 percent fewer health resources than fee-for-service insurers because of reduced use of specialists, hospitalization, and expensive ancillary services such as intensive care wards. Similarly, one study found that the rate of angioplasty was twice as high and the rate of prostatectomy three times higher among fee-for-service patients than among HMO patients (Leape, 1992). The study noted, however, that both these surgeries were often of questionable benefit and that the differences in rates of surgeries might reflect *overtreatment* of fee-for-service patients rather than *undertreatment* of HMO patients.

On the other hand, studies consistently have found that MCO members are more likely to use preventive services such as mammograms (Commonwealth Fund Commission on Women's Health, 1999; Herzlinger, 1997). Similarly, although some studies based on random samples have found that persons in managed care plans are more likely than are persons with fee-for-service insurance to believe that they sometimes have problems gaining access to needed treatment, other studies have found no differences (Commonwealth Fund Commission on Women's Health, 1999; Donelan et al., 1996). When differences have been found between MCOs and fee-for-service organizations, they often are small and are limited to persons with chronic illnesses; Table 8.1 shows the results for the eight variables out of the 64 studied by Donelan and his colleagues (1996) in which significant differences were found.

Neither research on rates of medical services nor on patient satisfaction can tell us whether MCOs or traditional insurers produce better health outcomes overall. Unfortunately, few data are available that directly compare outcomes. However, two studies published in the *Journal of the American Medical Association* suggest that HMOs can offer care as good as or better than that found in traditional fee-for-service medicine. The first study tracked almost 2,000 patients who had high blood pressure or adult diabetes over seven years and found no differences in outcomes between those treated in HMOs versus those treated on a fee-for-service basis (Greenfield et al., 1995). However, doctors working in HMOs hospitalized 40 percent fewer patients and ordered 12 percent fewer tests, thus resulting in lower costs overall for the HMO patients. The second study (Riley et al., 1999) looked at women over age 65 who had breast cancer and found that those who belonged to HMOs were more likely than those with fee-for-service insurance to have had their cancers diagnosed at an earlier stage and, if

Table 8.1 **Self-Reported Experiences with Health Care in Fee-for-Service and Managed Care Plans, Among Persons with Significant Illnesses**

PROBLEMS	FEE-FOR-SERVICE	MANAGED CARE
Problems getting treatment you and your doctor thought necessary	13%	22%*
Inappropriate or incorrect primary care treatment in last year	5	12*
Primary doctor did not explain when and how to take medicines	4	10*
Long waits for appointment with primary doctor in last year	7	17*
Primary doctor did not explain what he or she was doing	6	12*
Inappropriate or incorrect specialty treatment in last year	3	10*
Specialist exam not thorough enough	3	12*
Time specialist spent was not adequate	6	15*

*$p<.05$

Source: Donelan et al. (1996).

they had a lumpectomy, to have received radiation afterwards, as is currently recommended.

At any rate, current research provides a poor basis for predicting the impact of MCOs in the future. As the use of MCOs has spread (especially through Medicare and Medicaid), MCOs are attracting a more representative and less healthy population than in the past. For these less healthy patients, MCOs' emphasis on preventive, primary care rather than on interventionist and specialty care may be more problematic. In addition, as market pressures drive MCOs to behave more like traditional fee-for-service insurance, the differences in the impact of these two types of insurers on cost and quality of care necessarily will diminish.

Perhaps the more important issue is not the impact of managed care *per se* but, rather, the impact of the for-profit motive. A review of all comparative research conducted from 1993 to 1997 found no clear differences between HMOs and fee-for-service medicine in quality of care, use of medical or hospital services, or patient satisfaction. These findings suggest that the use of HMOs *per se* neither harms nor improves patients' health (Miller and Luft, 1997). However, a later study, using data collected in 1997 from most HMOs in the country, found that for-profit HMOs scored lower than nonprofit HMOs on all 14 indicators of quality of care, including rates of childhood immunization, mammograms, prenatal care, and appropriate treatment of persons who had diabetes or heart attacks (Himmelstein et al., 1999). (Box 8.1 similarly discusses the impact of profit incentives on pharmacists' services.)

Finally, questions have been raised regarding the impact of managed care on patient/doctor relationships. One of the requirements for a good relationship is that patients trust their doctors to choose the best available treatments regardless of the doctors' own financial interests. Under the fee-for-service system, doctors had an incentive to provide as much care as their patients' insurance or budget would cover, whether or not it was fully necessary. In part as a result, rates of elective and questionable treatments, like cosmetic surgery and hysterectomies, were far higher than could be justified on purely medical grounds (Leape, 1992). Yet because that system had been in place for decades, the public rarely questioned whether they might be receiving unnecessary and potentially dangerous overtreatment.

With the rise of managed care, the inherent financial incentives of the health care system have reversed, so that now doctors can increase their incomes by *restricting* the services they provide rather than by *increasing* them. Indeed, in some MCOs, doctors do not even have the option of prescribing certain drugs or recommending certain specialized forms of care. Because this system is so new, and because doctors have campaigned actively to alert the public to its potential dangers (which, not coincidentally, also threaten doctors' incomes and clinical autonomy), patients seem more aware of the dangers of undertreatment inherent in managed care than of the dangers of overtreatment inherent in fee-for-service medicine. More broadly, some patients now think of their doctors as "double agents," whose loyalties are split between serving their patients and serving the MCOs that employ them (Shortell et al., 1998). Such patients are less likely to trust their doctors and, as a result, more likely to decline treatment, participate in treatment only halfheartedly or with little faith in the outcome, or withhold needed information about their health from health care providers (Mechanic, 1999).

These concerns have led to numerous legislative proposals to limit the perceived excesses of managed care, such as early release of women from hospitals soon after giving birth, refusals of MCOs to allow persons with bone cancer to receive experimental treatment, and internal MCO regulations that forbid doctors from suggesting certain treatments. Various "Patients' Bills of Rights" have been proposed and in some cases adopted, and "any willing provider" laws, which require MCOs to contract with any doctors willing to meet their terms for participation, have been adopted or

---

Box 8.1   *Ethical Debate: Pharmacists and Conflicts of Interest*

In the same way that the interests of doctors and patients can clash when doctors have a vested economic interest in referring patients for particular tests at particular laboratories, increasing numbers of pharmacists now have a vested economic interest in selling certain drugs rather than others (Kolata, 1994).

In 1992, Merck Pharmaceuticals bought Medco, a nationwide drug supply company that buys drugs from manufacturers and sells them at discounts to its 38 million U.S. members through pharmacies. Since then, two other major pharmaceutical companies, SmithKline Beecham and Eli Lilly, have bought drug supply companies.

Since Merck bought Medco, Merck has offered cash commissions to pharmacists who can convince customers to buy Merck products rather than competing drugs. For example, if a customer who belongs to Medco brings in a prescription for an ulcer medication not produced by Merck, the pharmacist might tell the customer that, under their Medco coverage, they can purchase a similar and equally effective drug more cheaply. The pharmacist then offers to call the customer's doctor to request that the doctor approve switching drugs. What the pharmacist will not tell either the customer or the doctor is that Merck makes the recommended drug and the pharmacist will benefit financially from this switch.

Because in the past pharmacists had no financial links to pharmaceutical companies, doctors generally assume that pharmacists' suggestions are both educated and impartial. Doctors therefore agree to switch drugs in about 80 percent of cases (Kolata, 1994). After several such phone calls from pharmacists, doctors may begin routinely prescribing the

---

considered in the majority of states. These legal changes are also moving MCOs toward convergence with traditional insurers.

THE CRISIS IN HEALTH CARE

Whereas the rise of managed care has raised concerns about the quality of care now available in the United States, the increased cost of health care and the resulting decrease in access to it has challenged the very basis of the U.S. health care system.

Rising Health Care Costs

According to federal researchers, average costs per household in 1995 for all medical care, drugs, supplies, and insurance was $1,732 (U.S. Bureau of the Census, 1998:124). Expenditures have continued to rise since then, and are expected to consume 16.2 percent of the entire Gross Domestic Product—$2.2 trillion—by 2008 (www.hcfa.gov/stats/nhe%2Dproj/proj1998/hilites.htm). Although costs for health care rose more slowly in the second half of the

recommended drug instead of the drug that they used to prescribe.

Is it unethical for pharmaceutical companies to offer financial rewards to pharmacists who sell certain drugs or for pharmacists to accept those rewards? Those who participate in these arrangements, of course, consider them merely an extension of normal business practices. Because many of the most popular drugs on the market are virtually identical to competing drugs, they argue, customers lose nothing by switching drugs and gain if the new drugs are cheaper. Moreover, they claim, if drugs do differ significantly, it is the doctor's responsibility to know that and to protect his or her patients, not the pharmaceutical company's or pharmacist's. In essence, they argue, drugs are like any other consumer good and no ethical rules apply beyond the normal rules of the marketplace, such as not advertising a product's effects falsely.

Opponents of these arrangements, on the other hand, argue that such arrangements necessarily produce unethical conflicts of interest. A pharmacist who can earn extra money by recommending certain drugs over others is more likely to recommend that drug, whether or not it really is the best drug for the customer. Moreover, the entire transaction is grounded in dishonesty, for neither customer nor doctor knows that the pharmacist has a vested interest in selling certain products. Rather, both customer and doctor reasonably assume that pharmacists, as professionals, are bound by a code of ethics that restrains, even if it does not eliminate, any tendency to place their economic self-interest ahead of customer welfare.

1990s than in earlier years, those costs continue to be higher and to rise more quickly than in other industrialized nations, as Figures 8.1 and 8.2 show (Organization for Economic Cooperation and Development, 1999).

What accounts for the rising costs of health care? Sociologist Paul Starr has identified four "myths" popularly used to explain the rising costs of U.S. health care.

The first myth is that Americans expect more care than do citizens of other nations. Yet in some ways the reverse is true. According to Starr, "the annual rate of physician visits in the United States is below average for industrialized countries. In fact, the American rate of 5.5 visits per year is less than half that of Germany (11.5 visits) and Japan (12.9 visits), both of which have lower health care costs" (Starr, 1994:16).

The second myth attributes our high health care costs to our unique propensity for filing malpractice suits. Yet malpractice insurance accounts for less than 1 percent of total U.S. health care costs (De Lew et al., 1992). Even if we add the estimated costs of **defensive medicine**—tests and procedures doctors perform primarily to protect themselves against lawsuits—these expenses increase to only 4 percent of total health care costs.

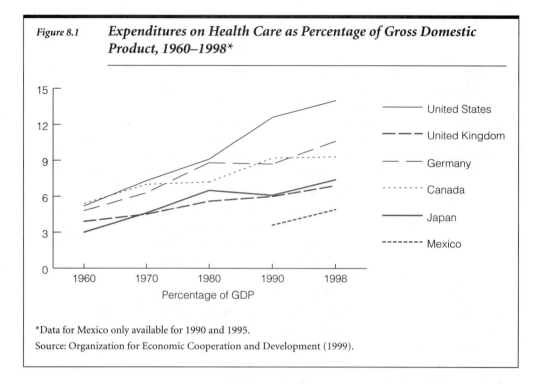

Figure 8.1   **Expenditures on Health Care as Percentage of Gross Domestic Product, 1960–1998***

*Data for Mexico only available for 1990 and 1995.

Source: Organization for Economic Cooperation and Development (1999).

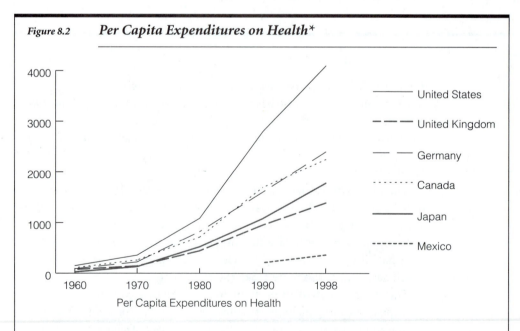

Figure 8.2   **Per Capita Expenditures on Health***

*Dollar amounts adjusted for purchasing power parity. This strategy controls for differences over time and across countries in the worth of a nation's currency by factoring in the number of units of a nation's currency required to buy the same amount of goods and services as $1 would buy in the United States. Data for Mexico only available for 1990 and 1997.

Source: Organization for Economic Cooperation and Development (1999).

Moreover, as Starr points out, those tests and procedures would offer doctors no legal protection if they were obviously unnecessary. Consequently doctors might do them even if not pressured by fear of lawsuits. Nor would health care costs necessarily decline if doctors stopped doing defensive medicine, because, to maintain their incomes, they might instead increase the number of other services they provide.

The third myth attributes our rising health care costs to our aging population. Yet the population of the United States is no older than that of any of the other top 27 industrialized nations (Population Research Bureau, 1998).

The fourth myth is that health care costs are so high in the United States because of our advanced technologies. Yet, as Starr points out, although advanced technologies do play a role in our health care costs, they are only a small fraction of all health care costs. Moreover, the same technologies exist in the other industrialized nations without producing equally high health care costs. Thus the mere existence of technology cannot explain these costs.

If patient demand, malpractice costs, the aging population, and advanced technology do not explain the rising costs of health care, what does? Canadian society is probably the society most similar to the United States, and comparing health care costs between these two countries can shed some light on this question. When the sources of health care costs in these two countries are compared, three factors emerge as significant: administrative costs, doctors' fees, and hospital costs (Starr, 1994). Administrative costs account for between one-third and one-half of the difference in costs, doctors' fees for another one-third, and hospital costs for the remainder.

The next chapter describes the Canadian health care system in detail. At this point, we need only note that Canadians receive their health insurance directly from the government. Hospitals receive an annual sum from their provincial government to cover all costs and therefore do not need an expensive administrative apparatus to track expenses and submit bills for each patient. Doctors, too, need submit their bills only to the national insurance system, rather than filing myriad different forms with different insurers. Meanwhile, no one need spend money on advertising or selling insurance, trying to collect unpaid bills, or covering the costs of any bills that remain unpaid.

The greater amount spent on doctors' fees in the United States results from the fact that profit-making remains at the center of our health care system. The rise of managed care has constrained doctors' incomes only slightly, as the primary goal of most MCOs is to increase their profits, not to restrict costs to consumers. The high costs of doctors' fees are also artificially maintained by the oversupply of specialists in the United States, which has a higher ratio of specialists to primary care providers than any other major industrialized nation (De Lew et al., 1992). Because health care consumers typically purchase whatever medical services their doctors recommend, when an oversupply of doctors increases competition for patients, doctors can protect their incomes by increasing the number of services they perform or increasing their charges for those services. As a result, areas with the greatest numbers of doctors per capita also have the highest rates of medical tests, surgeries, and other medical procedures and the highest prices for those services (Center for the Evaluative Clinical Sciences, 1996). In sum, whereas under the normal laws of the marketplace, greater supply leads to lower prices, in health care, greater supply leads to *higher* prices.

Similar processes have driven up the costs of hospital care. Hospitals, like doctors, can create demand for their services. Consumers have even less ability and opportunity to make decisions about hospital care than about medical care. This would not matter if the number of hospital beds matched the need. Yet, because historically the United States has placed few restraints on hospital construction, the nation has a surfeit of hospital beds, leaving approximately one-third of all beds empty at any given time. Similarly, no legal constraints keep hospitals from purchasing or developing expensive technologies such as heart transplant units or CT scan machines. Rather, hospitals are pressured to make such purchases as a way of attracting community doctors—and hence their patients—to use their facilities. For example, during the 1980s, the number of hospitals equipped to perform coronary artery bypass surgery rose sharply and the rate of such surgeries among men over age 65 quadrupled (*Economic Report of the President,* 1993:149–150). Yet a meta-analysis (combining results from multiple **controlled** studies) suggested that about 10 percent of these surgeries are clearly unnecessary and many more may be unnecessary (Leape, 1992).

Even when demands for their services expand, hospitals still typically own more high-technology equipment than they can use. Unfortunately, whereas in any other field, low demand would lead to lowered prices, the reverse is true in medical technology. As Starr explains (1994:25),

> Consider the case of early detection of breast cancer through the use of mammography. With fully utilized mammography machines, a screening mammography examination should cost no more than $55, according to studies by the GAO [U.S. General Accounting Office] and Physician Payment Review Commission. But because machines are typically used far beneath capacity, prices run double that amount [so that hospitals can recoup their investment]. With prices so high, many women cannot afford a mammogram. . . . In other words, *because* we have too many mammography machines, we have too little breast cancer screening. Only in America are poor women denied a mammogram because there is too much equipment. [Emphasis in original.]

Moreover, when equipment is underutilized, health care providers cannot maintain their skills, so rates of complications and death rise significantly.

In contrast, hospital costs are considerably lower in Canada even though admission rates are about equal and average stays are longer. As the next chapter will describe in more detail, Canada has succeeded far better than the United States at restraining costs by requiring hospitals to obtain approval from a central authority before they can add beds or purchase advanced technologies. Consequently, doctors in Canadian hospitals order fewer tests, use fewer resources, and provide care at a reduced cost.

In addition, Canada has succeeded at cost control better than the United States because attempts at cost control occur in a unified system where everyone shares the same goal. In contrast, those who have attempted in the past to control the costs of medical and hospital care in the United States have failed because they did not take into account the broader, and often hostile, profit-driven system in which those costs were generated. For example, faced with rising costs under the Medicaid and Medicare programs, the government since 1983 has used a system of **diagnosis related groups (DRGs),** which sets an average length of hospital stay and cost of inpatient treatment for each possible diagnosis. Under this prospective reimbursement system, the government determines in advance each year the amount it will pay hospitals per patient based on the average cost of treating someone with that patient's DRG. If the hospital spends less than this amount, it earns money; if it spends more, it loses money. Theoretically, then, the DRG system should have limited the costs of providing care under Medicaid and Medicare. Instead, and taking advantage of the fact that patients often have multiple illnesses and that the same symptoms often suggest more than one diagnosis, doctors and hospitals now sometimes use sophisticated computer software to help them determine the most remunerative, but still plausible, diagnosis for a given patient—a process known as "DRG creep." In addition, hospitals responded to the adoption of the DRG system by shifting services to outpatient units (where the DRG system does not apply) and by increasing the number of patients they admitted. As a result, the DRG system only marginally reduced government costs for hospital care. Similarly, when the government restricted the fees it would pay health care providers for treating Medicare and Medicaid patients, providers increased the fees they charged other patients.

Declining Coverage

Uninsured Americans

The rising costs of care have led directly to declining coverage. Approximately 44.3 million persons—16.3 percent of all Americans—lacked health insurance on any given day in 1998, more than at any time since the initiation of Medicare and Medicaid in 1965 (Campbell, 1999).

Because almost all Americans over age 65 are covered by Medicare, health care coverage is essentially a problem of the young and middle aged. As Figure 8.3 shows, lack of health insurance affects substantial portions of all age groups below age 65, but is especially acute among persons 18 to 34. During these years, most individuals are neither covered by their parents'

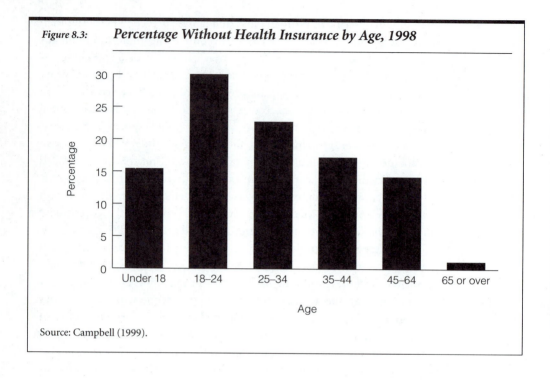

Figure 8.3:   *Percentage Without Health Insurance by Age, 1998*

Source: Campbell (1999).

insurance policies nor well-enough established financially to obtain insurance on their own. In addition, most are in good health and so do not feel it is worth paying for expensive health insurance.

As described earlier, insurance in the United States is typically linked to employment. This system is far from perfect, however. To keep profits high over the last decade or so, employers have reduced their expenses by cutting back on the benefits they offer to full-time employees and hiring more part-time workers who receive no benefits. Consequently, in 1998, 16.9 percent of all full-time, year-round workers lacked health insurance, as did an astounding 47.5 percent of poor full-time workers (Campbell, 1999). Similarly, almost one-quarter (23.2 percent) of all part-time or part-year workers lacked insurance, as did 44.7 percent of poor part-time or part-year workers. Not surprisingly, insurance coverage also varies by income level, with poorer persons substantially less likely than wealthier persons to have health insurance (Figure 8.4).

Size of employer also affects the likelihood of insurance coverage. The percentage who receive insurance from their employer rises from 29 percent among workers at firms with fewer than 25 employees to 66 percent among workers at firms with 500 or more employees (Campbell, 1999). Three factors explain why persons who work in small firms or are self-employed are most likely to be uninsured. First, whereas large firms can spread the administrative costs of insurance over many employees, small

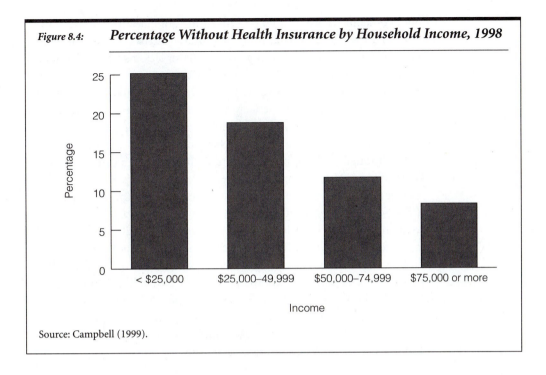

Figure 8.4:   *Percentage Without Health Insurance by Household Income, 1998*

Source: Campbell (1999).

firms and self-employed persons cannot. Consequently, although those costs pose a minor nuisance for large firms, they can make insurance prohibitive for small firms and self-employed persons. Second, large firms, unlike small firms, have enough ready capital to **self-insure**—putting aside a pool of money from which to pay all health care expenses for their workers rather than purchasing insurance from a commercial provider. By so doing, they can avoid paying the overhead costs of commercial insurers and often can avoid various state taxes and restrictions on insurance provision. As of 1999, 48 million Americans received their health coverage from employers who self-insured. Third, insurers are more willing to offer lower rates to large firms because they assume that any money they might lose paying for the health care of ill employees will be more than counterbalanced by the money they earn on the many healthy employees in the same firm.

Gender plays little role in insurance status: 17.3 percent of men were uninsured during 1998 compared with 15.3 percent of women (Campbell, 1999). Ethnicity plays a considerably more important role: more than one-third of Hispanics (35 percent) are uninsured, compared with 22.0 percent of African Americans, 21 percent of Asian Americans and Pacific Islanders, and 12.0 percent of non-Hispanic whites.

Insurance coverage also varies by state, with insurance less common in those states that provide less generous Medicaid coverage, have higher proportions of residents who work for small firms, or have higher proportions of poor residents. More than 20 percent of the population lack health insurance in Mississippi, Nevada, Texas, New Mexico, Arizona, and California (Campbell, 1999). Conversely, ten percent or fewer lack insurance in Rhode Island, Hawaii, Vermont, Iowa, Minnesota, or Nebraska.

Finally, insurance coverage varies by health status. Ironically, health insurance is hardest to get if one actually needs it. In most jurisdictions, insurers legally may reject any applicants for individual health insurance who do not pass a series of medical tests and have clean health records. Consequently, although most uninsured adults are healthy, a minority are much sicker than the rest of the population.

Paradoxically, not only have rising costs led to declining coverage, but declining coverage has led to rising costs. As the costs of coverage have increased, many relatively healthy people have concluded that they cannot afford insurance. Those who know they have health problems, however, more often decide that they must purchase insurance regardless of costs or financial hardship. Consequently, compared with the past, a higher proportion of insured Americans are ill. To maintain their financial stability, therefore, insurance companies must increase prices, driving away still more relatively healthy persons. Through this process, a **rate spiral** is created, in which increasing costs and declining coverage each foster the other.

Underinsured Americans

In addition to those who have no coverage, many more Americans have insurance that leaves them with more medical bills than they can afford to pay. These problems stem from required deductibles and copayments; long waiting periods before insurance covers preexisting conditions; caps on insurance reimbursement per treatment, per year, or per lifetime; and lack of insurance for certain costs, such as nursing home care and prescriptions. A report released in 1999 by the nonprofit Consumers Union concluded that 31 million Americans are underinsured, spending more than 10 percent of their income yearly on health care costs and risking bankruptcy should they have a major illness (Shearer, 1998). Because of these problems, many Americans who live near Mexico go there to purchase health care or prescription drugs more cheaply, and many who live near Canada fraudulently use the Canadian health care system (Rosenau, 1997; Vuckovic and Nichter 1997).

Other Americans face financial difficulties not because they lack sufficient insurance but because they cannot get their insurers to pay for their care (Light, 1992). In the past, once an individual had belonged to a plan for a given period of time (typically six months), their insurance generally would cover any medical bills for preexisting conditions. Now, however, insurers sometimes demand or employers seek new insurance contracts each year, with new lists of preexisting conditions.

In some cases, insurers also have avoided paying bills by adding near-impossible rules for receiving coverage (Light, 1992). When consumers fail to meet these rules, such as obtaining approval from the insurer within twenty-four hours after receiving emergency care or within twenty-four hours before an operation, the insurer legally can refuse to pay the claim, even if the insurer agrees that the procedure was medically justified. Similarly, insurers increasingly engage in **claims harassment:** establishing bureaucratic rules and systems that make it virtually impossible for consumers to file claims or to fight the insurer if a claim is denied (Light, 1992). For example, some insurers require that consumers file complex forms to receive reimbursement, provide insufficient telephone lines to their claims department, or deny claims on the slimmest grounds in the hopes that consumers will lack the time or energy to fight back. In these ways, insurers can delay payment (earning interest on the money for each day they keep the funds) or avoid paying claims altogether.

Precariously Insured Americans

Finally, in addition to the millions of Americans who are uninsured or underinsured, many more are precariously insured—liable to lose their insurance coverage at any time. This can happen in several ways. Those who receive Medicaid lose their coverage once their income rises above a specified ceiling. As a result, persons with chronic illnesses who become employed usually lose their coverage. If they become more seriously ill, they must first use up their savings paying their medical bills and then reapply for Medicaid.

Other problems can await those who hold private insurance. Because many adults receive their insurance as part of a family plan, they and their children can lose their insurance following divorce. Although courts can order a spouse or parent to continue paying insurance premiums, the courts do not have to do so and, at any rate, often cannot enforce such orders.

In the same fashion, because insurance is generally tied to employment, any change in work status can affect insurance coverage. Individuals who change their jobs sometimes find that their new jobs offer less coverage. In addition, even if their new jobs provide similar insurance, that insurance may have a long waiting period before it covers any health problems they developed while working at their previous jobs. The 1993 *Economic Report of the President* reported that 25 percent of American households include someone who cannot change jobs because of health insurance.

Individuals who lose their jobs face even more serious risks. Since passage in 1986 of the Combined Omnibus Budget Reconciliation Act (COBRA), such individuals have had the right to continue their insurance coverage for eighteen months, as long as they pay for that coverage themselves. (Their former employers may pay a portion but do not have to do so.) However, employers do not necessarily tell their former employees about COBRA and do not necessarily comply with COBRA requirements, assuming—often correctly—that their now-unemployed former workers cannot afford to take them to court. Similarly, many formerly employed workers cannot afford to pay the insurance premiums on their own, especially because after the first eighteen months they must purchase insurance at individual rates, which can be as much as 6 times higher than group rates (U.S. General Accounting Office, 1998).

Even those who remain in the same job for years can find themselves unexpectedly without insurance coverage. Firms that self-insure (and therefore are not bound by state insurance regulations) legally can drop coverage at any time for high-risk employees or employees who become ill. Individuals do not have much more guarantee of coverage if their employers purchase insurance. In the search for lower costs and greater profits, employers now often change their insurance provider yearly and insurers now sometimes require employers to renegotiate their contracts yearly. Insurance companies now sometimes agree to provide coverage to small firms only if certain individuals are excluded or fired (Light, 1992). (This issue does not arise with larger firms because any bills generated by a few high-risk individuals will be balanced by the low costs of providing care to their many low-risk co-workers.) In other cases, insurance companies agree to provide coverage to a firm, but charge exorbitant premiums because they anticipate that one or more employees will generate high bills. In all these circumstances, employers' only options are to fire their high-risk employees, retain them without coverage, or stop providing coverage to any employees.

To alleviate the problems of those who lose their group health insurance, Congress in 1996 passed the Health Insurance Portability and Accessibility Act. Unfortunately, the Act has had little impact, as it guarantees access to insurance coverage to only a small number of Americans and, more important, does not increase affordability for any (U.S. General Accounting Office, 1998).

The Consequences of Declining Coverage

The decline in health care coverage in the United States has directly affected the use of health care services and indirectly affected health outcomes among the uninsured and underinsured.

Individuals who do not have health insurance still sometimes can obtain health care. Federal, state, and some local governments provide clinics and public hospitals that offer low-cost or free care. In addition, governments sometimes provide low-cost or free vaccination, cancer screening, and well-child programs. These facilities and programs, however, are not always geographically accessible to those who need them. In addition, these facilities are continually underfunded, so individuals may have to wait hours for emergency care and weeks or months for nonemergency care. Uninsured persons are about twice as likely as insured persons to travel more than sixty minutes to receive care and to wait more than sixty minutes to receive care (Weissman and Epstein, 1994).

Uninsured persons also sometimes can obtain health care through the private sector. First, some individuals can find private doctors who will reduce or waive their fees and some live in communities where nonprofit hospitals offer inexpensive outpatient clinics. Second, uninsured persons can obtain care for both acute and chronic, emergency and nonemergency health problems from hospital emergency rooms; although emergency rooms legally can refuse care to anyone who is medically stable, many provide at least basic treatment to all who present themselves. As a result, emergency rooms around the country have become primary care providers for those who cannot afford care, even though the services they offer only poorly match the needs of these individuals and could be provided at far lower costs elsewhere. Finally, uninsured persons increasingly have volunteered for experimental trials of new drugs as a way of receiving sporadic treatment (Kolata and Eichenwald, 1999). Yet in such experiments some patients will receive **placebos,** some will receive drugs that prove ineffective, and some will receive drugs that prove harmful. Moreover, even if the drugs work well, patients receive only temporary benefit, as the drugs become unavailable once the experiments end.

Depending on where they live, therefore, uninsured persons may have some access to health care. However, this access is substantially less than that available to other Americans. According to a large national random survey conducted in 1995 (Donelan et al., 1996), 64 percent of the uninsured but only 26 percent of the insured had postponed seeking needed medical care, and 37 percent of the uninsured but only 9 percent of the insured had gone without needed care altogether. Similarly, uninsured persons are significantly less likely than others to receive basic preventive health care, such as physical examinations, blood pressure checks, pap smears, and mammograms. Because of these differences in access to care, the health problems of uninsured persons are usually worse and more difficult to treat when they do seek care. As a result, uninsured persons have higher hospitalization rates than others for health problems that generally do not require hospital care, including diabetes, hypertension, and conditions for which vaccinations exist, and uninsured persons are more likely to die once admitted to a hospital (Hadley et al., 1991).

When uninsured persons do seek health care, they typically receive less care than do insured persons (Hadley et al., 1991). For example, one study looked at a random sample of almost 30,000 sick newborns who needed at least six days of care in California hospitals during 1987 (Braverman et al., 1991). To isolate the effect of insurance status, the researchers controlled statistically for a broad range of demographic and medical variables. The researchers found that although on average the uninsured newborns were sicker at birth than the insured newborns, the former had an average hospital stay 16 percent shorter. In addition, the uninsured newborns received far fewer services than did the insured newborns, generating total charges averaging 28 percent less.

Finally, evidence suggests that when uninsured persons do receive care, the quality of that care is lower than that given to other Americans. In a major study in which a panel of doctors reviewed the records of more than 30,000 hospital patients (Burstin et al., 1992), researchers found that 20.3 percent of insured patients but 40.3 percent of uninsured patients received obviously negligent care and that uninsured patients were significantly more likely to die as a result of this negligence.

CONCLUSIONS

The United States stands virtually alone among the industrialized nations in lacking a rational and unified health care system and lacking any mechanism for guaranteeing health care to its citizens. Americans obtain their health care through a wide range of funding mechanisms, from publicly subsidized health care programs to private fee-for-service insurance to nonprofit HMOs. Although some Americans have nearly unlimited access to health care—including unneeded and potentially dangerous care—others lack access to even the most basic health care. As a result, the United States must cope simultaneously with economic and health problems caused by both overuse and underuse of health care services.

Whether we choose to tackle these dilemmas depends on how we—both individually and as a nation—define the situation. If we view obtaining health care as an individual responsibility, we are likely to oppose any attempts to extend government sponsorship of health care. However, if we view health care as a basic human right, we are likely to support extending health care to all. At the same time, regardless of whether we view health care as a right, we may support health care reform as a means of protecting the nation's economy; many corporations, for example, have begun lobbying for health care reform because they believe the money they spend on insuring their employees places them at a disadvantage compared with manufacturers in other nations that have national health care systems.

For those who believe reform is necessary, the question of *how* to reform the system becomes paramount. The next chapter grapples with this question.

SUGGESTED READINGS

Abraham, Laurie Kaye. 1993. *Mama Might Be Better Off Dead: The Failure of Health Care in Urban America.* Chicago: University of Chicago Press. A journalist's account of how our current health care system has abandoned poor, urban, minority communities.

Callahan, Daniel. 1998. *False Hopes: Why America's Quest for Perfect Health is a Recipe for Failure.* New York: Simon & Schuster. One of America's foremost ethicists argues that we need to emphasize providing primary care to all rather than high-technology care to the few.

Weissman, Joel J., and Arnold M. Epstein. 1994. *Falling Through the Safety Net.* Baltimore: Johns Hopkins University Press. Describes the health insurance crisis in the United States.

GETTING INVOLVED

People's Medical Society, 462 Walnut St., Allentown, PA 18102. (610) 770-1670. www.peoplesmed.org. A consumer organization that investigates the cost, quality, and management of health care; promotes self-care and alternative health care procedures; and represents consumer interests in health care.

REVIEW QUESTIONS

What is the nature of Blue Cross/Blue Shield insurance and how does it differ from commercial health insurance?

Why did the originators of health maintenance organizations believe these organizations would provide better health care at lower cost than their competitors?

What is managed care? How can it restrain health care costs and how can it harm individuals' health?

What are Medicaid and Medicare?

Why have health care costs in the United States risen?

Who are the uninsured?

Why do individuals who have health insurance still sometimes face financial difficulties in paying their health care bills?

How can individuals lose their health insurance?

How does lack of insurance affect health care and health status?

INTERNET EXERCISES

Find the Web site for the nonprofit Kaiser Family Foundation. Search the Web site for information on the characteristics of uninsured children.

Find the Web site for the nonprofit Consumers Union, then find its section on health care. What does Consumers Union believe are the most serious problems in the U.S. health care system? What sorts of strategies does Consumers Union propose for relieving those problems?

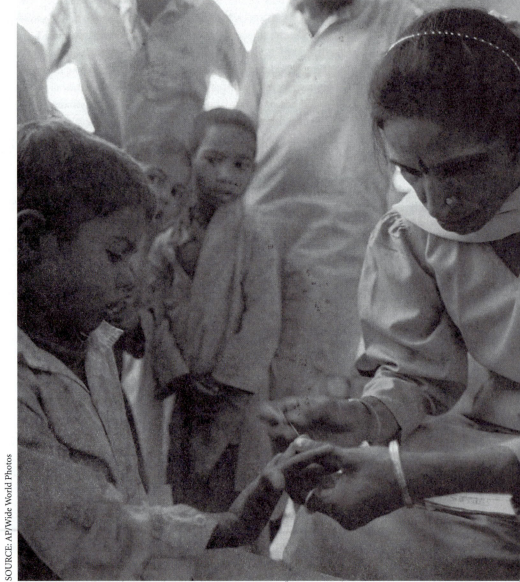

Health worker tests blood for malaria in Malab, India

CHAPTER 9

Alternative Health Care Systems

—

Joan Brooks, a 58-year-old grandmother, lives in Toronto. Her husband died a year and a half ago after suffering from cancer and kidney failure. He spent his last nine months in the hospital. The Ontario Health Insurance Plan covered all his medical expenses, leaving her no unpaid bills.

Brooks' only income is her husband's veterans pension—about $15,000 in U.S. dollars. But paying for medical care is not one of her worries. The Ontario health plan, to which every Ontarian belongs, covers those expenses.

She has severe arthritis and gout in both ankles and is unable to walk unless she takes prescription medicine. Not long ago, she was experiencing dizziness; her doctor suspected a drug toxicity affecting her liver and ordered a diagnostic ultrasound procedure. Brooks had the procedure within one week. She says it could have been done sooner but her schedule didn't permit an earlier appointment.

When the ultrasound revealed an enlarged liver, her doctor referred her to a specialist. Within days, the specialist admitted her to the hospital's outpatient unit and performed a needle biopsy. The Ontario plan paid the doctor about $54 for his work. Under the rules of the Canada Health Act, the doctor must accept the plan's payment, which is negotiated by the province and the provincial medical association. He cannot bill Brooks any additional amount.

Across Lake Ontario, in Buffalo, New York, insurance carriers pay doctors about $139 to perform the same procedure, and doctors can "balance bill"—that is, charge their patients more than they receive from insurers.

The biopsy shows that Brooks was suffering from excessive fat in her liver. . . . Her doctor has referred her to a nutritionist at the hospital who is helping her plan low-fat menus.

Right now she takes a drug for arthritis that costs $18 a month. The Ontario plan doesn't cover prescription drugs for Ontario residents *unless they are over 65 or on welfare. But many people who are employed have drug coverage through private insurance provided by their employers. . . .*

Because of the medication, Brooks' doctor checks her blood every three months to see if the dosage needs fine tuning. Ontario's health plan pays for the lab tests. When she needs to have her eyes examined, the plan pays for that checkup as well. The insurance plan covers the cost of one complete eye exam each year. Recently she had to return to the optometrist because the glasses he prescribed weren't adequate. The plan also covered the second visit, since it pays for subsequent visits if they are necessary.

Brooks . . . isn't interested in trading in Canadian health care for treatment in the U.S. "You can't buy the kindness and caring of this system," she says. "I have no dark tales to tell" (Consumer Reports, *1992:585*). *

On television, in newspapers, and in public discussions, we often hear that the United States offers the best health care in the world. Yet other countries— both Western and non-Western, rich and not so rich—provide far better access to care for their citizenry, at lower costs, with better health outcomes. In this chapter, I look at alternatives to the U.S. health care system. I begin by suggesting some basic measures for evaluating any health care system and then describe the systems in five other countries—Germany, Canada, Great Britain, China, and Mexico. The first three of these are often cited as possible alternative models for a restructured U.S. system, and the last two demonstrate how poorer countries have struggled to improve their nation's health despite limited resources. Finally, I look at the prospects for reforming the U.S. health care system.

EVALUATING HEALTH CARE SYSTEMS

Universal Coverage

The most basic measure of any health care system is whether it provides **universal coverage,** guaranteeing health care to all citizens and legal residents of a country. The United States is the only **industrialized nation** that does not do so. Instead, the U.S. government provides insurance to a small percentage of the population, and private insurers have nearly free rein to choose whom they will insure and at what prices. In contrast, any legal resident of Great Britain, Canada, or Germany, regardless of income, place of residence, employment status, age, or any other demographic characteristic, can obtain state-supported health care—although not necessarily everything they want when they want it.

In the absence of universal coverage, uninsured U.S. citizens must do without needed care, rely on charity, or hope that they can obtain government-funded health care. When individuals are not eligible for government-funded care, hospitals and doctors may provide care, but must make up the financial losses that they incur by raising the prices they charge others or "**cost shifting.**" Consequently, from the perspective of the system as a whole, it is more cost-effective to plan to provide care to everyone who needs it and budget accordingly than to have to find ways to pay for that care after the fact.

Portability

A second important measure of health care systems is whether they offer portable benefits. As described in the previous chapter, most U.S. citizens receive their health insurance through their jobs or their spouses' or parents' jobs. Consequently, individuals can lose their insurance if their family or work situation changes. Similarly, individuals who receive **Medicaid** can lose this coverage if they move to another state or if their income rises above the legal maximum. In contrast, in other developed nations individuals need not worry about losing their insurance no matter what changes occur in their personal lives.

Geographic Accessibility

Even those who have health insurance can face obstacles to receiving care depending on where they live. Both rural areas and poor inner-city neighborhoods in the United States typically have relatively few health care providers per capita. Meanwhile, other areas have an excess of doctors, which can pressure doctors to increase their prices or perform perhaps-unnecessary procedures to maintain their incomes. Consequently, for both economic and medical reasons, we should also evaluate health care systems according to whether they include mechanisms for encouraging an equitable distribution of doctors, such as providing low-cost loans to doctors who work in under-served areas or refusing permission for doctors to open practices in over-served areas.

Comprehensive Benefits

Another important measure of health care systems is whether they offer all the essential services individuals need. The difficulty, of course, lies in

Box 9.1 *Ethical Debate: Is there a right to health care?*

Every industrialized nation in the world other than the United States considers health care a basic right and provides all citizens with access to health care. In the United States, on the other hand, many question whether individuals have a right to health care, and no court has ever recognized such a right.

Those who argue against a right to health care draw on the language of autonomy and individualism, stressing the rights of individuals over any socially imposed rights accruing to all members of a society (Sade, 1971; Engelhardt, 1986). Those who take this position note that in asserting that individuals have a right to health care, we implicitly assert that health care workers have a duty to provide that care. In so doing, therefore, we restrict the rights of health care workers to control their time and resources. If we would not force a baker to give bread to the hungry, how can we force doctors to give their services away, or restrict what patients doctors see, what services they provide, and what charges they assess?

Similarly, in asserting a right to health care we implicitly assert that all members of a society have a duty to pay the costs of that care. When we subsequently use tax dollars to pay for health care, we restrict the rights of individuals to spend their money as they please. Some individuals, both rich and poor, might consider this a good investment, but many others would prefer to choose for themselves how to spend their money.

Moreover, according to those who take this position, asserting a right to health care fails to differentiate between *unfortunate* circumstances and *unfair* ones (Engelhardt, 1986). Although it is certainly unfortunate that some individuals suffer pain, illness, and disability, it is not necessarily unfair. Society might have an obligation to intervene when an individual unfairly suffers disability because another acted negligently, but society cannot be expected to take responsibility for correcting all inequities caused by biological or social differences in fortune.

Finally, if we assert that individuals have a right to demand certain social goods from a

defining what is essential. Although probably all informed observers would agree that comprehensive health care must include coverage for **primary care,** agreement breaks down quickly once we begin discussing specialty care. Some individuals, for example, consider coronary bypass surgery an essential service, whereas others consider it an overpriced and overhyped luxury. Similarly, some favor offering only procedures necessary to keep patients alive, whereas others support offering medical procedures like hip replacement surgery or dental care that improve patients' quality of life even though they don't help patients live any longer. By the same token, some support providing nonmedical services like home health care and hearing aids, whereas others consider purchasing such services an individual responsibility.

Any system that does not provide comprehensive benefits runs the risk of devolving into a two-class system, in which some individuals can buy

society, where do we draw the line? Do individuals have a right only to a minimum level of health care, or do they have a right to all forms of health care available in a given society? And, if we grant individuals a right to health care, how can we deny them a right to decent housing, education, transportation, and so on?

Those who argue in favor of a right to health care, on the other hand, draw on the language of social justice (Rawls, 1971). Believing each individual has inherent worth, they reject the distinction between unfortunate and unfair circumstances and the idea that health care is a privilege, dependent on charity or benevolence. Instead, they argue that each individual has a right to at least a minimum level of health care. Moreover, they argue, all members of a society are interdependent in ways that a rhetoric of individualism fails to recognize. For example, doctors who believe they should have full control over how and to whom they provide services fail to recognize the many ways they have benefitted from social generosity.

Medical training relies heavily on tax dollars, as do medical research projects, technological developments, hospitals, and other health care facilities. In accepting these benefits of tax support, therefore, doctors implicitly accept an obligation to repay society through the health care they provide.

Similarly, those who support a right to health care argue that to consider the decision to purchase health care as simply an individual choice misrepresents the nature of this decision, for it hardly makes sense to define something as a choice when the alternative is death or disability. Nor does it make sense to talk about the purchase of health care as a choice when individuals can do so only by giving up other essentials such as housing or food.

Finally, those who support a right to health care recognize that society could never afford to provide all available health services to everyone, but argue that this should not limit society's obligation to provide a decent minimum of care to all. Doing any less denies the basic worth of all humans.

more care than others. To those who believe health care is a human right, such a system seems ethically unjustifiable. Others object to such systems on more practical economic grounds, because, as described earlier, some who cannot afford care eventually seek care anyway, with the costs of that care shifted haphazardly to the rest of the population rather than budgeted in from the start.

Affordability

Guaranteeing *access* to health care does not help those who cannot afford to *purchase* it. Consequently, we also must evaluate health care systems according to whether they make health care coverage affordable, restraining the costs not only of insurance premiums but also of **copayments, deductibles,** and other health care services such as prescription drugs.

For health care to be affordable, individual costs must parallel individual incomes. As noted earlier, most insured Americans receive their insurance through employers. Employers typically pay a proportion of the costs for premiums and deduct the remainder from individuals' wages. To pay their share of the premiums, employers typically pass their costs on to their employees, dividing the costs equally among all employees and reducing salaries accordingly (Iglehart, 1999). As a result, low-wage and high-wage workers in essence pay (through reductions in salary) the same dollar amount for their health insurance, even though that dollar amount represents a much higher percentage of income for the low-wage worker. Consequently, paying for health insurance imposes a far heavier burden on poorer persons than on wealthier persons; having to pay $3,000 per year for health insurance, for example, might force wealthier persons to scale back their vacation plans but might force poorer persons to put off re-roofing their houses. In contrast, when, as in Great Britain and Canada, health coverage is paid for through graduated income taxes, poorer persons pay a lower percentage of their income for taxes and therefore for health care than wealthier persons. Either way—whether through taxes or lowered wages—citizens pay all costs of health care. The only difference is who pays how much.

Financial Efficiency

Another critical measure of a health care system is whether it operates in a financially efficient manner. Currently, the multitude of private and public insurers in the United States substantially drives up the administrative costs of the health care system. At the same time, the atomized and essentially entrepreneurial nature of our health care system makes it virtually impossible to impose effective cost controls. For example, when the federal government, as described in the last chapter, began paying hospitals prospectively for patient care based on **diagnostic-related groups** (**DRGs**) that established set fees for all patients with the same diagnosis, hospitals kept their profits high by shifting patient care from inpatient to outpatient settings where DRGs did not apply. Similarly, doctors have responded to financial limits on Medicare payments by raising the fees they charge to non-Medicare patients. For these reasons, true reform probably must include some mechanism for simplifying and centralizing control over the health care system and for restraining entrepreneurial elements.

Consumer Choice

We also need to evaluate health care systems according to whether they offer consumers a reasonable level of choice. Currently, wealthy Americans can purchase any care they want from any willing provider. In addition,

Americans who have **fee-for-service insurance** can seek care from any provider as long as they can afford the copayments and deductibles and, if their plan uses **managed care,** as long as their insurer approves the care. Those who belong to **HMOs,** meanwhile, can seek care only from providers affiliated with their plans, unless they have purchased additional coverage and can afford the extra charges. Finally, those who have Medicaid or Medicare coverage can obtain care only from providers willing to accept the relatively low rates of reimbursement offered by these programs, and those who have no health insurance can obtain care only from the few places willing to provide care on a charity basis.

No health care system can afford to grant all individuals full access to any willing provider. To be acceptable to Americans, however, an alternative health care system probably would need to provide at least the level of consumer choice that HMOs now offer and that many Americans have come to expect.

Provider Satisfaction

Finally, for a health care system to function smoothly, providers as well as consumers must feel satisfied with the system. Consequently, we must evaluate health care systems according to whether they offer health care providers an acceptable level of clinical autonomy, an income commensurate with providers' education and experience, and some control over the nature of their practices.

HEALTH CARE IN OTHER COUNTRIES

With these measures in mind, we can now look at the health care systems in Germany, Canada, Great Britain, China, and Mexico. The first four of these have guaranteed portable, affordable, and universal or near-universal health care coverage to their citizens, but the last, Mexico, continues to struggle to improve access to care.

Of these countries, Germany's system most closely resembles that found in the United States. Health care in the United States is primarily organized through an **entrepreneurial system,** based on the principles of private enterprise. Germany's system retains many entrepreneurial aspects, with health care (outside of hospitals) provided by private practitioners paid on a **fee-for-service** basis by a multitude of insurance companies. In contrast, Canadian primary care doctors, although also functioning as private practitioners paid on a fee-for-service basis, receive their payments through provincial government insurance programs. In Great Britain, meanwhile, most primary care doctors are paid through a mix of **capitation** and fee-for-service payments directly from the government, with insurance companies playing little role in health care. Finally, the Mexican and Chinese systems combine socialistic and entrepreneurial elements. (Table 9.1 summarizes the characteristics of these systems.)

Not surprisingly, each of these systems has changed over time. More interestingly, the changes seem to have moved these and other health care systems toward increasing convergence. This observation led David Mechanic and David Rochefort to propose a **convergence hypothesis,** which argues that health care systems become increasingly similar over time because of a combination of "scientific, technological, economic, and epidemiological imperatives" (1996:242).

First, Mechanic and Rochefort argue, doctors always seek the most current medical knowledge and technology, both to improve the services they offer and to increase their incomes and prestige. In recent decades, **globalization** has expanded access to such knowledge, as doctors increasingly use medical journals and Internet resources from around the world and as medical and pharmaceutical corporations market new technologies to doctors internationally. Thus, doctors in many different countries are adopting the same technologies and placing similar economic pressures on their health care systems. In turn, those systems have adopted similar strictures to limit both specialization and the use of technological interventions whose benefits do not justify their costs.

Broader economic shifts can also push health care systems inadvertently toward convergence. In countries with booming economies and largely capitalist health care systems, expenditures on health care typically rise steeply, eventually resulting in efforts to contain costs through restricting the role of the market in health care. Conversely, countries with weakening economies and largely socialistic health care systems find it increasingly difficult to support universal health care and typically respond by adopting measures designed to *increase* the role of the market, such as allowing wealthier persons to purchase health care outside of a national health care system. Thus, both sets of countries gradually move towards health care systems in which market forces play a restricted role.

Epidemiological changes can also promote convergence. As populations have aged around the world, health care systems have had to shift more toward treating chronic degenerative diseases rather than treating acute diseases. At the same time, the globalization of knowledge has increased people's expectations regarding health and health care because they now compare themselves not only to their neighbors but also to those they see in the mass media. This shift has forced health care systems to pay greater attention to patient satisfaction and choice, while providing further support for the parallel systems that allow the wealthy to buy care unavailable to others.

Table 9.1 **Characteristics of Health Care Systems in Various Countries**

CHARACTERISTICS	UNITED STATES	GERMANY	CANADA	GREAT BRITAIN	CHINA	MEXICO
Nature of system	Entrepreneurial	Social insurance	National Health Insurance	National Health System	In flux from national health system to entrepreneurial	Multiple options, providing unequal access
Role of private enterprise	Very high	High	Moderate	Low but rising	Moderate and rising	Moderate
Payment mechanism for primary care	Wide variety of mechanisms	Fee-for-service	Fee-for-service	Capitation	Primarily fee-for-service	Primarily salaried
Payment source for primary care	Mix of private, nonprofit, and government insurers	Primarily nonprofit insurers (sickness funds)	Government	Government	Primarily individuals	Primarily government
Universal coverage	No	Yes	Yes	Yes	No, but good access to primary care	No
Payment mechanism for hospital doctors	Salaried and fee-for-service	Salaried	Salaried	Salaried	Salaried	Salaried
Payment mechanism for hospital expenses	Varied	Fee-for-service and lump sum from government	Lump sum from government	Lump sum from government	Lump sum from government	Lump sum from government

Germany: Social Insurance for Health Care

Modern Germany is a product of a tumultuous twentieth-century history, including more than a decade of Nazism and the division of the country in two following its defeat in World War II. Yet despite the destruction wrought by two World Wars and the economic stresses that accompanied the reunification of East and West Germany in 1990, the nation is a stable constitutional democracy and now enjoys one of the strongest economies in Europe. Gross National Product (GNP) per capita is $28,870, just slightly more than the U.S. GNP per capita of $28,020 (Population Reference Bureau, 1998).

Structure of the Health Care System

Health care in Germany is based on a system of **social insurance.** Social insurance refers to the banding together of social groups, such as cities, occupations, or industries, to provide insurance for their residents or members. This system was adopted in 1883 by politicians who hoped that by offering workers accessible health care, as well as housing, unemployment, and retirement benefits, they could diffuse political forces that might otherwise have pressed for a more radical redistribution of power and wealth in German society (Lassey et al., 1997). Social insurance remains the center of the current German health care system.

Purchasing Care

As in the United States, nongovernmental insurance providers form the basis of the German health care system (Lassey et al., 1997). Whereas in this country, insurance providers (whether for-profit or not-for-profit) must compete to survive in a profit-driven market, in Germany, social insurance groups, motivated not by profit but by a commitment to providing care to members of their groups, provide about 90 percent of insurance. As of 1993, all Germans earning less than $40,000 must purchase insurance from one of the more than 1,000 social insurance groups known as "**sickness funds.**" Most blue-collar workers belong to funds organized along geographic lines, whereas most white-collar workers belong to funds organized along occupational lines. In addition, those who work for large corporations often belong to sickness funds attached to their corporations. Individuals can switch sickness funds at will, although most remain with the same employer and union and therefore the same sickness fund throughout their adult lives.

Members of a sickness fund pay a set percentage of their gross income toward the cost of their insurance premiums, which cover both workers and their dependents. Percentages are based on **community ratings** of risk rather than on **actuarial risk ratings,** so they vary from fund to fund but not by individuals' health statuses. The sickness funds set these percentages yearly based on estimates of how much income they will need to cover their anticipated expenses. By law, employers must match individuals' contributions; currently, the average total cost of premiums (including both employers' and employees' contributions) is 13 percent of individual income. Premiums cost anywhere from 8 to 17 percent of income, depending on the proportion of fund members who are elderly and the costs of care in that particular community.

When workers lose their jobs, the government pays their premiums, so individuals neither lose nor change insurance benefits. Similarly, to keep insurance premiums affordable for retired workers, pension funds pay about 60 percent of the cost of these premiums, with the remaining 40 percent taken from the premiums that younger workers pay. If a sickness fund has an unusually high proportion of retired workers, the national government partly subsidizes the fund to ensure that younger workers' premiums remain affordable.

Individuals earning more than $40,000 per year legally can purchase private insurance instead of joining a sickness fund. Like commercial insurers in the United States, private insurers base their premiums on individuals' actuarial risks at the time they purchase insurance. Unlike in the United States, however, German insurers can raise premiums only to counterbalance inflation and not because individuals' age or other health risks have increased.

Private insurance provides essentially the same coverage as the sickness funds plus various amenities such as private hospital rooms with televisions and telephones. In addition, because private insurers pay doctors about twice as much per procedure as the sickness funds, many Germans believe (although the data are inconclusive) that privately insured persons receive health care appointments sooner, receive more attentive care from doctors, and, if hospitalized, are more likely to receive care from experienced doctors rather than from medical **residents.** Despite these perceived benefits, however, only about 8 percent of Germans (as of 1999) choose to purchase private insurance. As a result, the sickness funds remain the core of the German health care system.

Paying Doctors

German doctors are sharply divided into hospital doctors and **ambulatory care doctors** (that is, those who work outside of hospitals). Hospital doctors are paid on salary, based on specialty and seniority, but can earn extra income by running outpatient clinics or performing outpatient surgery. In contrast, ambulatory care doctors, about 58 percent of whom are specialists, typically work in solo, fee-for-service practice and are paid indirectly by the sickness funds.

To pay doctors for their services, the sickness funds each year turn over most of the premiums they collect from individuals, employers, and the government to their regional medical associations. In turn, the associations each year set fee schedules and then reimburse the doctors in their region

Table 9.2	**Percentage of Consumers and Doctors Who Believe Their Health Care System Needs Fundamental Changes or Complete Rebuilding**	

COUNTRY	CONSUMERS*	DOCTORS**
Canada	79%	67%
West Germany	66	52
Britain	79	n.a.
United States	72	77

*German data from 1994, other data from 1998. Obtained from Organization for Economic Development and Co-operation (1999).
**1991 data, from Blendon et al. (1993).

according to those schedules. However, because the regional medical associations can reduce doctors' fees at the end of each quarter if they receive more bills from doctors than they expected, ambulatory care doctors cannot necessarily increase their incomes by increasing the services they perform. On the other hand, German doctors, unlike U.S. doctors, have retained almost complete autonomy in patient care. Whereas in the United States, as the last chapter described, the government and insurers increasingly have regulated medical practice through such policies as denying approval for certain treatments or denying reimbursement for patients hospitalized longer than their DRGs suggest is needed, the German system only recently and halfheartedly has instituted similar measures to limit clinical autonomy. In part as a result, and as Table 9.2 shows, German doctors—along with Canadian and British doctors—are significantly less likely than U.S. doctors to feel their health care system needs fundamental rebuilding (Blendon et al., 1993). Satisfaction with the system is also undoubtedly kept high by the fact that incomes among both ambulatory care and hospital doctors, although lower than a decade ago, remain higher than those of almost any other profession in Germany.

Paying Hospitals

To pay their operating expenses, hospitals receive reimbursements from the sickness funds for each procedure they perform. To pay for any capital expenses, such as purchasing expensive technologies or building a new hospital wing, hospitals receive a lump sum from the government yearly. Because the government controls the funds available for capital expenses, Germany has achieved greater success than the United States (although, as we will see, less than Canada) in concentrating medical technology in a limited number of hospitals. This has helped restrain costs by avoiding unnecessary duplication of technology, while ensuring that workers at hospitals with advanced technologies use them often enough to maintain their skills.

Access to Care

All Germans receive a comprehensive package of health care benefits. With the exception of minimal copayments, insurance pays all costs of dental care, maternity care, hospitalization, ambulatory care, prescription drugs, and preventive measures such as vaccinations. As a result, Germans have few incentives to put off obtaining needed care and see ambulatory care doctors on average more than twice as often as do U.S. citizens (Starr, 1994:16). Germans can see any ambulatory care doctors they like but must get referrals to see hospital-based specialists. In addition, health insurance replaces any income lost because of illness and provides funeral benefits to survivors. For these, among other reasons, Germans are less likely than Americans to believe their health care system needs fundamental changes (Table 9.2).

Controlling the Costs of Care

Concern about rising health care costs led Germany to adopt a series of changes, beginning in the late 1980s (Lassey et al., 1997). To stem the oversupply of doctors, German doctors no longer can receive reimbursement from the sickness funds after reaching the mandatory retirement age of 68 (although they still can practice medicine in other sectors). To control drug costs, the sickness funds educate both doctors and consumers about the relative costs and benefits of similar drugs and use various economic incentives to encourage doctors to prescribe cheaper drugs rather than equivalent, more expensive ones. To control the proliferation of expensive technology, neither clinics nor private doctors can buy any major capital equipment without prior approval from the government.

Several factors still hamper efforts to constrain costs. The sharp division between ambulatory care doctors and hospital doctors makes it difficult to shift care when medically warranted from hospitals to less expensive outpatient settings because of resistance from hospital doctors who fear their incomes will fall. At the same time, the oversupply of doctors in Germany has driven up the costs of care, as each doctor tries to perform enough medical procedures to maintain an income he or she considers adequate.

The nature of hospital financing also has kept costs high. Government control of budgets for capital improvements leaves those budgets particularly susceptible to political influence by legislators who seek special favors for local hospitals. Moreover, because those who establish capital budgets are not involved in establishing operating budgets, they need not remain informed about or concerned with overall hospital costs and so need not consider how increases in capital budgets might affect hospital operating costs.

Finally, the sheer complexity of the German system, with its more than 1,000 insurance providers, has kept administrative costs high. In addition, the existence of private insurance increases costs over time, as sickness funds must match at least some of the attractive and expensive benefits provided by private insurers if they are to retain their wealthier members (Wysong and Abel, 1991).

Health Outcomes

Whether because of its health care system or because of its higher standard of living and more general commitment to providing social services to its population, Germany enjoys a very high standard of health. Although conditions in the former East Germany remain poorer than in West Germany, those differences are rapidly disappearing. Life expectancy in Germany now averages 77, one year more than in the United States (and with far less variation among its citizenry) and infant mortality (at 5.6 per 1,000 live births, compared with 8.02 in the United States) is among the lowest in the world (Population Reference Bureau, 1998; U.S. Department of Health and Human Services, 1998).

Canada: National Health Insurance

Like the United States, Canada is an industrialized democracy made up of various provinces and territories more or less equivalent to U.S. states. Although its GNP per capita of $19,020 is almost one-third lower than in the United States, its economy is strong. In addition, because of steady immigration, Canada has a younger population than most industrialized nations have. These factors, combined with its ongoing commitment to social equity, including universal, comprehensive health care, have enabled Canada to achieve an overall health status rated first in the world by the United Nations in 1992 (Lassey et al., 1997).

Canada is also, however, the second largest country in the world, with vast social differences reflecting its vast geographic spaces. Its population is highly concentrated along its southern border, as are most health care personnel and facilities. Neither health status nor health access is as good in its northern regions, where many of the residents are poor Native Americans (known in Canada as "First Nations").

Structure of the Health Care System

Whereas in Germany individuals obtain their health insurance from any of a large number of sickness funds, in Canada all health insurance is obtained through one source: the government. For this reason, the Canadian system is referred to as **national health insurance,** or a **single-payer system.** In fact, however, the Canadian system is a decentralized one, with each province retaining some autonomy and offering a somewhat different health care system. Thus this brief discussion of the Canadian health care system necessarily obscures some of these differences.

The national health insurance system has evolved gradually since the late 1940s (Armstrong and Armstrong, 1998). Underpinning the system are payments that the federal government gives the provinces yearly to run their health care systems, as long as those systems meet four basic criteria:

> First, they must enroll virtually everyone in the province and eliminate essentially all out-of-pocket costs for covered services. Second, benefits must be portable from province to province—that is, if you are from Ontario and get sick in Quebec you must be covered. Third, the provincial program must cover all medically necessary services. . . . Fourth, the program must be administered through a public, non-profit agency (Himmelstein and Woolhandler, 1994:92).

Purchasing Care

Unlike patients in the United States, Canadian patients never see a medical bill, an insurance form, or any other paperwork related to their health care. Through a combination of federal and provincial taxes, the health insurance systems cover the costs of hospital care; medical (but not dental) care; long-term care; mental health services; and prescription drugs for the elderly. Some provinces provide additional benefits as well, such as prescription drugs for younger persons. Because the system is based primarily on graduated income taxes, it is **financially progressive,** placing the heaviest financial burdens on those who can best afford it: those who earn more money pay a higher proportion of their income in taxes and therefore pay more toward health care than do those who earn less money.

Paying Doctors

Most Canadian doctors are paid on a fee-for-service basis. The remainder—a small proportion in most provinces but a larger proportion in Quebec—work on salary for hospitals or clinics. Doctors submit their bills directly to the health insurance system using fee schedules negotiated annually between the provincial medical associations and provincial governments. Unlike in the United States, doctors who consider these fees too low cannot **balance bill** (that is, bill their patients for the difference between what the patients' insurance will pay and what the doctor wants to charge). In addition, some provinces control costs by setting annual caps on the amount they will reimburse each doctor. Others achieve the same goal by using diminishing reimbursement schedules under which doctors receive less for each service rendered as the number of services they bill for rises.

Data collected in the early 1990s suggests strong support among Canadian doctors for their country's health care system; unfortunately, more recent data are unavailable. Of a random sample of 3,387 Canadian doctors who participated in a nationwide survey in 1992, about 85 percent preferred the Canadian system to the U.S. system (Himmelstein and Woolhandler, 1994:265). Similarly, as Table 9.2 shows, Canadian doctors are less likely than U.S. doctors to believe their health care system needs significant changes.

Several factors account for Canadian doctors' support for their system. First, Canadian doctors have retained considerably more clinical autonomy than have U.S. doctors. In addition, most Canadian doctors work in solo private practice, free from the constraints of group settings or regulations. At the same time, doctors' workloads have remained essentially unchanged since the start of national health insurance and their incomes have remained high, rising at approximately the same rate as inflation (Marmor and Mashaw, 1994). Primary care doctors (about half of all doctors in Canada, compared with 13 percent in the United States) earn approximately the same net incomes in Canada and the United States, although specialists earn considerably more in the United States. Moreover, because medical education is highly subsidized by the Canadian government, Canadian doctors do not enter practice burdened by heavy debts, so their incomes go farther.

Paying Hospitals

To cover their operating costs, Canadian hospitals receive an annual operating budget from their provincial insurance system. Hospitals can spend their budgets as they like, with no controls imposed by the government, as long as they provide care to anyone in their region who needs services. In addition, and as in Germany, hospitals annually receive a capital expenditure budget. Because the government controls both operating and capital budgets, it can limit both unneeded hospital growth and the proliferation of high-cost technologies.

Access to Care

Despite having national health insurance, on average Canadians retain more control than most U.S. residents over their health care (Armstrong and Armstrong, 1998). Whereas most Americans can receive care only from the doctors affiliated with one particular health care plan, Canadians can choose any primary care doctor they want and theoretically can switch doctors at will (although often doctors will not accept new patients). However, as in HMOs in the United States, individuals typically must get a referral from their primary practitioner before seeing a specialist.

Nevertheless, access to care has decreased during the last few years, as budgetary pressure has led to reductions in federal subsidies for health care (Armstrong and Armstrong, 1996). As a result, some Canadian provinces now purchase certain high technology services from providers in the United States (for example, sending persons from Toronto to Buffalo for chemotherapy) and some now offer a more limited package of benefits than in the past. Consequently, some Canadians now purchase certain medical or surgical services out of pocket in Canada or in the United States. Meanwhile, provinces have closed some hospitals, sparking interest among U.S. investors in buying these hospitals and turning them into fee-for-service facilities.

These problems, though real, have been blown far out of proportion by the U.S. media, leading many U.S. citizens to conclude that Canadians have far less access to health care than do persons living in the United States (Brundin, 1993). This image is almost totally inaccurate. Although Canadians do sometimes seek care in the United States, the reverse is also true: U.S. citizens sometimes claim to be Canadian residents to obtain care that would otherwise be unaffordable, while U.S. HMO's, like Canadian health insurers, sometimes find it cheaper to send their patients across the border for services than to provide those services themselves (Lassey et al., 1997). Finally, Canada, like the United States, had in past years permitted the building of unneeded hospital beds, driving up the cost of health care. By closing some of these beds and centralizing services to a smaller number of locations where staff constantly practice their skills, Canada has both lowered costs and improved the quality of care. (However, decisions regarding which hospitals to close are partially shaped by political rather than by strictly health concerns.)

Finally, although it is true that U.S. citizens who have the most money and best insurance can obtain better (or at least more) health care than the average Canadian, when we look at the two populations overall, Canadians have the same or better access to care than do U.S. citizens on almost every measure, such as number of doctor visits per capita, number of hospital admissions per capita, and average length of hospital stay (Himmelstein and Woolhandler, 1994). Canadians do wait longer than Americans for high-technology care, but not in life-threatening situations (U.S. General Accounting Office, 1991). Canadians also are less likely to receive some (although not all) high-technology procedures, such as coronary artery bypass graft surgery. However, this more likely reflects *overuse* in the United States than *underuse* in Canada (Himmelstein and Woolhandler, 1994). Moreover, a 1994 national random survey (Donelan et al., 1996) found U.S. citizens slightly more likely than Canadians (or Germans) to report that they were unable to get needed medical care, had postponed getting needed medical care, or had serious problems paying their medical bills (although Canadians and Germans more often reported long waits to get appointments with specialists); Table 9.3 provides details. Finally, although in both Canada and the United States poorer residents have worse health and so

Table 9.3   ***Consumers' Self-reported Experiences with Health Care in the United States, Canada, and West Germany***

	UNITED STATES	CANADA	WEST GERMANY
Not able to get needed medical care	12%	8%*	6%*
Postponed needed medical care	30	16*	13*
Had serious problems paying medical bills	20	6*	3*
Had long waits to get appointments with specialists	20	34*	30*

*p<.05

Source: Donelan et al. (1996).

need more health care than do more affluent residents in the United States, affluent persons receive more surgical procedures than poorer residents, whereas the reverse is true in Canada (Armstrong and Armstrong, 1998:47).

Nevertheless, whereas less than a decade ago Canadians were more satisfied than U.S. residents with their health care system, they now are less satisfied (Table 9.2). The sharp drop in satisfaction among Canadians probably best measures how they feel about the changes in their health care system, rather than suggesting how they might feel about a U.S.-style system.

Costs of Care

In addition to improving access while maintaining quality of care, the Canadian health care system has at least partially restrained health care costs. As of 1995, the United States spent $3,701 per capita, or 14.2 percent of its gross national product, on health care, whereas Canada spent $2,049 per capita, or 9.6 percent of its national product, on health care (Armstrong and Armstrong, 1998:104).

How does the Canadian system restrain health care costs? Most important, a single-payer system dramatically reduces administrative overhead. In a single-payer system, no one need spend money on selling or advertising insurance. Nor is money spent on collecting funds to run the system, for those funds are collected from the public through already-existing taxation systems. Doctors, too, have fewer expenses because they only need to submit bills to one insurer using one standard form. Hospitals, meanwhile, need not spend money tracking or collecting bills for each patient because they receive a lump budget for the year regardless of how many patients they treat or what services those patients receive. As a result, in 1993, insurance overhead cost $212 per capita in the United States compared with $34 in Canada (U.S. General Accounting Office, 1991). According to the U.S. General Accounting Office (1991:3), "if the universal coverage and single-payer features of the Canadian system were applied in the United States, the savings in administrative costs alone would be more than enough to finance insurance coverage for the millions of Americans who are currently uninsured. [In addition,] there would be enough left over to permit a reduction, or possibly even the elimination, of co-payments and deductibles."

The single-payer system also saves money by centralizing purchasing power. As the sole purchaser of drugs in Canada, the Canadian government has substantial leverage to negotiate with pharmaceutical companies regarding drug prices. As a result, Canadians pay an average of 38 percent less than Americans for identical drugs (Himmelstein and Woolhandler, 1994:138). Similarly, Canadian doctors, like fee-for-service doctors in the United States, can increase their incomes by increasing the number of services they perform, but, as the sole payer, the Canadian government can control how much it will reimburse them per service. For example, Canadian doctors receive about $349 to remove a gallbladder, compared with $945 in Buffalo and $2,700 in New York City (Marmor and Mashaw, 1994). Finally, the single-payer system restrains costs by enabling Canada to implement efficient regional planning and to avoid unnecessary duplication of expensive facilities and technologies.

The Canadian system is not, however, free of problems (Armstrong and Armstrong, 1998:124–138). Payment of doctors on a fee-for-service basis makes it difficult for Canada to control medical costs. When, for example, the provinces banned balance billing, doctors responded by increasing the number of services they performed. Similarly, the provinces have instituted **utilization review** boards to identify any doctors who perform medically unjustifiable tests and procedures, but have given these boards little authority to sanction doctors. Finally, as noted earlier, declining budgets have led to declining benefits across the country.

Concern about costs, coupled with political pressure from corporations interested in profiting from health care provision, has resulted in some movement toward increasing the role of market forces in the Canadian health care system (Armstrong and Armstrong, 1998:138–142). Some hospitals have begun contracting with private corporations to provide certain services, usually those related to food and cleaning. Perhaps more important, Canadian hospitals, like U.S. hospitals, have begun shifting toward outpatient services and clinics and shorter patient stays, moving some of the burden of care from the health care system onto family caregivers. These changes have resulted in considerable public concern. Although support for the national health insurance system remains strong, that system nonetheless seems increasingly fragile and under attack.

Health Outcomes

Perhaps the most important question to ask about the Canadian health care system is how health outcomes compare with those found in the United States. The data suggest that outcomes are at least as good if not better in Canada. Canadians have lower infant and maternal mortality rates, live two years longer on average, and enjoy more years free of disability, even when ethnic differences are controlled (Armstrong and Armstrong, 1998:79–80). Of course, these health outcomes tell us more about social conditions than about the quality of health care. Nevertheless, studies that have looked more directly at health care have reached similar conclusions. For example, a study that compared matched populations of elderly persons who received surgery in Manitoba and New England found that long-term survival rates were higher in Manitoba for nine of the ten studied surgical procedures (Roos et al., 1992).

Great Britain: National Health Service

As the home of the Industrial Revolution, Britain for many decades was a leading industrial power. Along with its industrial strength came a strong labor movement, as workers united to gain political power within Britain's parliamentary government. As a result, a commitment to protecting its citizens, including a commitment to universal health care coverage, has long been a central part of Britain's identity. Recent years, however, have seen a gradual decline in its economy—GNP per capita is now only $19,600, just above Canada's—which has put Great Britain's national health insurance system in increasing jeopardy.

Structure of the Health Care System

Since 1911, Great Britain has provided low-income workers with subsidized care from general practitioners. As a result of the sacrifices made by the British people during World War II, however, popular sentiment increasingly held that all Britons had earned the right to a decent quality of living, including access to health care. This sentiment, coupled with other social forces, resulted in the creation of the **National Health Service (NHS)** in 1946. Two years later, the NHS came into existence, offering comprehensive health care to the entire British population.

Whereas Canada provides its citizens with national health *insurance*, Great Britain provides a national health *service*. In Canada, the government offers individuals insurance that allows them to purchase health care from private practitioners. In Great Britain, on the other hand, no individual need purchase health care or health insurance because the government directly pays virtually all health care costs. As a result, the two systems look quite similar to health care consumers, but differ substantially from the perspective of hospitals, health care workers, and the government.

Purchasing Care

As in Canada, British citizens can obtain comprehensive health care unburdened by bills or bureaucratic forms. The NHS uses tax revenues to pay virtually all costs for a wide range of health care services, including medical, hospital, dental, and optical care. In turn, the NHS receives its funds almost solely through general taxation, with small supplements from employers and employees. As in Canada, the health care system is paid for through taxes and is financially progressive.

To obtain care through the NHS, individuals first must choose a general practitioner. As in Germany and Canada, individuals can choose any general practitioner, but, once they have registered with one, can see only that general practitioner. Individuals can, however, change their general practitioner at any time, although few choose to do so. Individuals can see specialists only if referred by their general practitioner. In addition, individuals can obtain care through community service units, which provide a wide range of public health services, including visiting nurses for the homebound, homemakers for chronically ill persons, and long-term care programs for those who need more comprehensive care.

Paying Doctors

As in Germany, British doctors divide sharply into ambulatory care doctors (almost all of whom work in primary care) and hospital-based doctors (all of whom are specialists).

Britain's general practitioners work as private contractors, not as state employees. They are paid via **capitation,** receiving a set annual fee from the government for each patient in their practices regardless of how many times they see the patient or how many procedures they perform. In addition, general practitioners receive additional payments for each low-income or elderly patient in their practices to compensate them for the extra time and money required to care for such patients. Doctors also receive allowances to pay for office expenses and bonus payments if they meet targets set by the government for preventive services, such as immunizing more than a certain percentage of children in their practices. In addition, to encourage access to health care, the NHS offers financial supplements to doctors who practice in medically underserved areas. Finally, to restrain costs by increasing the role of market forces, general practitioners in group practices with more than

10,000 patients receive an annual lump payment from the NHS to pay for elective surgery for their patients at either NHS hospitals or private hospitals. If at the end of the year the doctors have money left, they can use that money to improve their facilities or to add health prevention services to their practices.

When the NHS began, most British general practitioners worked in solo practice. Over time, however, NHS administrators became convinced that working in group practices improved quality of care by enabling doctors to learn from each other and reduced costs by enabling doctors to share clerical and nursing staff. As a result, the NHS began offering financial incentives to those who practice in groups, and most general practitioners now work in groups of three or more.

To encourage doctors to enter primary care, the NHS increased capitation payments to general practitioners and added supplemental payments for house calls and other services offered by general practitioners. As a result, general practitioners now earn approximately the same incomes as specialists. In addition, senior general practitioners can have private patients, which can increase general practitioners' income by 10 percent or more (although few take advantage of this option).

Unlike general practitioners, specialists almost always work as salaried employees of the NHS at hospitals or other health care facilities. All specialists, regardless of field, receive the same annually negotiated salary from the NHS, adjusted for the local cost of living. Specialists can, however, earn merit awards (usually given toward the end of a person's career) which can increase their salaries by as much as 50 percent. In addition, the majority of senior specialists earn substantial extra income by reserving about 10 to 20 percent of their time for private patients.

Income for all doctors remains far higher than for other occupations in Great Britain. Moreover, those incomes go farther than they would in the United States because municipal governments pay most costs of medical training, so British doctors enter practice virtually debt free.

Paying Hospitals

Beginning with the Great Depression of the 1930s, both public and private hospitals in Britain experienced severe economic difficulties. As a result, when those who designed the NHS proposed nationalizing almost all British hospitals, they met little opposition. Since the start of the NHS, therefore, the vast majority of hospitals in Britain, including mental hospitals, chronic disease facilities, tuberculosis hospitals, and so on, have belonged to the government (although some NHS hospitals have retained beds for private patients). Until recently, hospitals received their funds in an annual budget allocated by their regional health authority. To control costs, the government increased the role of market forces by shifting to a system under which hospitals obtain their revenue by competing with each other for contracts with local Health Authorities and general practitioners to provide services to their patients. To fund new hospital construction, hospitals now have the authority to consolidate, sell land, or relocate to cheaper sites.

Access to Care

Under the NHS, individual financial difficulties no longer keep Britons from receiving necessary care. In addition, the NHS has reduced substantially the geographic inequities that for generations made medical care inaccessible to many rural dwellers, although serious deficits in access to care in inner-city areas remain.

Although Britons' access to primary care is excellent, their access to high-technology care is somewhat limited. Britain's economic decline during the last few decades has left few funds available for new hospital construction. Consequently, although the quality of health care offered in Britain remains high, both the quality of hospital facilities and the number of hospital beds have fallen. As a result, British citizens receive fewer days of hospital care per capita than do citizens of almost any other country in Europe. At the same time, the government has restricted the purchase of advanced technologies, leaving such technologies considerably more accessible in some areas than in others. Nevertheless, although individuals sometimes experience long waits before receiving elective surgery, no one must wait for emergency care. Still, Britons are more likely than U.S. residents to think that their health care system needs fundamental change (Table 9.2)

Controlling the Costs of Care

Great Britain spends less per capita on health care than does any other nation in Europe, and little more than a third of what the U.S. government spends per capita (Lassey et al., 1997). Like Canada and, to a lesser extent, Germany, Britain has made its health funds go farther than they otherwise would through national and regional planning. Because the government owns a large proportion of health care facilities and employs a large proportion of health care personnel, it can base decisions about developing, expanding, and locating high-technology facilities on a rational assessment of how best to use available resources. Thus, Britain can avoid unnecessary proliferation of expensive facilities that would otherwise increase costs.

Great Britain also has restrained government health care expenditures through a series of actions designed to increase the role of market forces and shift costs and services from the NHS to the private sector (Lassey et al., 1997). During the 1980s, the then-ruling Conservative Party refused to grant salary increases to specialists to encourage them to develop private practices. As specialists increased their private practices, they had less time for NHS patients, who soon complained of having to wait longer for specialized care. (However, it remains unclear whether patients had to wait longer or whether general practitioners had begun putting patients on specialists' waiting lists sooner.) As a result of these problems (whether perceived or real), a small number of Britons began buying private health insurance so they could buy their way more quickly into a specialist's office. The government further encouraged this shift by granting tax deductions to employers and employees who purchased private health insurance. Private insurance, however, never became popular, and less than 10 percent now have such coverage.

Similarly, during the 1980s the Conservative Party increased the number of beds set aside for private patients in NHS-owned hospitals. During those same years, private corporations, many owned by U.S. hospitals, began building private hospitals in Britain. Although these hospitals contain only a small fraction of all beds in the country, they have drained personnel from the NHS by offering higher salaries and better working conditions. As a result, staffing pressures and waiting times at NHS hospitals increased, contributing to dissatisfaction with the NHS. Nevertheless, popular commitment to the NHS remained strong, causing the Conservative Party to moderate its position on the NHS during the early 1990s—deciding, for example, to drop a proposal to institute tax incentives to encourage elderly persons to purchase private health insurance. Since the Labor Party returned to power in 1997, no major changes have occurred in the structure of the NHS.

Health Outcomes

Despite the problems in the NHS, health outcomes have remained good. Infant mortality (6.2 per 1,000 live births) is only slightly higher than in Germany, and average life expectancy, at 77 years, is identical.

China: Good Health at Low Cost

Although many observers have proposed using the health care systems of Germany, Canada, and Great Britain as models for a restructured U.S. health care system, few would seriously propose China as a viable model. China's culture differs greatly from that of the United States, and so its citizenry have very different values regarding what constitutes an acceptable health care system. In addition, China's GNP per capita of only $750 (Population Reference Bureau, 1998) severely limits its options, and the communistic underpinnings of its economy make some health care options more feasible and some less feasible than in the United States. Nevertheless, China's story provides useful clues regarding how to provide good health to the citizenry of a poor country.

China's health care system reflects its unique history and situation (Lassey et al., 1997; Liu and Wang, 1991). When, after many years of civil war, the Communist Party in 1949 won control of mainland China, it found itself in charge of a vast, poverty-stricken, largely agricultural, and densely populated nation of about one billion persons. Most people lived in abject misery while a small few enjoyed great wealth. Malnutrition and famines occurred periodically, life expectancies for both men and women were low, and infant and maternal mortality were shockingly high. In urban areas, only the elite typically could afford medical care, while in rural areas, where most of the population lived, Western medical care barely existed.

Structure of the Health Care System

In 1950, one year after winning control of mainland China, the Communist government announced four basic principles for the new nation's health care system. First, the primary goal of the health care system would be to improve the health of the masses rather than of the elite. Second, the health care system would emphasize prevention rather than cure. Third, to attain health for all, the country would rely heavily on mass campaigns. Fourth, the health care system would integrate Western medicine with traditional Chinese medicine.

These principles reflected both the political climate and the practical realities of the new People's Republic of China. The first goal—improving the health of the masses—stemmed directly from the communist political philosophy underpinning the revolution. The years of bloodshed were to be justified by a new system that would more equitably redistribute the nation's wealth and raise the living standards and health status of China's people. The second and third goals reflected unignorable facts about China's situation. Lacking either a developed technological base or an educated citizenry, China's greatest resource was the sheer labor power of its enormous population, which could be efficiently mobilized because of its now-centralized economy. Focusing on prevention through mass campaigns promised to deliver the quickest and biggest improvements in the nation's health. Finally, the decision to encourage both Western and traditional medicine similarly recognized the difficulties China would face in developing a fully modern and Westernized health care system, as well as the benefits of including traditional medicine in the new system. By encouraging traditional as well as Western medicine, China could take advantage of all its existing health care resources and gain the support of the peasantry, who remained skeptical of Western medicine. At the same time, incorporating traditional medicine into the new, modernized Chinese health care system offered a powerful statement to the world regarding the new nation's pride in its traditional culture. Simultaneously encouraging the growth of Western medicine, meanwhile, would help bring China into the scientific mainstream.

Given its large and poverty-stricken population and its lack of financial resources and medically trained personnel, China needed to adopt innovative strategies if it were to meet its goal of improving the health of the common people. Two of these strategies are the use of mass campaigns and the development of **physician extenders**—individuals (such as nurse practitioners and physician assistants in the United States) who can substitute for doctors in certain circumstances.

One of the more unusual aspects of China's health care policy has been its emphasis, especially in the early years of the People's Republic, on mass campaigns (Horn, 1969). For example, to combat syphilis, which was **endemic** in much of China when the Communists came to power, the government first closed all brothels, outlawed prostitution, and retrained former prostitutes for other work. Second, the government began the process of redistributing income and shifting to a socialist economy so that no young women would need to enter prostitution to survive. During the next decade, the government trained thousands of physician extenders to identify persons likely to have syphilis based on ten simple questions, such as whether one had ever had a genital sore. By so doing, the government made manageable the task of finding, in a population of one billion, the small percentage that needed to be tested and treated for syphilis.

To convince people to come to health centers for testing, these physician extenders posted notices in villages, performed educational plays in marketplaces, and gave talks around the country, explaining the importance of eradicating syphilis and attempting to reduce the stigma of seeking treatment for syphilis by defining syphilis as a product of the corrupt former regime rather than a matter of individual guilt. Those identified as likely to have syphilis were tested and treated if needed. These methods, coupled with testing, among others, persons applying for marriage licenses, newly drafted soldiers, and entire populations in areas where syphilis was especially common dramatically reduced the **prevalence** of syphilis in China.

Health Care Providers

The second innovative strategy for which China has won acclaim is its use of physician extenders. In urban areas, "**street doctors**" (sometimes known as "**Red Cross health workers**") offer both primary care and basic emergency care, as well as health education, immunization, and assistance with birth control. Street doctors have little formal training and work in outpatient clinics under doctors' supervision.

In rural areas, "**village doctors**" (formerly known as "barefoot doctors") play a similar role. Village doctors were first used in 1965 during China's Cultural Revolution, a political movement started by students and fostered by some members of the national government to uproot the last vestiges of the old class structure (as well as to eliminate political dissidents). Those who initiated the use of village doctors hoped these doctors would both alleviate the continued lack of health care providers in rural areas and reduce the political power of urban medical doctors, who remained a reminder of the precommunist elites. Novice village doctors were selected for health care training by their fellow workers based on their aptitude for health work, personal qualities, and political "purity." Following about three months of training (supplemented yearly by continuing education), village doctors returned to their rural communes, where they divided their time between agricultural labor and health care.

Since the end of the Cultural Revolution in 1976, the number of village doctors has declined substantially, although their quality has increased. Training is now more rigorous and individuals must pass two levels of certification before becoming qualified to practice. In addition, village doctors are now more closely supervised by more highly trained health care workers.

Above village doctors in the Chinese health care hierarchy are **assistant doctors.** These individuals receive three years of postsecondary training similar to that received by medical doctors, during which they learn both Western and traditional Chinese medicine. Finally, at the top of the hierarchy are medical doctors. Individuals must complete a minimum of five to eight years of postsecondary training to become doctors, plus a supervised residency program to become specialists. All doctors receive training in both Western and traditional Chinese medicine and may focus on either field, although relatively few choose traditional medicine.

Purchasing Care

As China's economy has changed from a largely socialized and centrally controlled system toward a more decentralized, economically heterogeneous model, so has its health care system (Lassey et al., 1997; Liu and Wang, 1991).

For urban residents, these shifts have brought few changes. As in the past, the government pays all costs of health insurance and health care for government employees, military personnel, and students (minus, in some circumstances, a small copayment). Other public industries and urban industrial collectives also pay for care for their workers.

About 78 percent of China's population, however, live in rural areas. For these individuals, recent years have dramatically changed the nature of health care. Before the 1980s, rural residents received their care at little or no cost through the agricultural communes where they lived and worked. Within these communes, members shared all profits and costs, including those for health care. Each commune had between 15,000 and 50,000 members and offered its own clinic staffed by assistant doctors (also commune members) who provided both primary care and minor surgery. In addition, communes were divided into production teams of 250 to 800 people, each including a village doctor.

Beginning in the early 1980s, most agricultural communes ceased functioning as such. Instead, these communal enterprises reverted to their original village structures, with each family receiving land to farm from the village. Families now keep for themselves any profits remaining after they pay taxes and other bills. By the same token, families now are responsible for their own welfare should their costs exceed their profits. As a result of this shift in financing, the former communes no longer earn sufficient revenues to continue providing health care. Many village doctors returned to full-time agricultural work, while, with government approval, rural assistant doctors gradually took over commune clinics and turned them into private fee-for-service practices. Almost all rural residents now receive their primary health care on a fee-for-service basis.

Paying Doctors

Currently, ambulatory care doctors in China work primarily on a fee-for-service basis, and hospital doctors work on salary. In addition, many townships (made up of six or more rural villages) have a clinic where doctors work on salary. As in many HMOs in the United States, however, these doctors can divide among themselves any profits generated by the clinic and not needed for new equipment or facilities, thus encouraging market forces to play a role in controlling costs.

Paying Hospitals

Unlike primary care, hospital care has remained largely a public enterprise. The federal government runs hospitals for military members and their families in all large cities, and local governments run virtually all other hospitals. All hospitals receive their operating budgets from the relevant government unit and receive capital budgets separately in the rare circumstances when such funds become available. In recent years, however, budgets have been cut and great pressure has been placed on hospitals to generate income through selling services.

Access to Care

As a result of the changes in China's health care system, prices for health care have risen and access has diminished, especially in rural areas, where fewer hospital beds and doctors are available per capita. Although primary care remains affordable, even for those who lack health insurance, hospital care is not. To equalize access to care, the government has established a national fund to supplement the health care budgets of poorer regions and an insurance program for childhood immunizations. Those who, for a small premium, purchase this insurance receive free immunization for children to age 7 and free treatment if a child develops one of the infectious diseases the immunization program is supposed to prevent. More than half of all children in the country belong to this program. Finally, a similar insurance program offers prenatal and postnatal care to women and infants; it is not known how many are covered by this program.

Health Outcomes

As a poor country, China spends only about 2 to 3 percent of its GNP on health care. Nevertheless, China's commitment to equalizing both income and health care has allowed it to attain health outcomes far greater than its economic status or investment in health care might predict. Although median income in China remains similar to those in many other **developing nations,** China boasts health outcomes only slightly below those of the industrialized nations. Between 1960 and 1998, infant mortality fell from 150 deaths per 1,000 live births to 31 per 1,000, and life expectancy increased from 47 years to 71 (Population Reference Bureau, 1998). Although large differences in health status remain between rural and urban dwellers, China now stands on the cusp of the **epidemiological transition.** In urban areas, all three leading causes of death are degenerative diseases. (Lung cancer, especially, is a growing problem, given tobacco's role as a major source of tax revenue and export dollars and the near-total lack of smoking prevention efforts within China.) However, infectious diseases remain the leading cause of death in rural China, where they are responsible for 23 percent of all deaths, and remain the fourth most common cause of death in urban China, responsible for 14 percent of all deaths (Republic of China, 1998). In addition, China continues to share with other developing nations such health problems as insufficient access to clean drinking water in some regions. Nevertheless, China's great accomplishments in improving the health of its people deserve recognition.

To find the key to China's successes, we need to look beyond the nature of its health care system. This topic has been investigated through a series of studies begun by the Rockefeller Foundation (Caldwell, 1993). These studies explored how China, along with Sri Lanka, Costa Rica, Vietnam, Cuba, and several other countries, has achieved substantially better health outcomes at lower cost than have countries that spend more and have higher per capita incomes. Three factors seemed to account for these outcomes. First, health outcomes in these countries improved somewhat when access to medical care improved. Second, and more important, health outcomes improved when nations encouraged education for men and emphasized family planning for both men and women. Finally, and as Chapter 4 explained, health outcomes improved most dramatically when nations made a commitment to educating women. Once this happened, women's status rose and women gained greater ability to control or delay reproduction. As a result, women's lives were less often cut short by childbirth and their babies were born healthier. A rise in women's status also brought a more equitable distribution of food between women and men, so that both women and the children who relied on them were less likely to suffer malnourishment and more likely to survive.

Mexico: Struggling to Provide Health Care Equitably

Understanding Mexico's health care system is particularly important because Mexico shares a long and permeable border with the United States. As a result, health issues in Mexico directly affect the United States, as people (and often diseases) travel across the border in both directions to seek work or pleasure (Skolnick, 1995). In addition, both Mexicans *and* U.S. citizens sometimes cross the border to the other country to seek health care, although Mexicans more often seek basic medical care for life-threatening health conditions, whereas U.S. citizens more often seek inexpensive cosmetic surgery, dental work, or pharmaceutical products.

Mexico stands on the cusp between being an industrialized and a developing nation. As industry has developed, more and more people have moved off the land, and now more than three-quarters of Mexico's population live in cities. Those cities contain both middle-class neighborhoods, which enjoy health and living conditions similar to those found in the industrialized nations, and impoverished slums that lack such basic facilities as running water and sewer systems. These slums are inhabited primarily by migrants from rural areas. Rural areas, especially those inhabited primarily by Indians, generally are poor, with only 53 percent of rural residents having access to safe drinking water and only 21 percent having sewer systems (Pan American Health Organization, 1998). Mean GNP per capita remains only $3,670—far higher than in China, but far lower than in any of the other nations discussed in this chapter (Population Reference Bureau, 1998).

Structure of the Health Care System

Unlike any of the other countries described in this chapter, Mexico has a three-tiered system for health care: private health care for the wealthy (used by as much as 10 percent of the population), high-quality government-provided insurance for the middle class (available to 55 percent of the population, but used by about 45 percent), and lower-quality government-provided services for most of the rest (Lassey et al., 1997; Pan American Health Organization, 1998). This three-tiered system is a product of Mexico's unique history, in which revolutionary fervor and conservative sentiments have always counterbalanced each other and in which the division between Indians (who now make up less than 10 percent of the population) and others (who are primarily a mix of Spanish and Indian) has remained socially and economically important.

Over the centuries, Mexico has experienced several revolutions—some violent and some at the ballot box. Throughout the twentieth century, these revolutions resulted in gradual improvements in the health care available to Mexico's citizens. In 1917, Mexico's new constitution first gave the federal government responsibility for health care. Simultaneously, many large estates were taken out of private control and divided into small cooperatives owned by the local peasantry. These rural cooperatives subsequently received funding from the federal government to establish local clinics, typically run by minimally trained health aides. Staffing improved during the 1930s, when, responding to the revolutionary spirit of the times, the federal government established a continuing program under which all new physicians must work for one year in a rural community.

The next major change in the health care system occurred in 1942, when the government established the Social Security program and opened a network of modern health clinics and hospitals around the country. However, only salaried workers employed by private industries in Mexico's cities were eligible for Social Security and allowed to use these facilities. Since then, the system has expanded to include government employees and salaried agricultural workers and now covers about 55 percent of the population (Lassey et al., 1997). In addition, as of 1997, other individuals can purchase Social Security insurance.

Social Security provides a comprehensive package of ambulatory and inpatient benefits. However, some Mexicans receive considerably more and better quality benefits than others because benefits are allocated through several separate Social Security organizations with separate clientele and budgets. For example, the Social Security organization responsible for the health care of workers in the oil industry spends twice as much per capita as does the organization responsible for the health of workers in the private sector.

Since 1973, most of those who do not receive health care under Social Security can receive care through programs run by the Ministry of Health. Like Social Security, the Ministry provides comprehensive coverage for both acute and chronic health problems, including both ambulatory and inpatient care. The Ministry has expanded access to health care steadily, building clinics and hospitals in both rural and urban areas. In general, however, these facilities are inferior to facilities run by Social Security. On the other hand, the Ministry also runs some of the country's best specialized hospitals, used by private patients as well as by Ministry patients. In addition, the Ministry and other governmental agencies have funded widespread improvements in living conditions—food subsidies, new school construction, fluoridation of water, home improvements, and sanitary water systems—which have improved the health of the population.

Despite the growth of these two health care systems, about 10.6 percent of Mexico's citizens, primarily Indians living in rural areas, still lack regular access to modern health care (Pan American Health Organization, 1998). Responding to this problem, the Ministry of Health instituted a program in 1996 to improve basic sanitation, care of pregnant women, vaccination rates for children, and treatment of common, dangerous infectious and parasitic diseases. In addition, the National Institute for Indigenous Culture currently provides some sanitation and health programs in rural Indian towns and villages.

At the same time, about 10 percent of the population choose to purchase private insurance or care from private doctors on a fee-for-service basis, even though they are covered by Social Security (Lassey et al., 1997). About one-quarter of Mexico's physicians work solely for private patients on a fee-for-service basis. The remainder work as salaried government employees, although most also take fee-for-service patients on a part-time basis. Because the government does not regulate the private purchase of medicine, little else is known about this sector of the health care system.

Purchasing Care

Individuals who purchase health care in the private sector have, of course, a wide choice of doctors and hospitals. Other Mexicans, however, must use the doctor or clinic to which they are assigned for primary care (although in theory they have some choice, and a new pilot program now allows some individuals to select their doctors in exchange for an extra charge). Copayments vary by source and type of service, but range from nominal to nonexistent.

To obtain specialty care, patients must first get referrals from their primary care doctors. Such referrals can be difficult to get, however, because of government cost controls that restrict the number of practicing specialists. For the same reason, patients who do get referrals typically have long waits before they can get appointments with specialists. As a result, many patients subvert the system by instead seeking specialty care at emergency clinics.

Mexicans' access to technologically intensive care remains limited. In addition, because no central organization controls the distribution of hospitals and high-technology specialty clinics, these services are poorly distributed, with considerably more services available in cities compared with rural areas and in northern regions of the country compared with the south. Consequently, some hospitals and clinics are underutilized whereas others are severely overburdened.

Health Outcomes

Although Mexico remains rife with social and economic inequities and resulting inequities in health, it has nevertheless achieved notable improvements in health outcomes for much of its population. Consequently, by some measures Mexico appears to have completed the epidemiological transition—cancer and heart disease now kill more Mexicans than do infectious diseases, and overall life expectancy is 72. On the other hand, the infant mortality rate remains high (17.5 per 1,000 live births). In addition, poor rural Mexicans still experience health conditions characteristic of developing nations, and rates of some infectious diseases, such as malaria and tuberculosis, are rising (Pan American Health Organization, 1998). Nevertheless, preventive health campaigns have improved health throughout the nation: a massive vaccination program eradicated polio in 1991, and, as of 1998, 88 percent of children receive all recommended vaccinations by age 1 and 96 percent receive them by age 4.

These health outcomes have been achieved at relatively little cost: as of 1994, Mexico spent 6.1 percent of its GDP—only a bit more than $100 per capita—on health care (Pan American Health Organization, 1998). However, almost half of these costs (49 percent) are paid out of pocket, primarily by affluent individuals who desire a higher quality of care or additional services not available to others, suggesting once again that inequities remain within the system. Employers pay 29 percent of health care costs and the federal government pays the remaining 22 percent.

REFORMING HEALTH CARE IN THE UNITED STATES

Having established some basic measures for evaluating health care systems and explored the systems in five other countries, we now have the tools we need to think about the prospects for reforming the U.S. health care system.

Concern about rising costs and declining coverage led to numerous Congressional proposals to reform the health care system during the late 1980s and early 1990s, culminating in the Health Care Security Act (HCSA), proposed by President William J. Clinton in 1993. The HCSA represented a liberal approach to health care reform. If adopted, the HCSA would have broadened access to care without threatening the basically entrepreneurial nature of the U.S. health care system or the power or wealth of any of the "big players" in health care—hospitals, health care conglomerates, pharmaceutical companies, insurance companies, or organized medicine. Most Americans would have received health insurance primarily paid for by employers (either theirs, their parents', or their spouses'), while self-employed or unemployed persons would have purchased coverage directly from proposed insurance-buying cooperatives in each state.

Because, as is currently the case, employers would most likely have recouped the money they spent on insurance premiums by reducing wages across the board, the HCSA likely would have retained our current **regressive financing system,** with low-income persons paying a higher percentage of their income for care than high-income persons (Himmelstein and Woolhandler, 1994). In addition, under the HCSA, all Americans would still have had to pay substantial amounts for health care out of pocket. Moreover, affluent Americans would have retained the right to purchase health care options unavailable to others. Consequently, the HCSA would have reinforced our current two-class health care system, along with much of its current administrative complexity and the resulting costs.

In the end, Congress rejected the HCSA without even a floor vote. Although it is perhaps too soon to explain its demise with any certainty—scholars are still sifting through thousands of documents, meeting minutes, and interview notes to try to figure out why—some contributing factors seem obvious (Rothman, 1997; Skocpol, 1996). First, the kind of bureaucratic procedures needed to support any modern democratic government necessarily make the complete overhaul of a massive health care system difficult. This is especially true given the strength and power of those with a vested interest in maintaining the existing system, ranging from the multibillion dollar insurance industry to anti-abortion activists who feared the HCSA would cover abortion. Second, the Clinton administration undoubtedly made numerous tactical errors in presenting the plan, most importantly ignoring concerns that the plan might lead to rationing care. Third, distrust both of government and of government spending has became a powerful political force in U.S. politics since the 1970s, making it difficult to generate public support for moving services from the private sector to the public sector, as the HCSA proposed. Meanwhile, these same anti-government sentiments created a catch–22 for those proposing broader health care coverage. Because the public had grown to think taxes an unfair burden, making it impossible to fund expanded health coverage with new taxes, the HCSA proposal instead tried to keep costs low through an intricate web of government regulations. The sheer complexity of the resulting proposal made it easy to convince a public already leery of "big government" to distrust it.

Even if it had passed, however, the HCSA would have increased access to health care only partially, without really changing the essential nature of our health care system. That system reflects and reinforces a conception of society as a strictly economic contract, in which individuals and corporations deserve any profits they can earn with little if any obligation to help others (including the workers and customers who make those profits possible) and in which individuals deserve only what they can afford to purchase (with no right to even the minimum needed to sustain life). As such, our health care system both reflects and reinforces basic inequities in American society between rich and poor, manufacturers and consumers, and large corporations and individual entrepreneurs.

The defeat of the HCSA has led many observers to conclude that, for the foreseeable future, any changes in the health care system are likely to be incremental rather than major. Thus new federal legislative proposals keep emerging to expand health insurance coverage for specific groups, such as children who are near-poor or disabled persons who fear losing Medicaid if they get paying jobs. These proposals have generally taken two forms: expanding eligibility for already-existing insurance programs, such as Medicaid, or offering tax credits or tax deductions to make insurance more affordable. The former proposals have the benefit of taking advantage of existing structures rather than requiring new bureaucracies, but run the risk of straining already over-burdened programs. The latter proposals present other problems. With tax credits, individuals can reduce their federal taxes by the amount they have spent on health insurance, up to a set limit. All the proposals so far, however, have set limits so low that they would only cover a small portion of the cost of health insurance. As a result, these proposals seem more likely to benefit those who already have health insurance than those who currently find insurance unaffordable. Proposals offering tax deductions, which allow individuals to deduct part of the cost of health insurance from their income before calculating their federal taxes, offer even less benefit, especially to poorer persons who are in low tax brackets anyway. Moreover, the existence of tax credits or tax deductions might make it easier for employers to justify not offering health insurance to workers, thus increasing rather than reducing the number of uninsured Americans. And if more people do start purchasing health insurance, insurance companies would likely respond to this increased demand by raising prices.

Incremental change in the health care system could also come about through state-level reforms. During the last decade, the federal government has become much more open to innovation at the state level, allowing states to develop their own programs to serve persons eligible for Medicaid and Medicare. Some of these programs could eventually serve as models for the nation as a whole.

Hawaii's program has generated especially great interest. In 1974, Hawaii's legislators passed the Prepaid Health Care Act, which required employers to pay at least 50 percent of the cost of health insurance for all full-time employees (Neubauer, 1997). Small businesses that cannot afford to pay their share of premiums can draw subsidies from a special fund established under the Act, although very few have done so. Because of Hawaii's booming economy and the resulting competition for workers, most employers voluntarily insure employees' families as well as their employees and pay more than their required 50 percent of the costs.

As in other states, elderly persons and very poor persons receive their health insurance from Medicaid or Medicare. To provide insurance coverage for the "gap group" of unemployed persons and part-time workers who earn too much to receive Medicaid but too little to purchase insurance on their own, Hawaii in 1989 established a State Health Insurance Program (SHIP), which purchases insurance from HMOs for these individuals. By closing the insurance gap, Hawaii secured health insurance for 90 percent of its residents; only four states had a better record as of 1998 (Campbell, 1999). Because such a high proportion of the state's population is insured, insurers can use community ratings rather than risk ratings—keeping rates affordable for all purchasers—and still remain financially viable.

In addition to ensuring a high level of coverage, the new system enabled Hawaii to achieve unusual success in restraining health care costs. In part, this success resulted from the unintended development of monopolistic, nonprofit insurance plans. About 70 percent of Hawaiians receive their insurance from one of two nonprofit insurers: the Hawaii Medical Service Association (a **Blue Cross/Blue Shield** plan) and Kaiser Permanente (an HMO that still uses a salaried staff). Because they control such a large share of the market, these two insurers exert considerable control over medical costs. Doctors who refuse to accept their reimbursement schedules or salaries can attempt to seek patients elsewhere, but will find few patients who do not belong to these plans.

More important, Hawaii restrained costs through reducing hospital use and costs. Neither of the major insurers charges deductibles, so individuals have less incentive to put off needed care. As a result, health problems more often are caught at early stages, when treatment is relatively inexpensive. In addition, and unlike most U.S. insurers, both of these insurers pay only for stays in hospital wards, not in semiprivate rooms. Finally, Hawaii has implemented a strict system for prospectively reviewing any hospital capital expenses. Hospitals cannot purchase major equipment or construct new facilities unless they can demonstrate need for those services. As a result, consumers need not pay the costs of maintaining unused hospital beds or duplicative technologies.

Conversely, the continued existence of Medicare and Medicaid has hampered Hawaii's ability to restrain health care costs. Because these plans do not reimburse hospitals at rates high enough to cover the actual costs of providing care, hospitals have shifted costs to patients with private health insurance. At the same time, Medicaid's and Medicare's low reimbursement schedules have produced problems in access to health care because many doctors will not accept patients who belong to these plans. To control costs, and to equalize the benefits available under SHIP and Medicaid, Hawaii in 1994 merged Medicaid into SHIP (now renamed "QUEST"). Nevertheless, costs have continued to climb (although not as steeply as in other states), largely because of nationwide economic shifts resulting in a larger pool of part-time workers who fall into the gap group.

These cost increases have forced Hawaii to reduce the benefits available through its insurance program.

In sum, the Hawaii experiment demonstrates both the advantages of moving toward a single payer, nonprofit system with strong centralized control and the problems when multiple payers—in this case, public and private insurers—continue to function in the same economic sphere. It also demonstrates the benefits available from a reasonably unified managed care system, and the difficulties of sustaining a strong system in the face of external economic pressures.

CONCLUSIONS

A critical approach to health care reform suggests that for meaningful reform to occur in the U.S. health care system, we must be willing to question and challenge the power dynamics underlying the current system. Once we do so, the way becomes clearer for us to learn from the experiences of countries that have reformed their health care systems. Canada's history, for example, suggests that eliminating private insurers—major power holders in the current system—can reduce costs substantially by eliminating the costs of selling, advertising, and administering the various insurance plans. Eliminating private insurers also eliminates the costs that accrue when doctors, hospitals, and other health care providers must track and submit bills for each client to each insurance company. Similarly, moving hospitals and other health care centers from private to public control, as Britain has done, and placing them under a single national authority (probably with some decision-making reserved for local authorities) would give the government control over both operating and capital budgets for these facilities. As a result, centralizing control of health facilities would allow the government to restrict the duplication of services and proliferation of technologies that have driven up the costs of the existing system. By the same token, establishing a national fee schedule for service providers, such as Canada uses, would enable the government to restrict the rise of those fees. Even more control is possible if the government, like Britain's, restricts doctors to salaried practices so that doctors cannot increase their incomes by increasing the number of procedures they perform. At the same time, mandating national health coverage would guarantee a large enough risk pool to make community rates feasible and affordable while eliminating the possibility of a **rate spiral.** Finally, using income taxes to pay for health care would more equitably distribute the costs of financing the system.

Of course, any proposals incorporating a critical approach would meet major opposition from those who benefit from the current system. Such a proposal, however, would be worth fighting for.

SUGGESTED READINGS

Himmelstein, David U., and Steffie Woolhandler. 1994. *The National Health Program Book: A Source Guide for Advocates.* Monroe, ME: Common Courage. A basic book for anyone interested in understanding the current debate over health care.

Lassey, Marie L., William R. Lassey, and Martin J. Jinks. 1997. *Health Care Systems Around the World: Characteristics, Issues, Reforms.* Upper Saddle River, NJ: Prentice-Hall. An excellent overview of 13 health care systems, covering industrialized nations, developing nations, and formerly Communist nations.

GETTING INVOLVED

Physicians for a National Health Program. 332 South Michigan Avenue, Suite 500, Chicago, IL 60604. 312-554-0382. www.pnhp.org. Organization of U.S. physicians for a Canadian-style health care system.

REVIEW QUESTIONS

Define the eight measures of health care systems and explain why each is important.

What is the convergence hypothesis? What evidence of convergence can be found in the histories of health care in Great Britain and China?

How are doctors and hospitals paid in Germany? in Canada? in Great Britain?

What is the difference between national health insurance and a national health service?

How does access to care—both primary care and hospital care—in Canada compare with access to care in the United States?

What aspects of the health care systems in Canada, Great Britain, and Germany have helped them to restrain costs? What aspects have kept costs high?

How has the rise of market forces affected health care in Great Britain?

What aspects of its health care system have enabled China to provide good health at low cost to its people?

In what ways is health care in Mexico a two-class system?

INTERNET EXERCISES

Choose a country you are interested in. Then use the Internet to see what you can find out about its health care system, looking for information comparable with that presented for other countries in this chapter.

Using the Web site for Health Hippo, an online archive of health law materials, find information on a current health policy issue of interest to you. What are some of the current proposals on this issue, and what are some of the arguments that have been offered for or against those proposals?

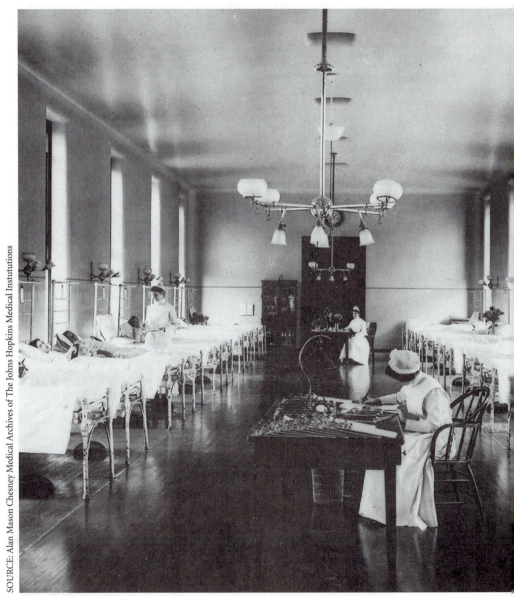

Even at Johns Hopkins, one of the most advanced hospitals in the country, patients in 1889 could not expect privacy, running water, or technologically intensive treatment.

CHAPTER 10

Health Care Settings

—

Timothy Diamond, a sociologist who spent several years working as a nursing aide in a variety of nursing homes, recounts the following experience:

Mary Ryan, like many others, spent all day in the day room, secured to her chair with a restraint vest. "How y' doin' today, Mary?" I once asked in passing.

She answered the question with a question. "Why do I have to sit here with this thing on?"

I responded automatically with a trained answer, "That's so you won't fall. You know that."

"Oh, get away from me," she reacted with disgust. "I don't trust anyone in white anymore."

Stunned by her rejection, and not completely confident of my own answer, I passed the question on to Beulah Fedders, the LPN [licensed practical nurse] in charge. "Beulah, why does she have to wear that thing all the time?"

Beulah accompanied her quick comeback with a chuckle. "That's so they don't have to hire any more of you."

We snickered together at the humor of her explanation, but an explanation it was, and more penetrating than mine to Mary. It posed a relationship between technology and labor, and in that connection Beulah explained that the use of one could mitigate the need for the other. A different kind of answer to the same question was given during our orientation [by the home's administrator]. "The restraint vests save on incidents. . . . "

Beulah's answer was more accurate than "so you won't fall" and "vests save on incidents," because she connected them both to a common denominator—available labor. If no nursing assistant was there to be with Mary, to walk with her or anticipate her dizziness, and if she sat in the chair without a restraint and without anyone to keep an eye on her, she might have fallen, thus generating an incident. Her restraint vest saved on incidents while it saved on labor costs.

As this story suggests, one of the central dilemmas of the American health care system is how to provide care in profit-driven institutions, as well as in nonprofit institutions that function within a broader, entrepreneurial system. In this chapter, I look at several settings where Americans obtain health care: hospitals, nursing homes, board and care homes, assisted living facilities, hospices, and family homes.

THE HOSPITAL

The Premodern Hospital

The hospital as we know it is a modern invention. Before the twentieth century, almost all Americans, whether rich or poor, received their health care at home, from friends, relatives, and assorted health care providers. Because these providers used only a few small and portable tools, hospitals were unnecessary.

Some form of institution, however, was needed for those Americans too destitute to pay for care at home and for those who had no friends or relatives who could provide care. For these individuals, the only potential source of care was the **almshouse.** Here they—along with orphans, criminals, the disabled, the insane, and other public wards— would receive essentially custodial care. Conditions in almshouses generally were appalling. Inmates often had to share beds or sleep on the floor, and rats often outnumbered humans. Hunger was common and blankets and clothing scarce. These conditions, coupled with the lack of basic sanitation, made almshouses ideal breeding grounds for disease (Rosenberg, 1987:31–32).

Wealthy Americans considered almshouse conditions quite acceptable for those they regarded as lazy, insolent, alcoholic, promiscuous, or incurable (categories they believed included all nonwhites). By the end of the eighteenth century, however, wealthy Americans began to view these conditions as unacceptable for those they considered the "deserving" poor—the respectable widow, the worker crippled by accident, the sailor struck by illness far from home. With such individuals in mind, philanthropists decided to develop a new form of institution, the hospital, devoted solely or primarily to inpatient care of the "deserving" sick. These hospitals would function as nonprofit, or **voluntary, hospitals,** so named because they reflected a spirit of voluntarism, or charity, rather than a profit motive. Such institutions would protect the morally worthy poor from the degradations of living in an almshouse and associating with the morally unworthy poor. The first two American hospitals were founded in the late eighteenth century, and a trickle of others appeared during the first half of the nineteenth century.

Reflecting their origins in social rather than medical concerns, these early hospitals accepted only patients certified as deserving. Hospitals often required those seeking care to provide letters of reference from their employers or ministers (Rosenberg, 1987:19–20). In addition, hospitals generally refused patients with chronic, contagious, or mental illnesses, making exceptions only rarely for the few who could pay for care.

Not surprisingly, given the essentially moral concerns of hospital founders, doctors played only a small role in hospital care and an even smaller role in hospital administration. Instead, hospitals relied on lay administrators or trustees, appointed more for their social status and charitable donations than for their medical knowledge (Rosenberg, 1987:47–68). From the beginning, though, hospitals partially justified their existence by pointing to their role in medical education, and the few elite doctors who worked in hospitals derived both status and financial profit from that association.

Early nineteenth century hospitals differed dramatically from modern hospitals. Until after the Civil War, the large ward remained the center of all hospital activity. Admissions, diagnostic examinations, surgical operations, the last moans of the dying, and the ministrations for the dead all occurred on the ward in full view of other patients and staff.

Although conditions in hospitals were better than in almshouses, they remained unpleasant. Throughout most of the nineteenth century, hospitals were chaotic and dirty places. According to historian Charles Rosenberg (1987:287),

> Nurses were often absent from assigned wards and servants insolent or evasive. Chamber pots [used for urinating and defecating] remained unemptied for hours under wooden bedsteads, and mattresses were still made of coarse straw packed tightly inside rough ticking. Vermin continued to be almost a condition of life among the poor and working people who populated the hospital's beds, and lice, bedbugs, flies, and even rats were tenacious realities of hospital life.

These conditions, coupled with the severe limitations of nineteenth century medicine, kept **mortality rates** high and taught the public to associate hospitals primarily with death rather than with treatment.

Hospitals functioned as **total institutions** (described in Chapter 7), in which patients traded individual rights for health care (Rosenberg, 1987:34–46). Hospital rules regulated patients' every hour, including mandating work schedules for all who were physically capable. Patients who did not follow the rules could find themselves thrown into punishment cells or frigid showers.

Beginnings of the Modern Hospital

Given the rigors of hospital life, the **stigma** of charity that accompanied hospital care, and the association of hospitals with death, early nineteenth-

SOURCE: Engraving from *Harper's Weekly*, 1860. Museum of the City of New York. Reprinted by permission.

A ward overrun by rats in New York's Bellevue Hospital. This woman's baby was eaten by rats.

century Americans entered hospitals only as a last resort. The Civil War, however, began to change this (Rosenberg, 1987:98–99). During the war, the need to care for sick and wounded soldiers exposed middle- and upper-class Americans to hospital care for the first time, as both patients and health care workers. Of necessity, during the course of the war, hospital organization and care improved, at least for the better-financed Union Army. These changes demonstrated that hospitals need not be either deadly or dehumanizing.

Following the war, widespread adoption of new ideas about the dangers of germs and the importance of cleanliness helped to make hospitals safer and more pleasant, as did technological changes, such as the development of disposable gauze and cheaper linens, which made cleanliness feasible (Rosenberg, 1987:122–141). Concurrently, demographic changes made hospitals more necessary. The tremendous spurt in immigration, the growth of cities, and the resulting overcrowding and dire poverty made it impossible for many Americans to recuperate from serious illnesses or injuries at home. Meanwhile, the growth of industry and technology fostered accidental injuries, while poor and crowded living conditions bred contagious diseases that required hospital treatment. Medical changes, too, made hospital care more necessary, as doctors came to value technologies and germ-free surgical conditions available only in hospitals (Rosenberg, 1987:149).

Yet affluent Americans remained by and large unwilling to tolerate the conditions on even the cleanest hospital wards. As a result, and to compete with the **for-profit, private hospitals** that began appearing during the second half of the nineteenth century, voluntary hospitals developed a class-based system of services (Rosenberg, 1987:293–294). Those who could pay for private accommodations received better heating and furnishings, exemption from many hospital rules, and privileges such as more anesthesia during operations. In addition, as hospitals increasingly became involved in medical education, private patients retained the right to treatment by their private doctors, while charity patients endured treatment by inexperienced medical students or **residents.** Through these changes, voluntary hospitals began to lose their ethos of service and to become increasingly like their for-profit competitors.

The Rise of the Modern Hospital

By the early twentieth century, the hospital as we now know it had become an important American institution and a major site for medical education and research. In the fifty years between 1873 and 1923, the number of hospitals increased from 178 to almost 5,000 (Rosenberg, 1987:341). These new hospitals also included **government hospitals,** established to provide services to those groups—the insane, the chronically ill, and the "undeserving poor"—that voluntary hospitals considered unworthy and for-profit hospitals considered unremunerative. However, African Americans still could obtain care only in a few segregated, poorly staffed, and poorly funded wards and hospitals; in municipal hospitals where medical students and residents could learn skills by practicing on African-American patients; and sometimes in other hospitals for emergency care (Stevens, 1989:137).

This hospital building boom reinforced the class division within voluntary hospitals. According to Rosemary Stevens (1989:112), the voluntary hospital of the early twentieth century "was like a multiclass hotel or ship, offering different facilities for different prices. The grade of semiprivate patients, tucked in between private patients and the wards, seemed the logical development of a new 'cabin class' between 'first class' and 'steerage.'" Thus by the 1920s, voluntary hospitals had abandoned much of their original charitable mission and become big businesses. As such, they had come to reflect the American ideology that individuals should get only what they pay for, in health care as in other areas (Stevens, 1989:112).

By this time, surgical admissions to hospitals far surpassed medical admissions (Rosenberg, 1987:150). The majority of patients went to a hospital to have their tonsils, adenoids, or appendix removed; their babies delivered; or their injuries treated (Stevens, 1989:106). The emphasis on technology as a defining aspect of modern hospitals further reinforced hospitals' tendency to focus on the care of **acute** rather than **chronic illness.**

This emphasis, coupled with hospitals' desire to maintain their image as proper middle-class institutions, created problems in the years following World War I, when hospitals proved extremely loath to deal with the chronic health problems of veterans (Stevens, 1989:126–128). Many veterans were poor and suffered from crippling or disfiguring problems not amenable to the acute or surgical care that hospitals emphasized. Yet Americans generally believed that veterans had earned the right to health care. As a result, in 1921, Congress voted to establish a national system of **veterans hospitals.**

Initiation of a federal system of veterans hospitals offered the government a chance to set national norms for health care overriding local norms of racial segregation (Stevens, 1989). Instead, however, the federal government bowed to local political pressure and decided to segregate its hospitals. As a result, a survey of veterans hospitals conducted as late as 1945 still could find that 17 percent accepted African Americans only in emergencies and 24 percent treated African Americans only in segregated wards. These percentages changed significantly only following the civil rights struggles of the 1960s (Stevens, 1989:222).

The number of hospitals continued to grow during the 1930s but did not increase dramatically until the passage in 1946 of the Hill-Burton Act, which provided funding for hospital construction. During the next fourteen years, 707 voluntary hospitals and 475 state and local hospitals were built, and the rate of hospital admissions increased substantially. As with the development of the veterans hospital system, however, the federal government did not use this opportunity to develop a rational and national health care system. Instead of tying funding to regional health needs, the government allowed hospitals to pursue their private financial interests: focusing on acute rather than chronic illness; discouraging nonpaying patients; reinforcing local norms of racial segregation; and buying expensive, esoteric technology even if it duplicated that owned by nearby hospitals (Stevens, 1989:200–232).

Hospitals Today

Federal subsidies for hospitals expanded substantially following the implementation in 1965 of **Medicaid** and, especially, **Medicare.** These plans dramatically increased the profits available to hospitals and spurred the merger of hospitals into large for-profit and voluntary hospital chains (such as Humana and Sisters of Charity, respectively). By 1993, chains controlled 42.6 percent of U.S. hospitals (American Hospital Association, 1994a:B3).

As hospital profits grew, so did costs to the federal government via Medicaid and Medicare. As a result, the government for the first time developed a vested interest in controlling hospital costs. Ironically, the resulting price control programs (described in Chapter 8) such as **diagnostic related groups (DRGs)** have pressured hospitals to pay more attention to the bottom line and therefore encouraged voluntary hospitals, which remain the center of the hospital system, to act more like for-profit hospitals (Stevens, 1989:305).

More recent cost-containment programs have especially squeezed funding for public hospitals. Under any circumstances, it is difficult for public hospitals to make ends meet, as about one-third of their patients cannot

Table 10.1   *Proportion of U.S. Non-Federal Hospitals Owning Various Technologies, 1984 and 1991*

TECHNOLOGY	1984	1991
Angioplasty	0.0%	20.5%
CT scanner	47.9	73.0
Magnetic resonance imaging (MRI)	3.1	20.8
Organ transplant facilities	4.6	11.2
Open heart surgery facilities	11.8	17.4

Source: American Hospital Association (1994b:xlviii).

pay their hospitals bills (Andrulis et al., 1996). Until recently, however, public hospitals could subsidize these patients through "disproportionate share funds" given by each state to hospitals that serve a disproportionate share of poor persons. In addition, public hospitals could subsidize nonpaying patients using grants received from the federal government for training medical residents. In the last few years, however, states instead have given some of their disproportionate share funds to **managed care** organizations in exchange for providing insurance coverage to Medicaid recipients. Meanwhile, the federal government has cut funding for medical residencies as a means of decreasing the oversupply of physicians. Taken in combination, these two changes have reduced budgets substantially at public hospitals, resulting in cutbacks and hospital closings.

Concern about costs and profits also has affected the mix of services offered by hospitals (Stevens, 1989:334). Because insurers (including Medicare under the DRG system) typically pay only preset amounts for inpatient surgery but give hospitals more leeway in setting prices for **outpatient** surgery (that is, surgery given without formally admitting the patient to the hospital or requiring an overnight stay), hospitals now offer outpatient surgery whenever technically feasible. As a result, outpatient surgery increased from 20 percent of all hospital surgeries in 1981 to 60 percent in 1996 (American Hospital Association, 1998). At the same time, the competitive market environment has encouraged hospitals to offer new, technologically intensive, treatments even if other nearby hospitals already do so. The result has been a proliferation of technology, as Table 10.1 demonstrates. Similarly, intensive care units, almost unknown as recently as the 1960s, were found in 69 percent of hospitals by 1994 (American Hospital Association, 1994b:231). As a result of these changes, hospitals now treat an older and sicker mix of patients, most of whom suffer from the acute complications of chronic illnesses.

Conversely, as hospitals have shifted toward providing more intensive care for middle-class Americans, some (especially government hospitals) have moved, if unwillingly, toward becoming **primary care** providers for the poor. Patients who have neither health insurance nor money to pay for care will sometimes turn to hospital outpatient clinics and emergency rooms not only for treatment of acute problems, such as gunshot wounds, but also for chronic problems, such as backaches. This "**emergency room abuse,**" as it is defined by hospitals, aggravates exhausted medical staff and worries hospital administrators concerned about budgets. In turn, it has fostered **patient dumping,** in which voluntary and for-profit hospitals place patients, sometimes in serious medical distress, in ambulances and deliver them to the emergency rooms of government hospitals—often without informing either the patient or the receiving hospital beforehand. In response to this problem, Congress in 1985 passed the Combined Omnibus Budget Reconciliation Act (COBRA), which made it illegal for hospitals to transfer physically unstable patients. This law, however, has not ended the problem: between March 1995 and September 1996, the Department of Health and Human Services confirmed reports implicating 256 U.S. hospitals in patient dumping (Public Citizen's Health Research Group, 1997). These numbers undoubtedly underestimate the problem, because the groups most likely to be dumped—the poor and the powerless—are the groups least likely to file complaints. For example, a review of patient charts at one Tennessee hospital found almost as many cases of patient dumping at that one hospital in 1988 as had been reported to the federal government by all hospitals in the country that year (Kellerman and Hackman, 1990).

The Hospital-Patient Experience

For many patients, a hospital stay is now a matter of only a few hours or days. For example, whereas women typically stayed in the hospital for two to three weeks following childbirth before World War II, they now stay for an average of 2.2 days (Graves and Owings, 1998). Similarly, the average stay for a patient admitted to a general hospital was 5.2 days in 1996, compared with 7.2 days in 1991 and 12.5 days in 1923 (American Hospital Association, 1994b; Graves and Owings, 1998; Starr, 1982:158).

Certainly hospitals no longer terrify and endanger patients as they did in the nineteenth century. Yet, a hospital stay often remains alienating and frightening. The bureaucratic nature and large size of modern hospitals, coupled with the highly technological nature of hospital care, often means that the patient as individual person and not just diseased body gets lost.

The reasons behind this are obvious and, to some extent, unavoidable. First, increasingly patients enter hospitals needing emergency care. Often, health care workers must respond immediately to their needs and have no time to talk with them to ascertain their preferences—which, at any rate, many are physically incapable of expressing. Second, the highly technical nature of hospital care encourages staff to focus on the machines and the data these machines produce rather than on the patient as a whole person. In the modern obstetric ward, for example, workers often focus much of their attention on the electronic fetal monitor rather than on the laboring woman (Martin, 1987:142–146). Third, medical training, as Chapter 11 will describe, encourages doctors to focus on biological issues much more than on patients' psychological or social needs. At the same time, short stays make it less likely that patients will develop a personal relationship with either hospital staff or other patients. Fourth, as large institutions necessarily concerned with economic profitability or at least stability, hospitals cannot afford to provide individualized care. Instead, hospitals rely on routines and schedules for efficiency. These routines and schedules leave little leeway for individual needs or desires, resulting in such ironies as nurses awakening patients from needed sleep to take their temperature or blood pressure.

Public dissatisfaction with the often-dehumanizing nature of hospital care, combined with market pressures, has led hospitals to make at least superficial changes in care. For example, since the early 1990s, the majority of U.S. hospitals have offered couples who consider the standard hospital labor and delivery rooms emotionally and physically uncomfortable the option of using a "birthing room," which offers a more home-like environment. Critics, however, note that these rooms are still filled with medical technologies, such as intravenous pumps, fetal monitors, and the like, whose very presence makes their use more likely.

NURSING HOMES

From the start, American hospitals focused on caring for acutely ill persons and assumed that families would care for chronically ill persons. During the course of the twentieth century, however, average life expectancy increased; families grew smaller, more geographically dispersed, and less stable; and women less often worked at home. As a result, more and more Americans needed to seek long-term care from strangers and **nursing homes**—facilities that primarily provide nursing and custodial care to groups of individuals over a long period of time—became part of the American landscape.

The number of nursing homes has skyrocketed in recent years. A survey conducted by federal researchers in 1995 identified 16,700 nursing homes, 66 percent of them run for profit (Strahan, 1997). These homes fall into two categories: skilled and intermediate care. Skilled nursing homes accept only patients under a doctor's care and provide both medical and trained nursing care, whereas intermediate care nursing homes provide bed, board, and a less intensive level of health-related care.

Who Uses Nursing Homes?

Researchers project that 39 percent of Americans who turned 65 in 1990 will enter a nursing home before they die (Murtaugh et al., 1997:213). Currently, about 1.5 million Americans, including 4 percent of those over age 65, live in nursing homes (Strahan, 1997). **Controlling** for differences in life expectancy, whites, unmarried persons, and women are more likely than others to use nursing homes (Dey, 1997). Women comprise 72 percent of nursing home residents, both because women live longer and thus more often eventually need assistance and because women less often have a surviving spouse who can and will care for them. Although illness and disability can force individuals into nursing homes at any age, nursing home residents overwhelmingly are elderly, with 74 percent age 75 or older and only 11 percent under age 65 (Strahan, 1997).

On average, current nursing home residents are sicker than were residents a decade ago. This change stems from the economic incentives built into DRGs, which have encouraged hospitals to discharge patients "sicker and quicker"—physically stable but still ill—once their bills and lengths of stay exceed the limits set by Medicare for hospital coverage. Those patients who cannot care for themselves at home often are discharged directly to nursing homes.

Although some people stay in nursing homes for only a few weeks, others stay for several years. A survey of nursing home residents conducted by federal researchers in 1995 found that the average length of stay for all persons over age 65 was 2.3 years (Dey, 1997).

Financing Nursing Home Care

In 1997, nursing home care cost an average of about $40,000 per year (*Consumer Reports,* 1997: 36). Few Americans have private insurance that will pay these costs. Although individuals can buy **long-term care insurance** to cover the costs of nursing or custodial care, its steep price and limited benefits make it unaffordable for most. Nor can most Americans rely on Medicare to finance nursing home care because Medicare pays only for skilled (rather than custodial) nursing care and only for the first 150 days.

In the absence of comprehensive coverage for long-term care, nursing home residents rapidly slide toward poverty. Those who survive long enough eventually reach the limits of any private or Medicare coverage. They may then obtain Medicaid or other public aid, but only after selling all their assets (minus their houses if they are married) and spending all their savings (minus the cost of burial expenses and minimum living expenses for their spouses).

Medicaid pays nursing homes directly, giving residents only a small monthly stipend from which to purchase all personal items, such as cigarettes, gifts, greeting cards, phone calls, or clothes. Moreover, because Medicaid will pay only a certain amount per month for care, as residents progress from Medicare to Medicaid, nursing homes often move residents to cheaper and lower-quality facilities either within a given home or in another home.

Working in Nursing Homes

Nursing home care is extremely labor intensive. To provide this care, nursing homes rely almost solely on **nursing assistants** (who often have no training) augmented by **licensed practical nurses** (who have completed approximately one year of classroom and clinical training).

Nationally, nursing assistants (half of whom work in nursing homes and one-quarter in hospitals) form one of the largest and fastest-growing health care occupations (U.S. Bureau of Labor Statistics, 1998:319–321). Almost all are women and most are nonwhite. Many come from Africa, Asia, or Latin America and are not native English speakers. Often they obtain their airfare to the United States as loans from nursing agencies in exchange for signing contracts obliging them to work for those agencies until they have repaid their debt (Diamond, 1992). These contracts leave them vulnerable to unscrupulous employers because, as essentially bonded laborers, these women have no legal grounds for requesting better wages or working conditions.

In some states, nursing assistants must complete a seventy-five hour course and pass a state examination before seeking employment, but in others, nursing assistants need neither training nor experience. In either case, those who work in nursing homes rarely earn much above the minimum wage, with median weekly earnings of $292 per week in 1996 (U.S. Bureau of Labor Statistics, 1998:321). To afford basic food, clothing, and shelter, therefore, nursing assistants commonly work two jobs or double shifts (Diamond, 1992).

To understand the life of nursing home residents and the nursing assistants who care for them, sociologist Timothy Diamond (1992) became certified as a nursing assistant and worked for several years in a variety of nursing homes. He soon concluded that the core of working as a nursing assistant is caregiving, but that those who train nursing assistants do not recognize this basic fact. Instead, his instructors taught him to recite biological and anatomical terms, measure vital signs, and perform simple medical procedures. Instructors divorced these skills from any social context or any sense that their patients were people rather than inanimate objects. Moreover, the skills he most needed he was never taught, such as exactly how do you clean an adult who has soiled a diaper in a manner that preserves the individual's sense of dignity? Only by labeling this caregiving mere physical labor could those who hire nursing assistants label them "unskilled" and treat them so poorly.

Life in Nursing Homes

Diamond's research underlines how the fates of nursing assistants and nursing home residents intertwine and how even in the best nursing homes the economics of a profit-driven system produce often-intolerable conditions for both. According to Diamond (1992:172), within nursing homes

> caregiving becomes something that is bought and sold. This process involves both ownership and the construction of goods and services that can be measured and priced so that a bottom line can be brought into being. It entails the enforcement of certain power relations and means of production so that those who live in nursing homes and those who tend to them can be made into commodities and cost-accountable units.

In this process of **commodification,** or turning people into commodities, "Mrs. Walsh in Bed 3" becomes simply "Bed 3." To keep down the price of this "commodity," only the most expensive homes provide private rooms or separate areas for residents who are dying, incontinent, smelly, or insane. Privacy, then, also becomes a commodity, which few residents can afford.

Nursing assistants, meanwhile, become budgeted expenses, which homes try to keep to an absolute minimum. Homes can do so by not recognizing the caregiving that assistants provide and residents need. For example, managers may hire only enough assistants to hurriedly spoon-feed residents rather than enough to allow assistants to chat with residents while feeding them or to help residents retain their dignity by feeding themselves. Similarly, managers can keep residents drugged, strapped to chairs, on a strictly regimented schedule, and in a single central room during the day so that a few assistants can supervise many residents; nationally representative studies have estimated that on any given day, nursing homes physically restrain between 20 and 38 percent of residents (Castle and Mor, 1998). The same logic frequently leads nursing homes to reward aides who work quickly and efficiently (even if the aides must bully or coerce patients to do so) and to penalize aides who spend the time needed to offer true caring (Foner, 1994).

Within this profit-driven system, managers constantly stress to staff that *providing* care is less important than *documenting* care. As a sign proclaimed in one nursing home where Diamond worked, "If it's not charted, it didn't happen." For example, state regulations where Diamond worked required homes to serve residents certain "units of nutrition" each day. Consequently, each day Diamond collected the cards placed on residents' food trays that named the foods and their nutritional content. Every few months, state regulators would inspect the cards and certify that the homes met state nutritional requirements. Yet these cards bore little relationship to reality, for the appetizing-sounding names given to the foods rarely matched the actual appearance or taste of the food. Nor did the cards note if a resident refused to eat a food because it was cold, tasteless, or too hastily served. Similarly, sanitation regulations required homes to shower residents regularly but did not require that the showers be warm. Nor did they require the homes to hire enough nursing assistants so that residents who used diapers could be cleaned as soon as needed or so residents could get the help they needed in using the toilet and avoid the indignity and discomfort of diapers.

Problems such as these led Diamond (1992:163) to conclude that

> It made a certain kind of sense . . . that in the schooling and textbooks there had been no vocabulary of caring. There was no place for it in the records. Words that concerned how to be gentle with Arthur, firm with Anna, delicate with Grace; how to mourn with Elizabeth and mourn for Frances; how to deal with death and dying, loneliness and screaming; how to wait in responding to someone else's slow pace—these constituted much of the work as it went along, but nothing of the job. In the documentation there was nothing relational, no shadow of the passion, only a prescribed set of tasks a doer gave to a receiver.

BOARD AND CARE HOMES

Nursing homes were developed to provide long-term care to individuals who did not need hospital care but who required too much medical or nursing care to live on their own. Other individuals, however, require neither medical nor nursing care but do need assistance in routine daily tasks such as bathing, dressing, and meal preparation. This group has grown substantially in recent years partly because of the aging of the American population, the increasing survival rates of severely disabled infants, and **deinstitutionalization** (described in Chapter 7). Recognition of this market has stimulated the growth since the mid 1980s of **board and care homes**—residential facilities, typically based in private homes with shared baths, that provide assistance in daily living but neither nursing nor medical care (Eckert and Lyon, 1992). Although some homes serve as many as twenty-five clients, many more are family homes with as few as one client.

Board and care homes remain largely unregulated, and licensure is not required in all states. As a result, few data on these homes are available. The absence of regulation coupled with the dependence of residents and the emphasis on profits increases the potential for physical as well as emotional abuse in board and care homes. However, at least one study has concluded that, despite the emphasis in the popular media on abuses within board and care homes, these homes more often provide safe, supportive, and family-like environments (Eckert and Lyon, 1992).

ASSISTED LIVING FACILITIES

During the 1990s, another institutional form, **assisted living facilities,** grew substantially in popularity (Mollica, 1997–1998). Assisted living facilities provide fewer medical and nursing services than nursing homes but more than board and care homes, and offer greater independence and privacy than either of these.

Unlike nursing homes, which typically consist of wards, assisted living facilities typically consist of small, private or semi-private apartments or kitchenettes. Like nursing homes, they provide help with basic tasks of daily living (such as meal preparation and housecleaning) and with routine nursing tasks (such as administration of medications). These facilities also typically offer some medical care, although most states forbid them from caring for persons who have unstable medical conditions or require around-the-clock nursing. In addition, assisted living facilities typically allow residents to "age in place" by offering local transportation and social activities for those who are reasonably healthy as well as the opportunity to transfer to nearby units with higher levels of care for those whose health deteriorates.

Assisted living facilities were first developed in response to market demand from upper-income persons (Mollica, 1997–1998). They have grown in number as states increasingly have looked to such facilities as a means of reducing the costs they pay for nursing home care, which account for about 35 percent of all state Medicaid expenditures. As of 1998, 26 states cover or have plans to cover the costs of assisted living facilities for those who otherwise would be placed in nursing homes at state expense (Mollica, 1997–1998).

HOSPICES

Origins of Hospice

Whereas nursing homes emerged to serve the needs for long-term care not met by hospitals and board and care homes arose to serve the needs not met by nursing homes, **hospices** emerged out of growing public recognition that none of these options provided appropriate care for the dying.

Only in the last few decades has institutional care for the dying become a public issue. At the beginning of the twentieth century, few individuals experienced a long period during which they or those around them knew they were dying. Instead, most succumbed quickly to illnesses such as pneumonia, influenza, tuberculosis, or acute intestinal infections, dying at home and at relatively young ages. Now, however, most Americans live long enough to die from chronic rather than acute illnesses. In addition, as doctors and scientists have developed techniques for detecting illnesses in their earliest stages, they now more often identify individuals as having a fatal illness long before those individuals actually die. Thus, dealing with the dying is to some extent a uniquely modern problem and certainly has taken on a uniquely modern aspect.

Although modern medical care has proved lifesaving for many, its ability to extend life can turn from a blessing to a curse for those who are dying (as Box 10.1 discusses in more detail). For a variety of reasons, including the **technological imperative** underlying medical care, legal concerns about restricting care, and financial incentives that encourage the use of highly invasive treatments, thousands of Americans each year receive intensive, painful, and tremendously expensive medical care that offers only a small hope of either restoring their quality of life or extending their lives. In nursing homes, on the other hand, the emphasis on profit-making and cost-cutting often results in dying persons receiving only minimal and depersonalized custodial care.

This lack of appropriate care for the dying led to the development of the **hospice** movement. The first modern hospice, St. Christopher's, was founded in England in 1968 by Dr. Cicely Saunders, specifically to address the needs of the dying and to provide an alternative to the often alienating and dehumanizing experience of hospital death (Mor, 1987). The hospice admitted only patients expected to die within six months and offered only palliative care (designed to reduce pain and discomfort) rather than treatment or mechanical life supports. The hospice provided care both in St. Christopher's and in patients' homes.

The hospice movement received a substantial boost with the publication in 1969 of Elizabeth Kübler-Ross' book *On Death and Dying,* which helped to make dying an acceptable topic for public discussion. The first American hospice, which closely resembled St. Christopher's, opened five years later in New Haven, Connecticut. Other hospices soon followed, emerging out of grassroots organizations of religious workers, health care workers, and community activists seeking alternatives to hospitals and nursing homes. Public support for hospices was so immediate and so great that in 1982, only eight years after the first American hospice opened, Congress (hoping that supporting hospices would both reduce health care costs and garner votes) approved covering hospice care under Medicare (Mor, 1987:12–14).

The Hospice Philosophy

The early hospice philosophy differed markedly from mainstream medical philosophy (Abel, 1986; Finn Paradis and Cummings, 1986; Mor 1987). First, the hospice philosophy asserted that patients should participate in their own care and control as much as possible the process and nature of their dying. Hospices strove to give clients choices over everything from

---

Box 10.1:   *Ethical Debate: A Right to Die?*

---

*In 1983, 26-year-old Elizabeth Bouvia, suffering near-total paralysis from cerebral palsy and near-constant pain from arthritis, presented herself for admission to Riverside General Hospital. In years past, and despite her physical problems, Bouvia had earned a degree in social work, married, and lived independently. However, after her efforts to have children failed, her husband left her, and the state stopped paying for her special transportation needs, she lost interest in living. Her purpose in coming to the hospital, she told the hospital staff soon after her admission, was to obtain basic nursing care and painkilling medication while starving herself to death, cutting short what might otherwise have been a normal life span. The hospital's doctors took her case to court and won the right to force feed her, on the grounds that although individuals have the right to commit suicide they cannot force health care workers to commit **passive euthanasia** (that is, to allow patients to die through inaction).*

*In 1990, Janet Adkins, 54 years old and suffering from Alzheimer's disease, killed herself with the assistance of Dr. Jack Kevorkian. A pathologist, Kevorkian had designed a machine that allowed people with severe disabilities to give themselves a fatal injection in the privacy and freedom of their homes. By 1999, Kevorkian had provided **doctor-assisted euthanasia** for more than 100 people. He has been charged with murder multiple times, but was first convicted in 1999, after administering a lethal injection himself rather than having his client do so and sending a videotape of the death to CBS-TV.*

*In the Netherlands, meanwhile, doctors legally can practice **active voluntary euthanasia**—ending a patient's life through action rather than inaction—as long as they follow established guidelines. Those guidelines restrict active euthanasia to cases in which mentally competent but incurably ill individuals, who suffer intolerable and unrelievable pain, authorize their doctors in writing to give them a lethal injection. According to national survey data collected in 1997, 68 percent of Americans favored passing a law modeled on this Dutch law (Harris Poll, 1997a).*

As of 1999, only one state, Oregon, has adopted a legal statute permitting doctor-assisted suicide, whereas twenty others have voted against such laws. Even in the absence of such a law, however, some U.S. doctors engage in euthanasia or physician-assisted suicide; in a nationwide random survey of oncologists

---

what they ate to where they would die. Most significantly, hospices allowed residents to decide when to receive pain medications, how much, and what kinds. To eliminate pain from the experience of dying, hospices used whatever drugs would work, including opiates such as heroin. In contrast, nursing home staff do not have the expertise to prescribe or supervise the drugs that dying patients need, and hospital staff often oppose using addictive drugs because their commitment to healing makes it difficult for them to

(physicians who treat cancer), 10.7 percent reported having done so at some point in their careers (Emanuel et al., 1998).

Those who support a "**right to die**" argue that competent adults have the right to make decisions for themselves, including the ultimate decision of dying. Supporters argue that death sometimes can be a rational choice and that forcing individuals to suffer extreme physical or mental anguish is unwarranted cruelty.

If we accept that death can be a rational choice, then harder questions follow. Why is it only rational if one's condition is terminal? Doesn't it make even more sense to end the life of someone like Elizabeth Bouvia, whose agonies could continue for another fifty years, than to end the life of someone who will die soon regardless? Why should this choice be forbidden to individuals simply because they cannot, either physically or emotionally, carry it out themselves? And why should we only allow individuals to choose death through passive euthanasia, leaving them to languish in pain while awaiting death, if instead they could be killed quickly and painlessly?

Opponents of this view argue that the duty to preserve life overrides any other values and that euthanasia is merely a nice word for suicide or murder. They question whether Elizabeth Bouvia would still want to kill herself if she once more had the resources she needs to live independently and wonder whether euthanasia is merely an easy way out for a society that wants to avoid responsibility for relieving the burdens imposed by illness and disability. Opponents who have studied the Netherlands suggest that doctors there in fact do not always follow the legal guidelines but instead sometimes end patients' lives without their consent and without first attempting to improve patients' quality of life (Hendlin et al., 1997). In addition, they question whether acceptance of euthanasia in the Netherlands explains why there are fewer hospices in the Netherlands than elsewhere in Europe and why Dutch doctors receive relatively little training in pain relief.

In sum, the use of euthanasia, whether active or passive, raises numerous difficult questions: What are the consequences of, in effect, declaring it reasonable for disabled people to choose death? What pressures does this place on individuals to end their own lives rather than burdening others? What responsibilities does this remove from society to make these individuals' lives less burdensome? Finally, given that social factors, such as age, gender, and social class, affect our perceptions of individuals' worth, how do we ensure that health care workers and courts will not be more willing to grant a right to die to those who belong to socially disvalued groups?

acknowledge that certain patients are dying and therefore cannot be harmed by addictive drugs.

Second, the hospice philosophy foreswore regimentation and stressed the importance of integrating hospice care into clients' everyday lives rather than integrating clients into hospice routines. Where possible, hospices would offer services in clients' homes. For those who needed care in the hospice, the hospice would offer a home-like environment, without the regulations regarding schedules, visitors, food, clothing, and so on that rule life in hospitals and nursing homes.

Third, the hospice philosophy emphasized a true team approach. Because hospices provided neither diagnosis nor treatment, doctors could claim little special expertise (Abel, 1986). As a result, within hospices, doctors had little more importance or influence than did social workers, nurses, ministers, psychotherapists, nutritionists, and others. Hospices explicitly worked to minimize the authority of doctors and to increase the role and status of nonprofessional volunteers.

Fourth, hospices focused not only on the dying person but also on his or her friends and relatives. Hospices attempted to involve these others in the process of dying and to meet their social and psychological needs. As a result, hospice care did not end with the client's death but extended to bereavement counseling for survivors.

Finally, hospices viewed dying "as a natural event rather than as technological failure" (Abel, 1986:7). Workers viewed dying as an important phase of life, suitable for and worthy of open discussion. Neither the dying process nor the disease was to be hidden.

The Cooptation of Hospice

The U.S. hospice movement has proved enormously successful, growing from one hospice in 1974 to more than 3,000 in 1999. As the movement has spread, however, it has undergone substantial **cooptation,** exchanging much of its initial philosophy and goals for social acceptance and financial support (Finn Paradis and Cummings, 1986; Mor, 1987:17).

The history of hospice resembles the history of many other reform movements and organizations. As various sociologists have observed, successful social movements over time often come to resemble the very institutions they sought to reform (DiMaggio and Powell, 1983; Kanter, 1972; McAdam, 1982; McCarthy and Zald, 1973). These changes evolve gradually and naturally. For a movement to survive, it must mobilize people and develop sources of funding. To do so, reformers typically must develop hierarchies and rules, abandon their grassroots and voluntaristic approach, and hire professional staff (McCarthy and Zald, 1973). Battered women's shelters, for example, initially established by feminists as a radical means of protecting women from violent men, now often rely primarily on social workers whose goal is restoring the family unit (Schechter, 1982).

The cooptation of hospice similarly derives from natural developments in that field, especially the need to develop a stable economic base. Initially, many hospice organizers, reflecting the countercultural values of the late 1960s and early 1970s, expressed little concern for financial stability (Abel, 1986:75). Very quickly, though, and despite qualms among some hospice organizers, hospices began to seek federal funds to support hospice development, as well as **third-party reimbursement** (that is, the ability to bill insurers for services rendered).

To gain support, organizers worked with the federal government and with the American Hospital Association to develop standards for hospice care and accreditation. The resulting standards legitimated hospice care and paved the way for Medicare and, later, Medicaid and private insurance reimbursement, but, not surprisingly, also made hospices more like hospitals.

Medicare funding and the associated federal regulations also have changed hospices and threatened the original hospice philosophy (Finn Paradis and Cummings, 1986). For example, Medicare will only reimburse hospices up to a set number of dollars and number of days of care per patient. In addition, it will reimburse hospices for the cost of inpatient care only to the extent that inpatient care comprises no more than 20 percent of all care given. These regulations encourage hospices to accept patients who have sufficient family support to stay at home rather than in the hospice, who are near death, and whose time of death can be predicted with reasonable accuracy. In addition, to obtain reimbursement, hospices must provide services that meet specified standards and must document these services. These requirements have made it difficult for hospices to maintain their commitment to individualized care and to patient control and participation.

Medicare and private insurers also have placed limitations on who can provide care, requiring hospices to reduce their reliance on volunteers, social workers, ministers, and the like, and instead to hire professionally trained health care workers and administrators. These latter individuals often bring with them traditional ideas about health care, about the health care team, and about dying itself. Former hospital nurses, for example, might resist allowing patients to refuse intravenous feeding because that seems an unacceptable admission that health care has failed and might resist allowing patients to choose when to receive medications because they prefer the ease of a hospital-like schedule (Abel, 1986:77).

Internal pressures have forced other changes in hospice care. The original hospices were freestanding units, unaffiliated with other health care institutions. This model has proved both financially and administratively unfeasible. Freestanding hospices lacked ready access to the support services available at hospitals and other health care institutions. In addition, their independent status hampered efforts to get funding and to get referrals of patients from hospitals. As a result, although most hospices remain nonprofit, the number of independent, community-based, and largely volunteer-run hospices has declined from almost all hospices initially to about 28 percent currently (National Hospice Organization, 1999). (Box 10.2, however, describes one inspiring exception.) Hospitals or home health care agencies own most of the rest. Yet despite these changes, studies find that hospice clients and their families feel more satisfied with their care than do those who receive care from conventional sources (Mor, 1987:150–156).

Box 10.2   *Making a Difference: The Human Service Alliance*
W. Bradford Swift with Kimberly Ridley

The Human Service Alliance (HSA) is an experiment both in delivering free health and social services and in voluntarism as a way of life. In addition to its Care for the Terminally Ill (CTI) facility, the organization runs . . . a weekend respite program for families with disabled children . . . and a health and wellness program for people with chronic illnesses. Even the administrative jobs here, from accounting to filing, are performed by HSA's twenty-four volunteer board members.

In 1996, HSA volunteers provided 70,000 hours of service work, the equivalent of thirty-five people working full time. They delivered an estimated $926,800 worth of services . . . on a total operating budget of just $80,000, which comes from individual contributions and a few grants from area corporations. In the eleven years since the organization's inception, its methodologies have drawn the attention of administrators from nursing homes and schools of medicine and public health. One visiting physician, a cancer specialist, remarked after perusing the caregiving charts and detailed notes on each patient, "In the hospital, we cannot come close to offering this kind of attention, and having the rapport that HSA's caregivers do."

Forty-seven guests spent their final days in the CTI wing, twenty-four families utilized the services of the Respite Care Program, and twenty individuals with chronic health problems were served by the Health and Wellness Project in 1996. . . .

Human Service Alliance began in 1986 when a handful of people in remote Boomer, North Carolina, started taking care of one terminally ill neighbor at a time in makeshift quarters in a refurbished trailer. . . . By 1988, a core group had evolved, incorporated HSA, and moved the organization to the outskirts of Winston-Salem. They committed to operating debt-free by recruiting volunteers and raising donations before spending money. Within a few years, they had raised $400,000 to build HSA's facility for the terminally ill, which opened in 1991. . . .

Perhaps among all of HSA's programs, the Care for the Terminally Ill unit is where some of the most intensive services are provided. The unit, which accommodates up to six terminally ill "guests" in private and comfortable rooms, helps fill an important gap by caring for dying individuals who don't require the medical services of a traditional hospice, but whose families are unable to care for them at home. . . .

All guests accepted onto the CTI unit are selected by a committee of board members based on the organization's ability to care for their specific needs, the guest's willingness to

---

Use of Hospice

As of 1999, slightly more than half (52 percent) of U.S. hospice patients are male and about 70 percent are 65 years old or older (National Hospice Organization, 1999). Whites, who make up about 75 percent of the general population, make up 83 percent of hospice users. Almost two-thirds (63 percent) of users are married. On average, clients use hospice services for only 51 days.

live out his or her final days at HSA, and the family members' willingness to be a part of the process of their loved one's death. . . .

Family members are expected to visit regularly and are encouraged to volunteer some of their time serving at HSA, not because more volunteers are needed, but because it's been found that volunteering is often therapeutic for the family. Using volunteer activity in a therapeutic manner has also worked well in HSA's Health and Wellness program for clients with chronic illness. Todd Thornburg, a board member who started the program in 1988, says volunteering seems to be some of the best medicine the organization has to offer. He describes one young woman who entered the Health and Wellness program a few years ago with the complaint that her physician had ruined her knee and her life [through botched surgery]. Volunteering allowed her to redirect her focus, Thornburg says, adding that when she completed the program approximately a year later, she had a new life before her, even though she still had a knee that didn't work properly. . . .

Inspired by their experiences at HSA, a few volunteers have begun developing their own projects back home. Two free, volunteer-run hospices have opened in Jamesville and Fredericksburg, Virginia, [while] in Blue Hill, Maine,

writer and former HSA volunteer Maggie Davis launched Neighborcare, a program in which local volunteers clean, cook, run errands, and provide other help for the sick, elderly, injured, or overwhelmed in their community. "We see ourselves as filling in the gaps where people don't have what they need," Davis says. At first, Davis had in mind a center for the terminally ill, but when she met with representatives from area hospitals and social service organizations, they described more basic needs like rides to and from appointments, companionship, and simple caring. Davis put out the word, and a year and a half later, approximately seventy volunteers are ready to assist their neighbors in a handful of surrounding towns. . . .

The board members and founders of HSA hope to inspire other efforts around the nation and in other countries. But how does the average person find time in a busy life for this kind of work? "Serve in a group," suggests board member Danziger. "If eight people get together and want to serve a respite child, each could do two hours of work a week to give the parents a substantial break. The important thing is to start small and start now." . . .

Source: "Where Care Is Free." *Hope Magazine.* November/-December 1997.

Because the early British hospices focused on cancer patients, American hospital staff from the start associated hospice care with cancer and therefore more often referred such patients to hospices. Hospices themselves are more likely to accept patients with cancer because doctors can predict their life expectancy fairly accurately, and thus hospices can assume that any cancer patient they accept will die within the six-month Medicare guidelines. Currently, 60 percent of hospice clients have cancer, and about 50 percent of all Americans who die from cancer receive care in a hospice (National

Hospice Organization, 1999). Conversely, the greatest unmet needs are found among dying patients who do not have cancer.

Costs and Financing

Hospices receive most of their revenue (65 percent) from Medicare. Another 12 percent comes from private insurance, 8 percent from Medicaid, and the remainder from a variety of sources (National Hospice Organization, 1999).

Whether hospice care saves money compared with other options remains unclear (Mor, 1987:177–212). Direct costs appear somewhat lower for hospital-based hospices than for traditional hospital care (Kidder, 1988a; Kidder, 1988b; National Hospice Organization, 1999). However, indirect costs are substantial. Currently, most hospice users (77 percent) receive care in their own homes, rather than in a hospice facility. In these circumstances, family members provide most care. As a result, family members often must take time off from work or drop out of the labor market altogether. Consequently, hospice care might not reduce the costs of caring as much as it shifts the costs from hospitals and insurers to families.

HOME CARE

As the discussion of hospices has suggested, most individuals who experience chronic or acute health problems—whether children, working-age adults, or elderly, and whether the problems are physical or mental—receive their care at home (Abel and Nelson, 1990). This is even more true now than in the recent past because of technical, demographic, and policy changes. Because of technological advances, babies born prematurely or with birth defects and persons who suffer severe trauma are increasingly likely to survive, although often with severe disabilities that require lifelong assistance. Much of this care is now given by family members in the home. Similarly, the rise in the numbers of frail elderly, many of whom suffer both multiple physical problems and cognitive impairments, has increased the number receiving care at home. At the same time, technological advances also have made it possible for families to provide at home treatments previously available only in hospitals, ranging from chemotherapy to respiratory ventilation to kidney dialysis. In addition, the movement begun in the 1960s (and described in Chapters 6 and 7) to **deinstitutionalize** disabled and mentally ill persons, combined with the lack of community supports for such individuals once deinstitutionalized, have shifted much of the burden of care from state institutions to the home. Finally, as described earlier, policy changes now encourage hospitals to discharge patients to their homes "sicker and quicker," in essence replacing paid hospital workers with unpaid family caregivers (Glazer, 1993).

Because of the limited public or private insurance funding for home care, about 85 percent of those who need long-term supportive care receive services only from family members and, less often, friends (American Medical Association, Council on Scientific Affairs, 1990:1243). The economic value of home caregiving has been estimated at between $113 billion and $286 billion per year, much greater than the $30 billion spent per year on paid home care and the $79 billion spent on nursing home care (Fisher, 1998). However, because of the substantial methodological difficulties inherent in trying to compare home with institutional care, no fully reliable data are available comparing the total benefits and costs (including loss of wages by family caregivers who can no longer hold paying jobs) of home versus institutional care. The existing data suggests that home care has little impact on the costs of care or the mental or physical functioning of ill or disabled individuals but can produce small, short-term improvements in their life satisfaction (Arno et al., 1995; Weissert, 1991).

The Nature of Family Caregiving

A national telephone survey conducted for the nonprofit National Alliance for Caregiving in 1996 gives us the best available data on caregivers of persons over age 65. (No comparable data are available for those providing care to younger persons.) The survey found that 24 percent of U.S. households include someone who is providing care for a person over age 65 and found that most of these caregivers (73 percent) are women (National Alliance for Caregiving, 1997). Ethnic minorities and poorer persons also are more likely to become caregivers, probably because these groups experience higher rates of illness and disability and have less access to formal services. These findings are similar to those from other studies that have looked at caregivers of both the elderly and the nonelderly (Kemper, 1992:448)

Those who care for the health needs of family members typically do so out of love and often reap substantial psychological rewards. Yet, care giving by family members should not be romanticized nor should the financial, physical, social, or psychological costs of caregiving be underestimated (Abel, 1990; Abel and Nelson, 1990; Arras and Dubler, 1995; Reinhard and Horwitz, 1996; Tessler and Gamache, 1994).

The financial costs of caregiving are substantial, if unquantified. The demands of caregiving force many to shift to part-time work or even abandon paid employment. In addition, caregivers must purchase, often out of pocket, both expensive drugs and technologies and many everyday items such as diapers and bandages. In addition, caregivers typically are responsible for purchasing a variety of services and therapies from a range of companies and health care workers.

The physical costs also can be high. Caregiving often includes exhausting tasks such as lifting physically disabled or mentally incompetent individuals, some of whom either cannot help or resist being moved. The time burdens of caregiving also can become physically draining. Caregivers for the elderly, for example, spend an average of 18 hours per week on caregiving and have done so for an average of four and a half years; eighteen percent work 40 hours or more. These hours quickly lead to exhaustion, especially for the 64 percent of caregivers who hold paid jobs, the 41 percent who have children at home, and the 31 percent who care for more than one person (National Alliance for Caregiving, 1997). In part as a result, caregivers experience worse physical health than do non-caregivers (Gallagher and Mechanic, 1996).

Taken together, the financial and physical burdens of caregiving often leave individuals with little time, energy, or money for social relationships. As a result, caregivers often report feeling almost totally isolated from the world outside the household (Abel, 1990; Abel and Nelson, 1990). Family relationships, too, can suffer. For example, a mother who spends hours each day caring for an ill child might feel guilty that she cannot spend more time with her other children, and those children might resent the attention given to their ill sibling. Problems are particularly acute when the person receiving care is mentally ill and throws family routines into chaos, embarrasses other family members, or physically threatens others' safety (Reinhard and Horwitz, 1996; Tessler and Gamache, 1994).

Family life also can suffer disproportionately when caregiving requires the use of high technology within the home. John D. Arras and Nancy Neveloff Dubler (1995:3) suggest that this

> invasion of the home by high-tech medical procedures, mechanisms, and supporting personnel exerts a cost in terms of important values associated with the notion of home. How can someone be truly "at home," truly at ease, for example, when his or her living room has been transformed into a miniature intensive care unit? . . . Rooms occupied by the paraphernalia of high-tech medicine may cease to be what they once were in the minds of their occupants; familiar and comforting family rituals, such as holiday meals, may lose their charm when centered around a mammoth Flexicare bed; and much of the privacy and intimacy of ordinary family life may be sacrificed to the institutional culture that trails in the wake of high-tech medicine.

Finally, caregiving brings with it numerous psychological costs. Caregivers can easily become depressed when their efforts cannot stop or even slow the disease process. This is especially true when caregivers must routinely inflict painful treatments on their charges or when the burdens of caregiving are unceasing, as when a parent must suction the lungs of a child with cystic fibrosis hour after hour, day after day, to keep the child from dying. Moreover, as this example suggests, caregivers also often bear the enormous psychological burden of being directly responsible for another person's life. In fact, family caregivers are now expected to manage in the home—often with little training or technical support—technology considered too complex for licensed practical nurses to manage in hospitals. Finally, caregivers of persons younger than themselves face anxieties about what will happen to their charges if the caregivers die first.

Summing up the burdens of caregiving, a woman whose husband has Parkinson's disease says,

> I need some help. I am burned out. I am locked in this house. I am used to going out to work and had to retire. I didn't plan to retire so soon. We had planned our retirement. We never did anything before because we didn't have the same vacation time. So you do all this and then bingo! . . . Two weeks ago I had a terrible pain in my ribs. But I can't run to the doctor for every little thing. How can I leave the house? I worry, what is going to happen to him, if I have to go to the hospital. . . . Medicare pays for only part of the things we need and doesn't pay for medications. That bottle of medication cost $130. . . . Sometimes he has to go to the bathroom just when I've finished eating. It is hard to get up at that instant to do it. You feel like everything [you just ate] is going to come up. You have all these things to contend with. People don't realize that unless they are in those situations themselves. . . . You have to really see it for yourself, be in it, to know what it is like (Corbin and Strauss, 1988:297).

Easing the Burdens of Caregiving

The problems faced by family caregivers have led to the development of new organizations, new organizational structures, and a new occupation to ease the burdens of caregiving. Two major organizations, the National Alliance for the Mentally Ill and the National Alliance for Caregiving, are now devoted to family caregiving. Both organizations work to increase assistance to family caregivers and improve access to community-based care, and the National Alliance for the Mentally Ill additionally fights to decrease the stigma of severe mental illness.

Both **respite care** (Montgomery, 1992) and **family leave programs** also were developed to ease the burdens of caregivers. Respite care refers to any system designed to give caregivers a break from their otherwise unrelenting responsibilities, including paid aides who provide care in the home for a few hours, day-care centers for elderly and disabled adults, and nursing homes that accept clients for brief stays. Unfortunately, only California and Pennsylvania offer formal programs for respite care, while in all other states, respite care is expensive and difficult to find; only 14 percent of those included in the National Alliance for Caregiving survey (1997) had ever used respite care.

The concept of family leave received considerable public attention with the 1993 passage of the federal Family and Medical Leave Act. This act gives employees the right to as many as twelve weeks of unpaid leave from work yearly to care for family members. Although the law has benefitted some family caregivers, its impact has been muted because only more affluent Americans can afford to take unpaid leaves and because the law does not apply to part-time workers, temporary workers, or employees of small firms. In addition, the law is problematic because it reinforces the idea that caring for ill and disabled persons is the responsibility of the family—which, in practice, usually means women relatives—rather than the responsibility of society as a whole (Abel, in press).

Finally, those who provide care to relatives or friends increasingly have turned for assistance to paid caregivers. Use of paid caregivers has grown both because increasingly family members cannot on their own provide the technological care needed by those released from hospitals "sicker and quicker" and because Medicare and Medicaid now offer some funding for paid caregivers as a means of decreasing the demand for far more costly nursing home care. As of 1996, on any given day about 1.7 million Americans received paid home care, most commonly help with bathing, dressing, and light housework (U.S. Bureau of the Census, 1998:142). Table 10.2 shows the tremendous growth in agencies providing such services and the shift from nonprofit to profit-making agencies, as government funding for these services became available.

Table 10.2	Home Health Agencies by Sponsor, 1967 and 1992*			
	1967		1992	
TYPE	N	%	N	%
Public	939	53.6%	1149	18.7%
Visiting nurse associations	642	36.6	604	9.9
Hospital-based (profit or nonprofit)	133	7.6	1688	27.5
Private, for profit	0	0.0	1953	31.9
Private, nonprofit	0	0.0	594	9.7
Other	39	2.2	141	2.0
Total	1753	100 %	6129	100%

*Table includes only agencies certified by Medicare.
Source: Cerne (1993:53).

Most home health care is provided by **home health aides,** who typically have no formal training, or **registered nurses,** who have received at least two years of nursing training and passed national licensure requirements. Aides are overwhelmingly minorities and women and are highly likely to be immigrants. Few receive any job benefits and most receive only minimum wage (Burbridge, 1993). Because the growth in paid home health care is so recent, little more is known regarding these workers or their work.

CONCLUSIONS

In this chapter, I suggested three difficulties inherent in the ways we provide care to those who are physically or mentally ill or disabled. First, I looked at some of the inherent contradictions of trying both to provide care and to make a profit. Health care workers, from medical students to home health aides, laboring long hours under often brutal conditions to keep their employers' costs low, cannot provide the quality of care they might like, and even institutions such as hospices, for whom profit-making is not a primary motive, must contend with the demands of a wider system that emphasizes cutting costs and generating profits. Meanwhile, other institutions, such as nursing homes, board and care homes, and home health agencies, have emerged specifically to make money, relegating caregiving to a secondary priority.

Second, I demonstrated the difficulty of providing individualized care in institutional environments. Almost by definition, large institutions must provide care *en masse,* ignoring individual preferences and desires. Patients must follow rules, schedules, and regimens established for the sake of efficiency, regardless of the impact on patients' quality of life. This tendency to ignore the individual is further reinforced by the fact that it is far cheaper to provide regimented rather than individualized care.

Third, I suggested some of the inherent difficulties of treating health care as an individual or family responsibility rather than a social responsibility. As we have seen, the burdens of caregiving can be enormous. Yet the United States offers little support to those who take on this responsibility. In contrast, other industrialized nations provide far more assistance; both Sweden and Finland, for example, allow parents of sick children to leave work for several months while still receiving most of their salaries and provide free or inexpensive assistance with domestic chores to elderly persons who might otherwise have to turn to family members for assistance (Swedish Institute, 1997, 1999; Zimmerman, 1993).

In sum, the data presented in this chapter regarding the virtual social abandonment of ill and disabled individuals and of their caregivers suggests the low priority this society places on caring for those who are weak or ill, especially if they also are poor. Only when these social values change can we expect the lives of ill persons, disabled persons, or their caregivers to improve significantly.

SUGGESTED READINGS

Annas, George J. 1989. *The Rights of Patients: The Basic ACLU Guide to Patients' Rights.* Carbondale, IL: Southern Illinois University Press. Written by one of America's foremost experts on health law.

Diamond, Timothy. 1992. *Making Gray Gold: Narratives of Nursing Home Care.* Chicago: University of Chicago Press. A riveting account of life in nursing homes, describing the experiences of both residents and nursing assistants.

Rosenberg, Charles E. 1987. *The Care of Strangers: The Rise of America's Hospital System.* New York: Basic. An excellent history of America's hospitals.

Stevens, Rosemary. 1989. *In Sickness and in Wealth: American Hospitals in the Twentieth Century.* New York: Basic. Continues where Rosenberg's book leaves off.

GETTING INVOLVED

National Citizens Coalition for Nursing Home Reform. 1424 16th Street, N.W., Suite 202, Washington, DC 20036-2211. 202-332-2275. www.nccnhr.org. Citizens' action group seeking reform of nursing homes and board and care homes.

National Alliance for Caregiving. 4720 Montgomery Lane, Suite 642, Bethesda, MD 20814. 301-718-8444. www.caregiving.org. Provides information and support to family caregivers of the elderly and to health care providers working in the field. Also collects and disseminates information about the value of family caregiving and the burdens borne by caregivers.

REVIEW QUESTIONS

In what ways were nineteenth century hospitals total institutions?

What led to the development of voluntary hospitals? veterans hospitals? government hospitals? the modern hospital as we know it?

What was the original philosophy of hospices, and why and in what ways has it changed?

What is patient dumping, and why does it occur?

Who uses nursing homes?

What is the difference between nursing homes, board and care homes, and assisted living facilities?

How does the process of commodification affect nursing assistants and nursing home residents?

Why has home care grown? What are the difficulties faced by family caregivers?

INTERNET EXERCISES

Using deja.com or the newsgroups feature on your Internet browser, access postings from the alt.support.alzheimers discussion group. Read a few "threads"—queries and the answers posted to them—to identify some of the issues faced by those who care for persons with Alzheimers. Do discussion groups seem to be effective means of helping family caregivers?

Using the Internet, find three Web sites advertising alternative living facilities. What information would you want if you needed to place a relative in such a facility? Do these Web sites provide the information you would need? What do they leave out? How does each Web site encourage you to believe that their facility would be the best one for your relative?

Health Care Providers and Bioethics

CHAPTER 11
The Profession of Medicine

CHAPTER 12
Other Mainstream and Alternative Health Care Providers

CHAPTER 13
Issues in Bioethics

In this final section, we shift our perspective to health care providers. In Chapter 11, I provide an overview of the history of medicine as a profession and describe how the social position of doctors has changed over time. I also describe how one becomes a doctor, including the nature of medical education and medical culture and the steps involved in building a medical career. Finally, I look at how medical education and medical culture, as well as broader social and cultural factors, affect relationships between doctors and patients.

Although doctors typically are the first persons who come to mind when we think of health care, they form only a small percentage of all health care providers. In Chapter 12, I discuss some of these other providers both within and outside the mainstream health care system, including nurses, pharmacists, midwives, and acupuncturists.

The final chapter in this section and in this book provides a history of bioethics, as well as a sociological account of how bioethics has become institutionalized and of its impact on health care and health research. I will show how issues of power underlie ethical issues and why we need a sociological understanding of bioethics.

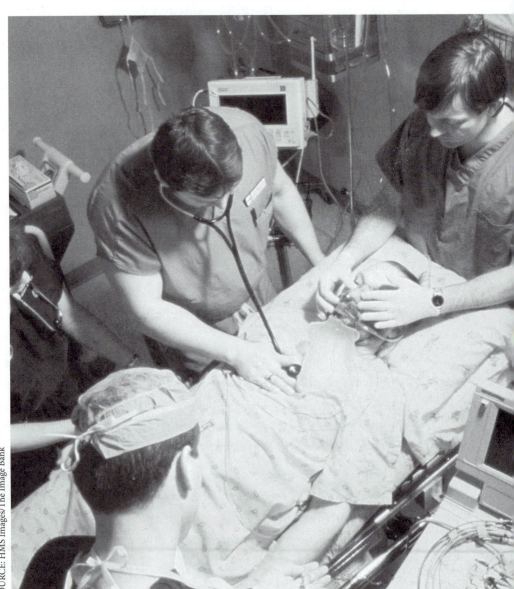

CHAPTER 11

The Profession of Medicine

—

One of the more traumatic events medical students can face early in their training is dog lab. The official purpose of the lab is to demonstrate the physical principles underlying human physiology by working on live dogs. Pressure from medical students who believe these principles could be more ethically taught using laboratory apparatuses has led many schools to drop dog lab or make it optional, as it was at Harvard Medical School when Perri Klass attended. Klass describes how her reactions to the prospect of dog lab evolved during her first weeks in medical school:

> Dog lab is a lab we are all supposed to do in connection with what we have been learning in physiology about the cardiovascular system. Essentially, dog lab means that a group of four medical students will spend six hours performing various experiments on the heart and circulatory system of an anesthetized dog. At the end of the afternoon, the dog is dead, and the medical students, in theory, have a much fuller appreciation of the functioning of the heart. And so I finally do have a decision to make: am I going to do this lab? . . .
>
> "The idea of doing the lab bothers me, that's all." That's what I end up saying to my classmates, to my parents, to friends who don't go to medical school. These friends are invariably horrified by the very idea of the lab ("You dissect it while it's still alive?") and yet somehow I feel their horror is not the same as mine. They make me wonder: have I crossed some dividing line already, that I can consider doing this thing?
>
> I think it is very ugly that this lab should come along so early in my medical education. Why should I be beginning my training with the prolonged killing of an animal? I worry so much, as it is, about the unpleasant and painful things I will see later on in my training and in my career. I am scared that in order to handle them, I will have to become "toughened" in ways I won't like. Isn't there something wrong with starting off by causing pain without any intention to cure?

*And then I worry that this is all hysterical nonsense, that I am pass-
ing up an incredible learning experience because of absurd scruples
about killing a completely anesthetized dog which comes from the pound
and would be killed anyway. I wake up in the middle of the night
dreaming about dogs or surgery; I keep waiting for a dream in which
dogs are operating on me. . . .*

*Meanwhile, at the medical school, the students who are definitely
planning to do the lab are beginning to say to the people who think they
won't, "How are you ever going to be a doctor if you're too sensitive to do
dog lab? How are you going to take it when you're in the hospital?"
Some of the women in my class are particularly vulnerable to this kind
of reasoning; they worry that they aren't tough enough, that they cannot
afford to pass up any opportunity to prove, to themselves and to everyone else, that they have what it takes. In some ways, it is this kind of
thing that finally decides me not to do the lab. It's a confirmation of all
my worst suspicions—this lab is intended to toughen me, to divide me
from ordinary normal people (Klass, 1987, 30–32).*

Klass' story illustrates several of the basic elements of modern-day medicine—its emphasis on scientific experimentation, on emotional detachment
or "toughness," and on doctors as different in some basic way from other
mortals. This chapter looks at how these and other aspects of medical culture
and training evolved and at the consequences for both doctors and their patients, as well as at the history and current status of medicine as a profession.

MEDICINE AS A PROFESSION

American Medicine in the Nineteenth Century

Confronted by disquieting illness, most modern-day Americans seek care
from a doctor of medicine. Little more than a century ago, however, that
would not have been the case. Instead, Americans received most health care
from family members. If they required more complicated treatment, they
could choose from an array of poorly paid and typically poorly respected
health care practitioners (Starr, 1982:31–59). These included both **"regular"**
doctors, the forerunners of contemporary doctors, and such **"irregular"**
practitioners as patent medicine makers, who sold drugs they concocted
from a wide variety of ingredients; botanic doctors, who offered herbal
remedies; bonesetters, who, not surprisingly, specialized in dislocated joints
and fractured bones; and midwives. Regular doctors were also known as
allopathic doctors, or allopaths (from the Greek for "cure by opposites"),

because they sometimes treated illnesses with drugs selected to produce symptoms *opposite* to those caused by the illnesses. For example, allopaths would treat patients suffering the fevers of malaria with quinine, a drug known to reduce fevers, and treat patients with failing hearts with digitalis, a drug that stimulates the heartbeat. Their main competitors were **homeopathic doctors,** or homeopaths (from the Greek for "cure by similars"). Homeopaths treated illnesses with drugs that produced symptoms *similar* to those caused by the illnesses—treating a fever with a fever-producing drug, for example. Although in retrospect the homeopathic model might seem odd, it drew on the same logic as smallpox inoculation, the one successful inoculation available at that time: people who were inoculated with a small quantity of cowpox cells, and who therefore developed a mild form of cowpox, somehow became immune to the related but far more serious smallpox. Homeopaths therefore concluded that patients who received a small quantity of a drug that mimicked the symptoms of a given illness would become better able to resist that illness. At any rate, although homeopathy probably did not help patients except through a **placebo** effect, it also did not directly harm them.

That Americans before the twentieth century placed no greater trust in allopathic doctors than in any others who claimed knowledge of healing should not surprise us. Although, by the nineteenth century, science—the careful testing of hypotheses in **controlled** experiments—had infiltrated the curricula of European medical schools, where many of the wealthiest or most dedicated Americans trained, it had gained barely a foothold in U.S. medical schools. Moreover, the United States licensed neither doctors nor medical schools (Ludmerer, 1985). Instead, and until about 1850, most doctors trained through apprenticeships lasting only a few months. After that date, most trained at any of the multitude of uncertified medical schools that had sprouted around the country, almost all of which were private, for-profit institutions, unaffiliated with colleges or universities and lacking any entrance requirements beyond the ability to pay tuition (Ludmerer, 1985). Nor were standards stricter at the few university-based medical schools. Thus, for example, in 1871, Henry Jacob Bigelow, a Harvard University professor of surgery, could protest a proposal to require written graduation examinations on the grounds that more than half of Harvard's medical students were illiterate (Ludmerer, 1985:12). Training averaged far less than a year and depended almost entirely on lectures, so that almost no students ever examined a patient, conducted an experiment, or dissected a cadaver. Any student who attended the lectures regularly received a diploma. This situation only began to change significantly in the better university schools in the 1890s, while other schools with weaker standards continued to function.

Lacking scientific research or knowledge, allopathic doctors developed their ideas about health and illness either from their clinical experiences with patients or by extrapolating from abstract, untested theories. The most popular theory of illness, from the classical Greek era until the mid 1800s, traced illness to an imbalance of bodily "humors," or fluids. Doctors had learned through experience that ill persons often recovered following episodes of fever, vomiting, or diarrhea. From this, doctors deduced—in part correctly—that fever, vomiting, and diarrhea helped the body restore itself to health. Unfortunately, lacking methods for testing their theories, doctors carried these ideas too far, often inducing life-threatening fever, vomiting, purging, and blood-letting. Consider, for example, the following description of how Boston doctors in 1833 used what was known as **"heroic medicine"** to treat a pregnant woman who began having convulsions a month before her delivery date:

> The doctors bled her of 8 ounces and gave her a purgative. The next day she again had convulsions, and they took 22 ounces of blood. After 90 minutes she had a headache, and the doctors took 18 more ounces of blood, gave emetics to cause vomiting, and put ice on her head and mustard plasters on her feet. Nearly four hours later she had another convulsion, and they took 12 ounces, and soon after, 6 more. By then she had lapsed into a deep coma, so the doctors doused her with cold water but could not revive her. Soon her cervix began to dilate, so the doctors gave ergot to induce labor. Shortly before delivery she convulsed again, and they applied ice and mustard plasters again and also gave a vomiting agent and calomel to purge her bowels. In six hours she delivered a stillborn child. After two days she regained consciousness and recovered. The doctors considered this a conservative treatment, even though they had removed two-fifths of her blood in a two-day period, for they had not artificially dilated her womb or used instruments to expedite delivery (Wertz and Wertz, 1989:69).

As this example suggests, because of the body's amazing ability to heal itself, even when doctors used heroic medicine, many of their patients survived. Thus, doctors could convince themselves they had cured their patients when in reality they either had made no difference or had endangered their patients' lives.

By the second half of the nineteenth century, most doctors, responding to the public's support for irregular practitioners and fear of heroic medicine, had abandoned the most dangerous of their techniques. Yet medical treatment remained risky. Allopathic doctors' major advantage over their competitors was their ability to conduct surgery in life-threatening situations. Unfortunately, until the development of anesthesia in the 1860s, many patients died from the inherent physical trauma of surgery. In addition, many died unnecessarily from postsurgical infections. Dr. Ignaz Semmelweis had demonstrated in the 1850s that because midwives (whose tasks included washing floors and linens) had relatively clean hands, whereas doctors routinely went without washing their hands or surgical instruments from autopsies to obstetrical examinations and from patient to patient, more childbearing women died on medical wards than on midwifery wards. Yet it took another 30 years before his ideas became incorporated into standard medical practice.

Until well into the twentieth century, then, doctors could offer their patients little beyond morphine for pain relief, quinine for malarial and other fevers, digitalis for heart problems, and, after 1910, salvarsan for syphilis—each of which presented dangers as well as benefits. According to the 1975 edition of *Cecil's Textbook of Medicine*, one of the most widely used medical textbooks, only 3 percent of the treatments described in the first edition, published in 1927, provided fully effective treatment, whereas 60 percent were harmful, of doubtful value, or offered only symptomatic relief (Beeson, 1980). Doctors' effective pharmacopeia did not grow significantly until the development of antibiotics in the 1940s.

Beginnings of Medical Dominance

Despite the few benefits and many dangers inherent in allopathic medical care, doctors had eliminated most of their competitors and gained control over health care by about 1900 (Starr, 1982:79–112). In this section, we will see how this change came about.

From its inception in 1847, the **American Medical Association (AMA)** had worked to restrain the practices of other health care occupations. State by state, the AMA fought to pass laws outlawing their competitors or restricting them to working only under allopathic supervision or to performing only certain techniques, such as spinal manipulation.

Most of these efforts met with little success initially, for nineteenth century Americans considered health care an uncomplicated, domestic matter, unrelated to science and not requiring complex training (Starr, 1982:90–92). By the beginning of the twentieth century, however, as improvements in public health and in living conditions ended scourges such as cholera and typhoid and as Americans began reaping practical dividends from scientific advances such as electric lights and streetcars, public faith in science swelled. As a result, Americans increasingly defined health care as a complex matter requiring expert intervention, assumed the superiority of "scientific" medicine, and turned to allopathic doctors for care (Starr, 1982:127–142).

Like the public, homeopaths and botanic eclectics (allopathic doctors' two major groups of competitors) also had come to recognize the benefits of science and therefore to realize that a lack of scientific foundation would soon doom their fields. However, they still received considerable popular support. Moreover, because, like allopaths, most were white men, homeopaths and botanic eclectics by and large held social statuses similar to allopaths. As a result, homeopaths and botanic eclectics retained sufficient influence to pressure allopaths to accept them into medical schools and licensing programs, eventually merging their fields into allopathic medicine.

Other health care workers could bring far less power to their dealings with legislators and with allopathic doctors. Newly emerging occupations such as chiropractic (described in the next chapter) lacked the longstanding history of popular support that had allowed homeopaths to push for incorporation with allopathy. Older occupations, meanwhile, such as midwives and herbalists, lacked the social status, power, and money needed to fight against doctors' lobbying because most practitioners were women or minorities and were therefore assumed to be incompetent by both legislators and doctors (Starr, 1982:117,124).

The Flexner Report and Its Aftermath

These differences between allopathic doctors and other health care practitioners increased during the early years of the twentieth century. Since the 1890s, the better medical schools had begun tightening entrance requirements, stressing higher academic standards, emphasizing research and science, and offering clinical experience. These changes placed pressures on the other medical schools to do the same. Those pressures increased following the publication in 1910 of the **Flexner Report** on American medical education (Ludmerer, 1985:166–190). The Report, which was written by Abraham Flexner and commissioned by the nonprofit Carnegie Foundation at the behest of the AMA, shocked the nation with its descriptions of the lax requirements and poor facilities at many medical schools. The Flexner Report increased the pressures on all medical schools to improve their programs and accelerated the process of change that was already underway. In the next few years, responding to pressure from both the public and the AMA, all fifty states adopted or began enforcing stringent licensing laws for medical schools (Ludmerer, 1985:234–249). These laws hastened the closure of all proprietary and most nonprofit schools, many of which were already suffering financially from the costs of trying to meet students' growing demand for scientific training. As a result, the number of medical schools fell from 162 in 1906 to 81 in 1922 (Starr, 1982:118, 121).

The Flexner Report, in conjunction with the changes already underway in medical education, substantially improved the quality of health care available to the American public and paved the way for later advances in health care. However, these changes in medical education also had some more problematic results. The closure of so many schools made medicine as a field even more homogeneous. Only two of the seven medical schools for African Americans survived and only one of the seven schools for women (Ludmerer, 1985:248; Starr, 1982:124). In addition, because the university schools set stricter educational prerequisites than had the defunct proprietary schools, few immigrants, minorities, and poorer whites could meet their entrance requirements. Even fewer could afford the tuition required by scientifically oriented university programs. Moreover,

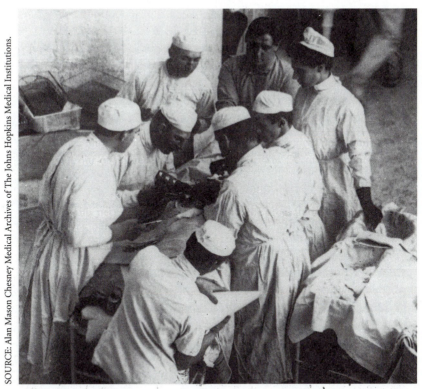

SOURCE: Alan Mason Chesney Medical Archives of The Johns Hopkins Medical Institutions.

In the 1890s, even surgery at a major hospital such as Johns Hopkins required no advanced technologies.

until well into the 1900s, many programs openly discriminated against women, African Americans, Jews, and Catholics. As a result, even though the technical quality of medical care increased, fewer doctors were available who would practice in minority communities and who understood the special concerns of minority or female patients. At the same time, simply because doctors were now more homogeneously white, male, and upper class, their status grew, encouraging more hierarchical relationships between doctors and patients.

Doctors and Professional Dominance

By the 1920s, doctors had become the premiere example of a **profession** (Parsons, 1951). Although definitions of a profession vary, sociologists generally define an occupation as a profession when it meets three characteristics. First, it must have the autonomy to set its own educational and licensing standards and to police its members for incompetence or malfeasance. For example, doctors, rather than consumers, make up the licensing boards that judge doctors accused of incompetence. Second, it must have its own technical, specialized knowledge, learned through extended, systematic training. Unlike nurses, for example, whose knowledge base largely comes from research conducted by others, doctors are generally believed to make decisions about care based primarily on medical research, theory, and experience. (In fact, much of this research is actually conducted by persons with science rather than medical degrees.) Third, it must be believed by the public to follow a code of ethics and to work more from a sense of service than a desire for profit. Such beliefs characterized public views of medicine for most of the twentieth century.

As the leading profession in the health care world, doctors enjoyed—and to some extent still enjoy—an unusually high level of **professional dominance.** Professional dominance refers to a profession's freedom from control by other occupations or groups and ability to control any other occupations working in the same economic sphere. This concept has been most fully analyzed by Eliot Freidson (1970a, 1970b, 1994). As Freidson has noted, for much of the twentieth century, most doctors worked in private practice (whether solo or group), setting their own hours, fees, and other conditions of work. Those who worked in hospitals or clinics were typically supervised by other doctors, not by nonmedical administrators. Although doctors often supervised members of other occupations, the reverse has started to happen only in the last two decades or so. Similarly, both in the past and currently, doctors often served on boards charged with judging the education and qualifications of other health care occupations, but members of other occupations played little role in setting standards for medical education and licensing. This high level of professional dominance by doctors—otherwise known as **medical dominance**—stemmed from the public's great respect for doctors' claims to a scientific knowledge base and service orientation. This respect in turn was bolstered by active lobbying by organized medicine.

The Problem of Medical Errors

The extent of medical dominance is perhaps best demonstrated by the treatment of medical errors. Doctors' actions can mean the difference between life and death for patients. Yet, despite the obvious vested interests of the public and the state in evaluating the quality of medical care, responding to doctors' errors has remained almost solely a prerogative of doctors.

Although doctors' explicit emphasis on a code of ethics and service orientation might lead one to assume that doctors would strive to ferret out their incompetent peers, such is only rarely the case. In his observations of doctors in group practices, Eliot Freidson (1975) found that few doctors know enough about their colleagues' actions to identify incompetent colleagues. Doctors rarely see each other's patients and typically communicate about patients only through perfunctory notes or telephone calls. Moreover, when doctors do communicate about errors, professional etiquette demands that they **normalize** those errors, emphasizing how such problems could happen to anyone. This tendency is heightened by the economics of group and specialty practice (which together now account for the vast majority of doctors). In such practices, doctors obtain their business more through referrals from colleagues than through word of mouth among patients. As a result, doctors have more of a vested interest in maintaining good relations with colleagues than with patients. In addition, medical values reinforce doctors' discomfort with judging their colleagues' actions by stressing that doctors should base their medical decisions more on their clinical experience than on results from scientific studies (Becker et al., 1961; Bosk, 1979; Freidson, 1970b; Knafl and Burkett, 1975). (This emphasis on clinical experience and judgment is discussed more fully later in this chapter.)

Both Charles Bosk (1979) and Marcia Millman (1976) reached conclusions similar to Freidson's from their observations of hospital **residents** and **attendings.** Residents (sometimes known as **house staff**) are doctors who have graduated medical school but now must continue their training at a hospital for several more years before they can enter practice; attendings are doctors who supervise residents and who are either employed by hospitals or in private practice. Unlike those studied by Freidson, residents and attendings worked together closely and therefore could not help learning if a colleague was incompetent. Yet professional etiquette, coupled with the need to maintain good relations with colleagues, almost always led these doctors to dismiss residents' errors as a natural result of inexperience and to dismiss attendings' errors as an unavoidable part of the job. Doctors might suggest subtly to a colleague that he or she lacked the necessary experience to do a particular operation or that the risks of an operation outweighed the potential benefit for a given patient, but—except in occasional circumstances when attendings restrained residents—they would do nothing further to stop a doctor from operating.

Similarly, both Bosk and Millman found through their observations that medical mortality conferences (weekly sessions in which doctors discuss the treatment of selected hospital patients who have died) seem designed more to downplay medical errors than to prevent them. As Millman describes it:

> A mortality and morbidity conference for doctors bears some resemblance to a wedding or a funeral for members of a family. In all these ceremonies there is some feeling among those who attend that tact and restraint must be exercised if everyone is to leave on friendly terms. . . . As one after another of the staff testifie[s] about how they were led to the same mistaken diagnosis, a convincing case for the justifiability of the error is implicitly presented and the responsibility for

the mistake is spread so that no one doctor is made to look guilty. . . . Responsibility for the error is also neutralized by making much of unusual or misleading features of the case, or showing how the patient was himself to blame, because of uncooperative or neurotic behavior. Furthermore, by reviewing the case in fine detail the doctors restore their images as careful, methodical practitioners and thereby neutralize the actual sloppiness and carelessness made obvious by the mistake (Millman, 1976:96–98).

The doctors further safeguard the cordiality of the conferences by inviting only the medical staff and not the family or other hospital staff and by selecting for review only cases in which the patients would have died regardless of the doctor's actions or in which there is no consensus nationally regarding how best to treat a given condition. In addition, the doctors downplay the seriousness of errors by framing the conferences as educational rather than investigative or punitive. Finally, the rituals of the medical mortality conferences require those who have made errors to admit their errors publicly and declare that they have learned from them, while requiring other doctors to respond sympathetically that these things happen to everyone. All these strategies help doctors to normalize medical errors (Bosk, 1979; Millman, 1976).

The Decline of Medical Dominance

As the example of medical error shows, for much of the twentieth century, doctors enjoyed an unusually high level of professional dominance. One of the most heated debates currently within the sociology of health and illness is the extent to which this professional dominance has declined (Freidson, 1994; Light and Levine, 1988; Starr, 1982:379–393).

Foremost among those who argue that professional dominance has declined are Marie Haug, John McKinlay, and John Stoeckle (Table 11.1). They differ, however, in where they locate the sources of this decline, with Haug (1988) focusing on changes in public sentiment and access to medical knowledge and McKinlay and Stoeckle (1989) on changes in health care financing and organization.

Changing Patient Attitudes and Deprofessionalization

In her writings, Marie Haug has focused on how the civil rights and feminist movements of the 1960s and 1970s increased popular emphasis on rights rather than duties and on questioning rather than obeying authorities (Starr, 1982:379–393; Haug, 1988). At the same time, Haug argues, the general rise in educational levels and in public access to medical information has helped patients to evaluate their symptoms and treatment for themselves and to challenge their doctors' diagnoses and decisions about care. These changes, coupled with growing public awareness of how un-

Table 11.1	*Divergent Views on Medical Dominance*
THE PROFESSIONAL DOMINANCE MODEL	DECLINE OF DOMINANCE MODELS
	A. *Deprofessionalization*
High level of prestige	Decline in public confidence and respect
Public defers to medical judgment and feels loyalty to their doctors	Public questions medical judgment and feels little loyalty to doctors
	B. *Proletarianization*
Doctors hold strong economic position	Doctors become economically vulnerable and AMA power declines.
Doctors set own working conditions	Doctors' working conditions set by corporate employers
Only doctors supervise doctors	Doctors supervised by nonmedical administrators and review boards
Doctors supervise and control other health care occupations	Other health care occupations gain considerable independence from medical control
Doctors act solely or largely based on their clinical judgment	Doctors' clinical autonomy constrained by corporate or governmental guidelines

questioning obedience to doctors sometimes can harm patients' health, helped foster both the feminist health movement and the patients' rights movement. These movements both reflected and created more egalitarian ideas about how doctors and patients should interact.

These new popular health movements have stimulated major changes in medical practice, ranging from the sharp decrease in use of general anesthesia during childbirth to the routine use of informed consent forms before patients receive experimental drugs. More broadly, through publications such as *Your Medical Rights: How to Become an Empowered Consumer* (Inlander and Pavalon, 1990), *Medical Self-Care* (Ferguson, 1980), and the best-selling *Our Bodies, Ourselves for the New Century* (Boston Women's Health Book Collective, 1998), these movements have encouraged consumers to take charge of their own health, to use practitioners other than doctors, and to obtain second opinions when they do go to doctors. The rise of the Internet has added impetus to this movement, giving consumers instant access to vast numbers of others who share their concerns and to vast quantities of medical literature, including literature on alternatives to allopathic medicine. Even the federal government is supporting this trend: in 1997, the U.S. Department of Health and Human Services opened a Web site (www.healthfinder.org) to give consumers online access to publications, clearinghouses, databases, other Web sites, self-help groups, government agencies, and nonprofit organizations related to both allopathic and alternative medicine; more than three million consumers used the service during 1998 (personal communication, U.S. Department of Health and Human Services, 1999).

The explosive rise during the 1980s in malpractice suits against doctors further suggests the public's loss of confidence in doctors. A recent survey of obstetrician-gynecologists (the specialty with the second-highest rate of lawsuits) found that 73 percent had been sued for malpractice at least once during their careers, with an average of 2.3 lawsuits per person (American College of Obstetrician-Gynecologists, 1998). In turn, these lawsuits have encouraged health care institutions, fearing expensive legal battles, to assert more control over the doctors who work for them. Fear of lawsuits has also encouraged doctors to change their own behaviors and to subordinate their clinical decision-making in favor of **defensive medicine**—doing tests or procedures solely or primarily to reduce their risk of a malpractice suit (American College of Obstetrician-Gynecologists, 1998; Tussing and Wojtowycz, 1997). For example, a recent study found that, after researchers controlled for a variety of maternal, physician, clinical, and other characteristics, doctors practicing in counties with high rates of malpractice suits (in which doctors' fear of malpractice suits would likely be greater) had significantly higher than average rates of cesarean deliveries. Fear of malpractice explained 24 percent of all cesarean deliveries in the study (Tussing and Wojtowcyz, 1997).

Taken together, these changes led Haug (1988) to conclude that doctors are becoming **deprofessionalized,** or losing the public confidence that defines professions. This concept gains credence from national polls showing that the proportion of Americans who place a "great deal of confidence in people in charge of running medicine" dropped from 73 percent in 1966 to 42 percent in 1976 and to 29 percent in 1997 (*Harris Poll,* 1997b:2).

The Changing Structure of Medicine and Proletarianization

In contrast, McKinlay and Stoeckle (1989) agree that medical dominance has declined but trace that decline to changes in health care financing. They maintain that doctors have lost substantial control over the most important professional prerogatives: deciding who may enter the profession and how; setting the conditions under which one works; owning one's own tools and workspace; and maintaining an individual relationship with freely chosen patients. As a result, McKinlay and Stoeckle conclude, doctors are becoming workers (or "proletarians") rather than autonomous professionals. They refer to this shift as **proletarianization** and trace it to three factors: the rise of corporatization, the growth of government control, and the decline of the AMA.

The Rise of Corporatization

McKinlay and Stoeckle (1989) begin their argument by noting that, before the 1960s, nonprofit or government agencies owned most hospitals and other health care institutions. With the initiation of **Medicare** and **Medicaid**, however, the potential for profitmaking in health care expanded tremendously, encouraging for-profit corporations to enter the field, as Chapter 8 described (Starr, 1982:428–32). During the last three decades, investor-owned corporations have purchased or developed a growing number of health care institutions. Moreover most of these corporations own not one institution but several, distributed regionally or even nationally; for example, in 1994 Columbia/HCA Healthcare bought both Healthtrust Inc. and Medical Care America, giving Columbia/HCA control of 311 hospitals and 125 outpatient centers, and 1999 Aetna Inc. bought Prudential Health Care, giving Aetna control over insurance benefits of almost 10 percent of all Americans. In addition corporations increasingly have shifted from **horizontal integration** (owning multiple institutions providing the same type of service) to **vertical integration** (owning multiple institutions providing different types of services, such as both **nursing homes** and pharmaceutical companies).

This growth of corporate medicine or **corporatization**, occurred at a time when doctors were experiencing increasing economic vulnerability (McKinlay and Stoeckle, 1989; Starr, 1982:446–448). Since the early 1960s, the supply of doctors has grown rapidly, more than doubling between 1970 and 1998 and far surpassing the ratio in most industrialized nations (American College of Physicians, 1999). The supply of doctors now exceeds demand in the most desirable communities and specialties. For example, among doctors who completed residencies in 1996, 56 percent of those in critical care medicine and 47 percent in anesthesiology (but only 7 percent in family practice) reported difficulty finding employment (Bodenheimer, 1999). Supply is expected to continue to increase until 2020, despite recent federal legislation to reduce funding for specialty training.

Because of the current oversupply, newly graduated doctors sometimes find the competition too great to enter private or small group practice, especially because most doctors feel compelled to earn substantial incomes immediately to pay off their medical school debts. Consequently, economic motivations have encouraged an increasing proportion of doctors during the last three decades to accept employment with hospitals, large group practices, **managed care** organizations (MCOs), or other corporate institutions. Others, especially women with children, have more freely chosen corporate employment because they prefer its more relaxed lifestyle and shorter, more predictable hours. A total of 43 percent of doctors worked as paid employees in 1997, compared with 24 percent in 1983 (Bodenheimer, 1999).

As employees of salaried or group practices, whether by choice or necessity and whether in small groups or in corporate-owned hospitals, doctors' autonomy has diminished. Fearing that, left to their own devices, doctors will overuse available resources and drive up costs, administrators now make many decisions formerly made by individual doctors. According to McKinlay and Stoeckle (1989:192), "doctors have slipped down to the position of middle management . . . , [while administrators are] organizing the necessary coordination for collaborative work, the work schedules of staff, the recruitment of clients to the practice, and the contacts with third-party purchasers, and determining the fiscal rewards." In addition, administrators now may such basic conditions of work as how many patients a doctor must see hour. Although administrators are often themselves doctors, many of view management as a career track from the start, obtaining management credentials and accepting management positions directly after medicine without ever entering medical practice (Montgomery, 1992). Consequently, the loyalties and identities of these physician administrators lie more with administration than with medicine.

Even those who not work directly for corporations now often find that the only way can get patients is to sign contracts with MCOs. These contracts also substantially limit doctors' autonomy by controlling the fees they may charge for their services and, most important, by scrutinizing their clinical decision-making and interactions with patients. Perhaps most important, some MCOs require doctors to sign contracts that forbid them from discussing with their patients any treatments that the MCO does not approve; laws against this practice are currently under discussion in several states and at the federal level. Prospective reviews, too, restrain doctors' autonomy. A recent study based on a national **random** sample of doctors found that MCOs prospectively reviewed length of hospital stay for 59 percent of the surveyed doctors' patients, site of care for 45 percent, and appropriateness of treatment for 39 percent (Remler et al., 1997). Although the MCOs rejected doctors' recommendations in less than 6 percent of cases, and half of these rejections were rescinded following appeal, the knowledge that their decisions will be reviewed leaves doctors feeling less autonomous and trusted and can lead some to avoid making recommendations they expect will be denied. Moreover, doctors often resign themselves to accepting the MCOs' initial decisions because they feel they cannot afford the time and energy it would take to appeal. Recently, however, doctors' dissatisfaction with prospective review, coupled with increasing doubts about its cost-effectiveness, has encouraged MCOs to rely less on prospective review and more on subtler means of controlling doctors' behavior.

Similarly, doctors increasingly find their autonomy constrained by "**practice protocols**," which establish norms of care for particular medical conditions under particular circumstances based on careful review of clinical research. There are now hundreds of these protocols, developed both by

MCOs concerned about the costs of unnecessary treatment and by physician groups concerned about the rise in malpractice suits and premiums (Good, 1995; Millenson, 1997). As early as 1995, a study found that 16 percent of patients seen by a national random sample of doctors had conditions for which care was set by practice protocols (Remler et al., 1997).

Moreover, both within corporate institutions and under managed care contracts, the nature of the doctor-patient relationship, and thus the power of doctors within that relationship, has changed. Doctors no longer have "their" individual patients, but now must see whatever patients their employers or MCOs assign to them. Conversely, even patients who continue to have a primary caregiver feel less loyalty to that doctor because they often see whatever doctor happens to be available when they need care.

McKinlay and Stoeckle (1989) additionally argue that doctors' power relative to other health care occupations has declined. Many health care institutions, including MCOs, now believe they can limit costs without limiting quality by hiring cheaper, allied health personnel (such as radiation technologists or nurse practitioners) to perform specialized tasks once performed by doctors. In many states, specially trained nurses can prescribe certain drugs, and in some states pharmacists can. Similarly, patient management now officially belongs to the health care team, in which allied specialists often have more knowledge of specialized tasks than do doctors. As a result, they argue, doctors' power to control the work of ancillary personnel has declined.

The Growth of Government Control

Increasingly, too, government regulations restrict doctors' professional autonomy. The same federal programs, such as Medicaid and Medicare, that made health care a lucrative field for corporations made it an expensive one for the government, as doctors discovered that no matter how many tests or procedures they performed, the government would pay. As a result, the government now has a large vested interest in controlling health care costs. Consequently, the government has established various programs aimed at limiting costs without limiting quality, such as the **DRG** system (described in Chapter 8), which establishes pre-set fees according to diagnosis for hospital care under Medicare (and, in some states, Medicaid). To protect their financial security, hospitals may cut the wages or terminate the contracts of doctors who consistently exceed DRG limits, thus pressuring all doctors in their employ to stay within those limits (Dolenc and Dougherty, 1985). As a result, doctors sometimes conclude that they have only two choices: to misreport a patient's diagnosis on the DRG forms so they can justify more expensive treatments or to ignore their clinical judgment about the treatment a patient needs so they can stay within the DRG limits.

Even doctors who do not work *for* hospitals are constrained by DRGs whenever their patients are hospitalized and the doctors must treat them *in* hospitals. In such situations, these doctors also face reviews by DRG boards, although the sanctions against them come from peer disapproval rather than from economic pressure. In addition, insurance companies can use publicly available DRG data on individual doctors' billing practices to decide which doctors to contract with (although, as Chapter 8 described, in many states doctors have won passage of laws requiring insurers to contract with any doctors willing to abide by the insurers' regulations).

Doctors' autonomy to set their own fees also has been constrained since 1992 by the federal **resource-based relative value scale (RBRVS).** Whereas DRGs were designed to control Medicare spending on hospital care, RBRVS is designed to control spending on doctors' bills.

RBRVS is a complex formula for determining appropriate compensation under Medicare for medical care, based on estimates of how much it actually costs to provide specific services in specific geographic areas (Dunn et al., 1988; Hsiao et al., 1988; Sigsbee, 1997). In addition, RBRVS provides extra funding to those who work in designated, underserved areas. Under this system, incomes of most specialists have declined, while those of generalists (other than pediatricians, who receive no Medicare funds) have increased. Its impact has been far broader than one might expect, given that Medicare provides only a small portion of most doctors' income, because most other public and private insurance plans also have adopted RBRVS. As a result, RBRVS, in the words of one observer, has become a "de facto national fee schedule" (Sigsbee, 1997).

The Decline of the AMA

Finally, McKinlay and Stoeckle (1989) argue that doctors' professional dominance has declined because the power of the AMA has declined. Although the AMA remains one of the most powerful lobbying groups in the country, both internal and external changes threaten its position. Externally, the AMA now often finds its power counterbalanced by the power of health care organizations. Evidence for this can be found in the spending patterns of the various political action committees (PACs)—federally recognized organizations that solicit contributions from individuals, associations, and corporations and distribute this money to candidates for election who support the PACs' political agenda. The AMA still controls one of the largest pools of PAC lobbying money in the country and the largest among the health care PACs (Center for Responsive Politics, www.opensecrets.org/home/index.asp, 1999). During the 1997–1998 election cycle, the AMA contributed more than $2 million to federal candidates. This sum is dwarfed, however, by the $25 million contributed by PACs representing other health professionals, pharmaceutical companies, and hospitals, all of whose legislative interests sometimes compete with those of the AMA.

Similarly, whereas in the past the AMA and the doctors it represented had nearly free rein to set both admissions criteria and curricula of medical schools, this freedom has eroded substantially. Legal changes and social pressures stemming from the civil rights and feminist movements forced medical schools beginning in the 1960s to acknowledge the rights of women and minorities to enter medicine and of foreign-trained doctors to gain access to U.S. licensing. Ironically, over time medical schools came to value having a more diverse student population, but recent court decisions have forced them to change admissions procedures in ways that could restrict minority enrollment. In addition, increased government, corporate, and foundation financing of medical training beginning in the 1960s has given these outside groups increased power to direct the nature of training, through choosing which educational programs to fund.

At the same time, the AMA has suffered internally from declining support among doctors. As of 1999, 43 percent of doctors belong to the AMA, compared with about 82 percent in 1962; membership is especially low among women, who as of 1997 comprised 43.5 percent of medical school students (Barzansky et al., 1998; Bodenheimer, 1999). Instead, some doctors have chosen to join more liberal organizations that often actively oppose the AMA, such as Physicians for Social Responsibility and Physicians for a National Health Plan, and others have abandoned the AMA for specialty organizations like the American College of Obstetrician-Gynecologists.

For all these reasons, then, McKinlay and Stoeckle (1989) argue that doctors are experiencing proletarianization. This conclusion is supported by the AMA's 1999 decision to end its long-standing opposition to unionizing doctors in order to increase doctors' bargaining power relative to MCOs (Greenhouse, 1999). About 40,000 doctors now belong to large, cross-occupational unions such as the American Federation of State, County and Municipal Employees.

The Continued Strength of Medical Dominance

Not all sociologists, however, agree that medical dominance has declined significantly. Some, such as Eliot Freidson (1984, 1994), argue that even though professional dominance has declined since its high point in the middle of the twentieth century, it remains strong. As Paul Starr notes, health care corporations depend on doctors both to generate profits and to control costs (1982:446). As a result, these corporations retain a vested interest in maintaining good relationships with the doctors who work for them. Consequently, the corporations continue to give doctors considerable autonomy in day-to-day clinical matters. As noted earlier, MCOs have moved away from prospective review and rarely reject doctors' treatment recommendations anyway (Remler et al., 1997). Similarly, although corporations increasingly hire professional managers as chief executive officers of health care institutions, they often also hire doctors as medical directors to work directly under these managers, as well as in a wide range of other administrative positions.

Freidson (1985, 1986) refers to this process as the **restructuring** of the profession of medicine into specialties organized not by clinical territory (for example, oncologists to treat cancer, pediatricians to treat children) but by functional sector: the producers who work in clinical practice, the knowledge elite who work in research or academia, and the administrative elite. Through restructuring, Freidson argues, medicine has retained control of critical areas of professional status (such as setting licensure regulations and practice standards) and thus preserved the dominance of the profession as a whole, even if the autonomy of individual physicians has eroded.

Moreover, Freidson argues, although individual doctors working in specific situations have lost some professional prerogatives, the power and dominance of doctors relative to other health care occupations have remained largely intact; the rhetoric of health care "teams," for example, greatly understates the power of doctors on these teams. By the same token, the use of medical technology by ancillary occupations tells us little about the relative power of those occupations, for medical innovations always have moved down the occupational scale over time. For example, nurses for some time have used stethoscopes and blood pressure cuffs without this either reflecting or causing any increase in their power relative to doctors. Similarly, although the rise of practice protocols could decrease the autonomy of individual doctors, supporters of protocols argue that only through such self-regulation can medicine preserve public faith and, in the end, its professional autonomy (Good, 1995).

Finally, although the environment within which physicians now practice medicine has changed considerably, they retain considerable ability to manipulate this new environment and thus retain much of their clinical autonomy. For example, and as mentioned earlier, many doctors now match the DRG system to their clinical decisions rather than changing their decisions to match the DRG system. Similarly, evidence suggests that physicians in certain circumstances have sufficient power simply to ignore bureaucratic directives that they find onerous. For example, interviews conducted with a random sample of hospital-based doctors in 1989 suggested that hospital policies regarding how doctors should obtain and use HIV tests for patients were only weakly correlated with doctors' actual behaviors (Montgomery, 1996). By the same token, doctors have proved surprisingly adept at maintaining their incomes because doctors, not consumers, largely control demand for medical services. Thus, doctors with shrinking patient pools can order more tests or treatments for their remaining patients or can expand the areas they define as suitable for medical intervention, as the discussion in Chapter 5 on **medicalization** explained. In addition, doctors have found that they can maintain their incomes and autonomy by performing elective procedures outside of hospitals, especially procedures typically paid for out of pocket such as cosmetic surgery, because such procedures are largely free of oversight by insurance, government, or hospital bureaucrats. Partly as a result, 26 percent of all surgeries took place in doctors' offices in 1999 compared with 5 percent of surgeries in 1981 (Zuger, 1999).

The continuing power of medicine as a profession is also demonstrated in the ongoing struggles, described in Chapter 8, to place legal limits on MCOs' control over doctors. Some of the legislative proposals typically described as "patients' bills of rights" might be more accurately characterized as "doctors' bills of rights." For example, during 1999 Congress discussed proposals to allow women to use obstetrician-gynecologists as their primary care doctors rather than the internists or family doctors that MCOs prefer. Although many women are indeed accustomed to using obstetrician-gynecologists for primary care, logic would suggest that women would be better served by using doctors who have trained broadly in primary care rather than doctors whose training has focused on surgical management of one set of bodily organs.

For these reasons, although Freidson's earlier model of professional dominance certainly needs modification, it remains a useful starting point for understanding the current status of medicine as a profession.

BECOMING A DOCTOR

Despite the assaults on medical dominance, becoming a doctor remains an attractive option, offering public prestige, the emotional rewards of service, and financial rewards far greater than most other professions (Gonzalez, 1999). Although applications to medical school declined during the 1980s, they have risen steadily for the last few years, from 26,721 applicants in 1989 to 46,968 applicants in 1996 (Barzansky et al., 1998). In the remainder of this chapter, we look at the process through which individuals become doctors and the consequences of medical training and medical culture for both doctors and patients.

The Structure of Medical Education

Becoming a doctor is not easy. Prospective doctors first must earn a bachelor's degree and then complete four years of training at a medical school. Before they can enter practice, however, and depending on their chosen specialty, they must spend another three to eight years as residents. (The term "intern," referring to the first year of a residency, is no longer commonly used.) As a result, most do not enter practice until age 30.

For most students, becoming a doctor entails incurring substantial financial debts. Most medical students (83 percent) acquire debts during medical school, with an average debt in 1998 (in addition to any debts incurred during their undergraduate training) of $85,619 (Cohen, 1998).

Becoming a doctor also carries tremendous time costs. An official investigation conducted in 1998 found that 37 percent of medical residents in

New York State and 94 percent in New York City work more than 85 hours per week, while 20 percent work more than 95 hours a week; hours for surgical residents are even higher (Kennedy, 1998). Yet statewide regulations adopted following the death in 1989 of a patient treated by exhausted residents supposedly limit residents to working "only" 80 hours per week. Even after graduation, doctors work an average of 58.3 hours per week (Gonzales, 1999). These time pressures, coupled with the financial pressures of training, encourage novice doctors to defer marriage, children, and other personal pursuits and to choose specialties requiring less training over those they otherwise might prefer.

Because of the financial costs, medicine has remained largely a privileged field, with most medical students coming from the middle and upper classes. However, medicine increasingly has opened to women and non-whites (Barzansky et al., 1998), although recent court decisions restricting affirmative action may reverse this trend. As of 1998, minorities constitute about one-quarter of the U.S. population but only about one-third of first-year medical students (Table 11.2). Women comprise 43.5 percent of first-year students, up from 28.9 percent in 1980. Gender does not affect acceptance rates of medical school applicants, but acceptance rates are somewhat higher among Native Americans and Hispanics than among non-Hispanic whites or African Americans (American Association of Medical Colleges, 1999).

Learning Medical Values

During their long years of training, doctors learn both a vast quantity of technical information and a set of **medical norms**—expectations about how doctors should act, think, and feel. As this section describes, the most important of these norms are that doctors should value emotional detachment, clinical experience, intervention, and working with rare or acute illnesses.

Emotional Detachment

Undoubtedly most doctors enter the profession in large part because they want to help others. Yet perhaps the most central medical norm is to maintain emotional detachment from patients. As Perri Klass discovered in her encounter with dog lab, medical culture values and rewards "strength" and equates emotional involvement or expression with weakness (Hafferty, 1991).

Given doctors' daily confrontations with illness, trauma, and death, some emotional detachment is a necessary coping mechanism. Sociological research suggests, however, that doctors develop emotional detachment not only as a natural response to stress but also because their superiors teach them to (Hafferty, 1991).

Professional socialization refers to the process of learning the skills, knowledge, and values of an occupation. According to sociologist and med-

Table 11.2   *Ethnicity and Gender of First Year Medical Students, 1997/98*

ETHNICITY	MEN		WOMEN		TOTAL	
	N	%	N	%	N	%
White non-Hispanic*	6549	38.9%	4585	27.2%	11134	66.1%
African-American	525	3.1	822	4.9	1347	8.0
Native American	62	0.4	64	0.4	126	0.7
Mexican American	262	1.6	202	1.2	464	2.8
Puerto Rican	172	1.0	174	1.0	346	2.1
Other Hispanic	161	1.0	126	0.7	287	1.7
Asian or Pacific Islander	1788	10.1	1352	8.0	3140	18.6
Total	9819	56.5	7325	43.5	16844	100.0

*Includes foreign citizens regardless of race.

Source: Barzansky et al. (1998).

ical school professor Frederic Hafferty (1991), who spent several years observing and interviewing medical students, this socialization typically begins even before students enter medical school. At some point during their undergraduate training, most pre-medical students volunteer in hospitals. Through observing the behavior of hospital doctors, students quickly learn the value placed on emotional detachment. This norm can be further reinforced during admissions interviews at medical schools. Currently enrolled students often take prospective students to see the most grotesque-looking, partially dissected, human cadaver available in the school's anatomy lab. Although officially they do so to display the school's laboratory facilities, their true purpose seems to be to elicit emotional reactions from prospective students. The laughter and snickers these reactions evoke in the medical students demonstrate to prospective students that such behavior is shameful while demonstrating to the current students how "tough" they have become.

The emphasis on emotional detachment is reinforced often during medical school, as faculty and students implicitly or explicitly ridicule those who display emotions and question the ability of such students to serve as doctors (Haas and Shaffir, 1987:85–99; Hafferty, 1991). Women often find this particularly problematic, for outside the profession society rewards women for showing emotional sensitivity. Women students do seem to remain more emotionally sensitive than their male peers during the medical school years, although by the time they enter practice, most women have adopted attitudes and behaviors virtually indistinguishable from those of their male colleagues (Hafferty 1991:143, 198–200; Martin et al., 1988).

Experiences during residency continue to reinforce the expectation of emotional detachment. During daily rounds of the wards, faculty members grill residents on highly technical details of patients' diagnoses and treatments. Except in family practice residencies, however, faculty members rarely ask about even the most obviously consequential psychosocial factors. Rounds and other case presentations also teach residents to describe patients in depersonalized language. Residents learn to describe individuals as "the patient," "the ulcer," or "the appendectomy" rather than by name. As Renee Anspach (1990:328) has described, using language like "the vagina and the cervix were noted to be clear" rather than "I noted that Mrs. Simpson's vagina and cervix were clear" reinforces the impression "that biological processes can be separated from the persons who experience them." The use of medical slang, meanwhile, which peaks during the highly stressful residency years, allows students and residents to turn their anxieties and unacceptable emotions into humor by using terms such as "crispy critters" for severe burn patients. Medical slang also enables doctors and residents to avoid emotionally distressing interactions with patients and their families by using terms that lay persons cannot understand, such as "adeno-CA" for cancer (Coombs et al., 1993).

The structure of the residency years largely prevents residents from emotionally investing in patients (Mizrahi, 1986). Long hours without sleep often make it impossible for residents to provide much beyond the minimum physical care necessary (Christakis and Feudtner, 1997). When combined with the norm of emotional detachment, such long hours can even encourage doctors to view their patients as foes. As Phillip Reilly (1987:226) explains in his autobiographical account of medical training: "At 3 o'clock in the morning as I stood over [a comatose patient's] bedside staring at his IV, he was an enemy, part of the plot to deprive me of sleep. If he died I could sleep for another hour. If he lived, I would be up all night." According to Terry Mizrahi, who spent three years observing, interviewing, and surveying residents in internal medicine, by the end of their training, most held "attitudes towards patients ranging from apathy to antipathy" (Mizrahi, 1986:122). These attitudes are reflected vividly in the many slang terms residents use (sometimes within earshot of patients) to describe those they dislike treating, including "trainwrecks" (seriously ill or injured patients who might not seem worth spending resources on), "scumbags" (dirty, smelly patients), and "negative wallet biopsies" (patients with neither money nor health insurance). Such terms help doctors vent their frustrations regarding the difficulties of their situation and maintain needed emotional distance, but also implicitly reinforce disparaging attitudes toward patients (Coombs et al., 1993).

More subtly, the language doctors use in formal case presentations similarly reflects an image of patients as untrustworthy enemies (Anspach, 1990). Doctors more often say a patient "denies" using alcohol than say a patient "does not" use alcohol. In case presentations, "physicians 'note,' 'observe,' or 'find'; patients 'state,' 'report,' 'claim,' 'complain of,' 'admit,' or 'deny.' The first verbs connote objective reality (i.e., only concrete entities can be noted or observed); the second verbs connote subjective perceptions" (Anspach, 1990:331). Such language questions patients' credibility as witnesses regarding their own experiences and feelings. Medical students and novice residents who observe such case presentations quickly learn to use this language and, perhaps, to adopt the attitudes embedded in it (Coombs et al., 1993; Hafferty, 1998; Hafferty and Franks, 1994).

Not surprisingly, given these structural factors and the resulting attitudes, the doctors Mizrahi studied sometimes appeared to care more about getting rid of patients than about providing care. The centrality of this motive to residents' lives is evidenced by the numerous slang terms for this process. For example, a resident who has "taken a hit" (received an unwanted patient on his ward) can "buff" a patient's record (making the patient seem ready to move on to another form of care) so that the resident can "turf" (transfer) the patient elsewhere (Coombs et al., 1993). Among those Mizrahi observed, the desire to get rid of patients grew as residents came to realize that many patients suffer from illnesses or social problems that medicine cannot cure. Doctors reserved their most negative attitudes for such patients, as well as for those the doctors deemed morally or socially unworthy of their time. The latter include patients whose illnesses seemed linked to self-destructive behaviors; who sought treatment for minor illnesses; who were poor, nonwhite, female, or old; or who suffered from common illnesses that the doctors, trained in research-oriented medical schools, found uninteresting.

Clinical Experience

In addition to teaching doctors certain attitudes toward patients, medical culture also teaches, at a more abstract level, a set of attitudes toward medical care, illness, the body, and what makes humans truly human. Ironically, given that doctors' prestige rests partly on their scientific training, medical culture values clinical experience more than scientific research and knowledge (Bosk, 1979:84–94; Ludmerer, 1985; Millenson, 1997). The structure of medical training unintentionally reinforces this notion. During the first two years of medical school (the preclinical years), students take basic science courses taught by professors who hold doctorates in fields such as biochemistry or physiology. Students spend the next two years training in hospitals and clinics under professors who are themselves doctors. This division between scientific training taught by scientists in the early years and clinical training taught by doctors in the later years teaches students that scientific training is something to be endured before the "real" work of medical training begins.

Once students begin their clinical training, they also learn to base treatment decisions primarily on their personal experiences with a given treatment rather than on scientific research (Becker et al., 1961; Ludmerer, 1985: 260–271). For example, Knafl and Burkett (1975:100) describe the following incidents observed during surgical rounds at a hospital they studied:

> After the residents finished presenting the case to the audience, one of the attendings asked, "What 'bout doing a cup arthroplasty on him?" Morrison replied, "There's some *literature* to back it up but it's *my experience* that 'cups' just aren't that successful on young people. [Emphasis in original.]

Similarly:

> The second case is presented by Dr. Lee, a 4th-year resident. He shows slides of a 13-month-old girl whose one leg is shorter than the other. The reason for presenting the case is to discuss whether or not the leg should be surgically lengthened. In presenting the case, Dr. Lee quotes from a source in favor of such a procedure. Dr. Eddy, an attending physician, interrupts with, "I know that's what he says, but that's not the way we do it here."

In this way, residents learned to value their own intuition and idiosyncratic clinical experience over scientific research. This factor partially explains why standard clinical procedure varies enormously from community to community and from doctor to doctor, producing high rates of medical error as well as rates of lumpectomies, prostatectomies, and back pain surgery that are as much as 33 times higher in some states than in others (Center for the Evaluative Clinical Sciences, 1996; Leape, 1994).

Recent events, however, suggest that scientific research may be growing as a value within medical training and the medical world in general. The practice protocols described earlier are part of a broader push for "**evidence-based medicine.**" Evidence-based medicine refers to the idea that doctors need to test therapies with randomized, controlled clinical studies and need to fully evaluate and critique the available research evidence before deciding how to treat patients with a given condition. It reflects not only the concerns about cost control mentioned earlier but also the growing recognition that less than half of modern medical treatments have good scientific support (Naylor, 1995). Even if evidence-based medicine is not widely adopted by currently practicing doctors, the growing use of practice protocols in medical training could eventually shift medical culture toward valuing scientific knowledge over personal experience.

Intervention

Medical culture also teaches doctors to distrust natural bodily processes. Doctors learn to view pregnancy and menopause as diseases, to stop the effects of aging if possible, to use drugs to control minor fevers (the body's natural process for fighting infection), and so on (e.g., Martin, 1987; McCrea, 1983). Thus, for example, when Perri Klass became pregnant at age 26, her classmates expressed horror that she had not had amniocentesis, a test designed to identify certain chromosomal abnormalities in fetuses. Yet for women in their twenties, who have extremely low rates of fetal abnormalities detectable by amniocentesis, the test more often causes miscarriage than detects abnormal fetuses. Klass' fellow students, however, had learned so well to distrust pregnancy and the natural body that they could not evaluate her situation objectively.

As this example suggests, learning to distrust natural processes is intimately interwoven with learning to value medical intervention. During the preclinical years, doctors receive only minimal instruction in using tools such as nutrition, exercise, or biofeedback to prevent or treat illness; during the rest of their training, such tools are rarely, if ever, mentioned. Meanwhile, those medical specialties that rely most heavily on intervention historically have received the most prestige and financial rewards (although RBRVS is starting to change at least the financial balance). For example, surgeons (known in medical slang as "blades"), earn twice the median net income of general and family practitioners (known in medical slang as "fleas"). Similarly, a study using a national random sample of fourth-year medical students found that fewer than one percent believe that faculty encourage good students to enter primary care and only 11.7 percent believe that specialists respect primary care doctors (Block et al., 1996). (Box 11.1 describes a recent program designed to change this situation.) Taken together, these forces support the **technological imperative**—the belief that technological interventions should always be used if available. The results are sometimes tragic, as Box 11.2 illustrates.

Emphasis on Acute and Rare Illnesses

As a natural corollary of valuing intervention (and a natural result of locating medical training within research-oriented universities), medical culture teaches doctors to consider **acute** illness more interesting than **chronic** illness. This is not surprising, for doctors often can perform spectacular cures for acute illnesses (such as appendicitis) but can do little for chronic illnesses (such as lupus). Similarly, medical culture teaches doctors to consider common diseases less interesting than rare ones, for the latter require complex and well-honed diagnostic skills even if no treatments are available. In sum, during the course of their training, doctors learn to value emotional detachment more than emotional involvement or expression; clinical experience more than scientific knowledge; intervention more than natural physiological processes; and working with rare or acute illnesses more than working with chronic illnesses.

Consequences of Medical Values

These values work against the provision of high-quality health care. Emotional detachment can lead doctors to treat patients insensitively and to

---

Box 11.1   **The American Medical Student Association**

The American Medical Student Association (AMSA) is an independent national association (not related to the AMA) of about 30,000 medical students. Since 1950, it has worked to improve the quality of health care and medical education, as well as to protect the welfare of medical students and residents. Recognizing some of the traditional limitations of medical education, AMSA promotes programs designed to encourage medical students to enter primary practice in underserved areas and to develop the skills needed to work effectively with minority populations.

With these goals in mind, AMSA recently obtained a four-year grant from the U.S. Public Health Service to develop and pilot two model programs. The programs, known jointly as "Promoting, Reinforcing, and Improving Medical Education" (PRIME), are designed both to provide students with necessary technical and interpersonal skills and to encourage students' idealism and commitment to working with underserved populations. The first program is expected to include such topics as recognizing bias and stereotypes, learning how doctors' behaviors can make patients from other cultures become defensive, recognizing and coping with communication difficulties, and learning how to use interpreters. In addition, the program will include training on some of the specific health care concerns of various minority populations and some of the diverse ways in which minority cultures explain health and illness and approach health care. The second program will include training in understanding the underserved and their particular health care needs, in the philosophy of primary care and the importance of community public health work, and in communicating effectively with underserved populations. This program also will include discussion of such issues as the impact of managed care on primary care practices and how to repay student loans and earn a living while working in underserved areas.

Both programs will stress experiential, service learning rather than lectures and demonstrations. This structure, it is hoped, will both make it easier for students to gain confidence and skills and make it more likely that students will incorporate what they learn in these classes into their personal values and career plans.

PRIME also will provide funding to help AMSA continue two of its successful current programs: an annual leadership training program for students interested in primary care and a survey and "scorecard" used to evaluate how well medical schools are promoting primary care.

---

overlook the emotional and social sources and consequences of illness. Meanwhile, the emphasis on clinical experience, although sometimes useful, can lead doctors to adopt treatments that have not been tested through controlled clinical trials and that lack scientific validity, such as treating ulcers (which are now known to be caused by bacteria) with a bland diet and training in stress reduction (Millenson, 1997). At the same time, the emphasis on working with rare illnesses (coupled with the financial incentives of specialty practice) has contributed to the oversupply of specialists

Box 11.2   *The Story of Baby Andrew**

The true story of the short life and harrowing death of Baby Andrew, as told by his parents (Stinson and Stinson, 1979), illustrates the problems that can arise when doctors, trained in emotional detachment, adopt and implement a technological imperative within a framework of hierarchical doctor-patient relationships:

Andrew was a baby born 15½ weeks prematurely, weighing only 1 lb. 12 oz., and in a state of painful deterioration almost from the start. [Knowing that at the time no babies like Andrew had ever survived], we wanted him to be allowed to die a natural death. Andrew's story is the story of what can happen when a baby becomes hopelessly entrapped in an intensive care unit where the machinery is more sophisticated than the codes of law and ethics governing its use. . . .

The sad list of Andrew's afflictions, almost all of which were iatrogenic [i.e., caused by his medical care], reveals how disastrous this hospitalization was. . . . He was "saved" by the respirator to endure [heart problems, oxygen deprivation, blindness, painfully invasive procedures] . . . , numerous infections . . . , fractured bones . . . , and, finally, seizures of the brain. He was, in effect, "saved" by the respirator to die five long, painful and expensive months later of the respirator's side effects. . . .

We think the question must be raised as to whose interests were really served by this six months of imposed hospitalization. Certainly not Andrew's. He had the misfortune of being declared "salvageable" . . . by people who knew neither how to "salvage" him nor when or how to stop. . . . It seems clear to us that all the benefits in this case went to Pediatric Hospital and its staff. The medical residents got a chance to

broaden their education by working with a baby with malfunctions of virtually every system of his body, the specialists took part in some "interesting consults" and gathered some data, and the hospital collected the mind-boggling sum of $102,303.20 from our insurance company. . . .

From the very first, we were treated as wholly external to the case. Our wishes, judgements, and thoughts were rarely of interest to . . . medical staff, who abrogated decisions to themselves as though we did not exist. . . . When we objected to the decision [to place Andrew on a respirator], Dr. Farrell accused us of wanting to "play God" and to "go back to the law of the jungle." . . . He reduced the issue to its most absurd level. "I would not presume," he told us, "to tell my auto mechanic how to fix my car." . . .

Many times we tried to give attending physicians, residents, nurses, and business office clerks a sense of how financially destructive this experience was. . . . Since December, when Andrew entered PHC, we have not been able to make ends meet and will do no better in the foreseeable future. . . .

What sort of memories or thoughts could we have of Andrew? By the time he was allowed to die, the technology being used to "salvage" him had produced not so much a human life as a grotesque caricature of a human life, a "person" with a stunted, deteriorating brain and scarcely an undamaged vital organ in his body, who existed only as an extension of a machine. This is the image left to us for the rest of our lives of our son, Andrew.

*For a similar, more recent account, see *Lost Lullaby*, by Deborah Golden Alecson (University of California Press, 1995).

Source: Stinson, Robert, and Riggy Stinson. 1979. "On the Death a Baby." *Atlantic Monthly* 244(1):64–72. Reprinted with permission.

and undersupply of **primary care doctors,** or **primary practitioners**—those doctors in family or general practice, internal medicine, and pediatrics who are typically the first doctors individuals see when they need medical care (Stimmel, 1992). As of 1996, 70 percent of U.S. doctors were specialists, although only about 20 percent of the problems patients bring to doctors require specialty care (Bodenheimer, 1999; Light, 1988:308). Similarly, emphasizing acute illness leads doctors to consider patients with chronic illnesses uninteresting. Finally, emphasizing intervention can lead doctors to act when inaction might be best. An individual who has a cold, for example, will likely recover regardless of treatment. Often, however, doctors will prescribe antibiotics either because they psychologically need to intervene or because their patients pressure them to do something. Yet, antibiotics cannot cure colds but can cause unpleasant or even life-threatening health problems. Moreover, in the long run, and as described in Chapter 2, unnecessary treatment can foster the development of drug-resistant bacteria.

Although medical values can decrease the quality of medical care for all patients, the impact is especially great for women, minorities, the poor, and the elderly. Doctors carry into medical practice the same attitudes toward these groups that most Americans hold. As a result, doctors are considerably more likely to use disparaging labels for such people compared with other groups and more often adopt paternalistic interaction patterns with them (Mizrahi, 1986; Scully, 1994).

Similarly, the value placed on emotional detachment can cause doctors to feel disdain for patients whom doctors consider too emotional. How much emotion one shows, however, and how one does so, depends partly on one's cultural socialization. In contemporary America, women and members of certain ethnic minority groups (such as Jews and Italians) are more likely than are men and nonminorities to display emotion openly (Koopman et al., 1984). Consequently, these groups are more likely to bear the brunt of doctors' disdain.

Finally, the value placed on clinical experience and responsibility has a greater impact on female, minority, poor, and elderly patients. Of necessity, medical students and residents must learn skills through practice. The desire for clinical experience sometimes encourages them to perform procedures, from drawing blood to doing surgeries, even if they cause unnecessary pain or lack sufficient training or supervision. Doctors are most likely to do so if they can define a patient as "training material" rather than as an equal human being. This is most likely to happen when a patient differs significantly both from the doctors and from the patients on whom those doctors assume they will someday practice.

Probably all these values, and the problems they create, are stronger during medical training than afterward. Once doctors enter practice, economic pressures encourage them, willingly or unwillingly, to show at least somewhat more sensitivity to patients' needs. In addition, those who consistently work with the same pool of patients—a situation, that, as described earlier, has become less common—can develop more meaningful relationships with them. Thus, over time, doctors may recoup some of their initial, more positive, attitudes toward patients and patient care (Mizrahi, 1986). These changes cannot, however, help the millions of Americans who lack either health insurance or the ability to pay for medical care and who therefore must rely on public clinics or hospitals for their care. These patients pay the highest costs for the medical value system.

Building a Medical Career

Not all medical careers are created equal. Some specialties and practices offer considerably more status, income, and autonomy than others. As a result, new doctors face greater competition for some residencies and jobs than others.

A new doctor's ability to enter a prestigious medical field or type of practice depends largely on **sponsorship** (Hall, 1949). Sponsorship refers to the process through which successful professionals in a given field actively help new members to establish their careers. This process is not an egalitarian one, for established members typically choose whom to sponsor based not only on **achieved statuses,** or earned qualifications such as medical school grades, but also on **ascribed statuses,** or innate characteristics such as ethnicity and gender.

Judith Lorber's longitudinal research on the careers of men and women doctors vividly shows the impact of sponsorship. For example, one young man tells how his residency supervisors sponsored him:

> Dr. _____ made a conscious effort to interest me in gastroenterology, and he had the support of the chief of medicine. I found the two of them both excellent researchers and clinicians. They made it seem very exciting and interesting, and to some extent, they also wooed me just a little bit. Dr. _____ took me to a meeting in Boston in the fall of that year. They took me to the national GI [gastrointestinal] meeting in Philadelphia in May and I loved it. The meetings were excellent, very stimulating. I had a good time, and that's when I decided to go into gastroenterology. I also had them behind me pushing me and guiding me into my choice of fellowships. I was starting late to look for fellowships, and it would have been difficult, but I had the two of them assisting and making entrés (Lorber, 1984:34–35).

In contrast, the women Lorber studied lacked such sponsorship. Although they rarely experienced overt discrimination, they endured constant covert discrimination. Their professors typically assumed that women would be happiest in traditionally female, low-status fields such as psychiatry, public health, and pediatrics. These professors therefore discouraged them from entering other fields and withheld the experience, recommendations, encouragement, and other forms of sponsorship needed to enter them. As a result, the women eventually found themselves in less prestigious and remunerative fields than their male peers, despite approximately equal academic grades, research records, and desire to enter high-status fields. (Table 11.3 shows median net salary and the percentage of doctors who are female for several medical specialities.) Meanwhile, women who do enter more male-dominated fields typically face continual disadvantage; this problem was highlighted when Dr. Frances Conley, the first female full professor of neurosurgery in the United States, resigned her tenured position at Stanford University in protest against years of discrimination.

Although little recent research is available on the topic, these same processes undoubtedly hinder the careers of those who differ from most doctors in ethnicity or class. Indeed, the many Catholic and Jewish non-profit hospitals around the country were founded earlier in the twentieth century because most hospitals refused to hire Catholic or Jewish doctors. Over time, religious discrimination within medicine all but disappeared, and we can hope that other social barriers eventually will fall as well.

Table 11.3 **Median Net Salary and Percentage of Doctors Who Are Female, by Specialty**

SPECIALTY	MEDIAN NET SALARY*	PERCENTAGE WHO ARE FEMALE**
Pediatrics	$110,000	45.5%
Obstetrics/gynecology	182,000	30.9
Psychiatry	120,000	27.6
Family/general practice	110,000	24.6
Internal Medicine	150,000	24.6
Anesthesiology	200,000	19.7
Radiology	220,000	19.1
General surgery	219,000	9.1
Total	$150,000	18.5

*1994 data.

**1996 data

Source: Gonzalez (1999:45); Roback et al. (1999:106).

Patient-Doctor Relationships

From the beginnings of Western medicine to the mid-twentieth century, medical culture has stressed a paternalistic value system in which only doctors, and not patients or their families, are presumed capable of making decisions about what is best for a patient (Katz, 1984); Box 11.3 gives an example of such a situation. Often, this **paternalism** is reinforced by patients who prefer to let their doctors make all decisions; indeed, at least part of doctors' efficacy comes simply from patients' faith in doctors' ability to heal. Paternalism is also reinforced by the structure of medical practice, in which doctors by their own (probably optimistic) estimates spend an average of only 19 minutes per patient per office visit and 15 minutes or less with 67 percent of their patients (Woodwell, 1997). As a result, doctors often do not have the time to inform patients fully or to assess patients' needs or desires.

Unfortunately, doctors' inclination to make decisions for patients is sometimes bolstered by doctors' racist, sexist, or classist ideas. Doctors are exposed to and sometimes adopt the same stereotypical ideas about minorities, women, and lower-class persons common among the rest of society, believing, for example, that African Americans are unintelligent, women flighty, and lower-class persons lazy. Doctors who hold such ideas sometimes make decisions for patients belonging to these groups, rather than involving the patients in the decisions, because these doctors believe it is easier and less time-consuming to do so. For example, medical residents in obstetrics and gynecology interviewed by Diana Scully (1994) made such comments as "I don't like women that think they know more than the doctor and who complain about things that they shouldn't be complaining about" and "I think the main thing is that the patient understands what I say, listens to what I say, does what I say, believes what I say." Similarly, "I don't care for the patient that gives you a fight every time you try to give them a drug. I don't care for the patient that disagrees with me" (Scully, 1994:92).

Finally, doctors' inclination to make decisions for patients can be reinforced when cultural barriers make it difficult for doctors' to gain patients' cooperation or to understand patients' beliefs or wishes. Those cultural differences are probably greatest when western-born doctors treat immigrants from nonwestern societies. In these circumstances, even the smallest gestures unintentionally can create misunderstanding and ill will. For example, in her observations of Hmong patients who had immigrated from Laos and their American doctors, Anne Fadiman (1997:65) found that

> when doctors conferred with a Hmong family, it was tempting to address the reassuringly Americanized teenaged girl who wore lipstick and spoke English rather than the old man who squatted silently in the corner. Yet failing to work within the traditional Hmong hierarchy, in which males ranked higher than

---

Box 11.3   *Ethical Debate: Truth Telling in Health Care*

Jeffrey Monk, an unmarried, 26-year-old accountant, goes to see Dr. Fisher because of recurrent headaches that have made it difficult for him to concentrate at work. Jeffrey generally enjoys good physical health, although he has experienced bouts of severe depression since his mother died a few months ago.

Dr. Fisher runs a series of tests and soon discovers that Jeffrey has an inoperable brain tumor, which will probably kill him within the year. Because no treatments are available, telling Jeffrey of his diagnosis would seem to serve little purpose at this point. Jeffrey has no dependents, so he need not make a will or other financial arrange-

ments immediately. Moreover, telling him might cause his health to deteriorate more rapidly, spark another depressive episode, or even lead him to commit suicide. Anyway, Dr. Fisher believes, few patients truly want to know they have a fatal illness. He therefore decides to defer telling Jeffrey his diagnosis. Instead, Dr. Fisher tells Jeffrey that the headaches are not serious and prescribes a placebo, counting on the fact that placebos significantly reduce patient symptoms in about 30 percent of cases.

Do doctors have an obligation to tell their patients the truth? Answering this question requires us to look at several significant ethical issues. The

---

females and old people higher than young ones, not only insulted the entire family but also yielded confused results, since the crucial questions had not been directed toward those who had the power to make the decisions. Doctors could also appear disrespectful if they tried to maintain friendly eye contact (which was considered invasive), touched the head of an adult without permission (grossly insulting), or beckoned with a crooked finger (appropriate only for animals).

In these circumstances, doctors conclude that collaboration with patients is impossible and that paternalistic decision-making is their only alternative.

Nevertheless, doctors only rarely have complete control over treatment decisions and interactions with patients. As Thomas Szasz and Marc Hollander (1956) explain, three **models of doctor-patient interactions** exist. Only in the first model, **activity-passivity,** is the doctor totally active and the patient totally passive. Emergency surgery performed on an unconscious patient would fall into this category, as would drugging a psychiatric patient against his or her will. In the second and most common model, **guidance-cooperation,** the doctor offers guidance to a cooperative, but clearly submissive, patient, such as one suffering from a cold. In the third model, **mutual participation,** both doctor and patient participate equally. This model occurs most often with chronic illnesses such as diabetes or multiple sclerosis, in which much of doctors' work consists of helping patients discover what works best for them.

most central ethical issues in this case are autonomy versus paternalism. According to the principle of autonomy, each rational individual is assumed capable of making his or her own choices if given sufficient information and each health care worker has the obligation to provide that information. Consequently, each individual has the right to decide what is in his or her own best interest and to act on those decisions without coercion from others. Counterbalancing this is the principle of personal paternalism—the idea that some individuals (in this case, doctors) have the expertise needed to decide what is in the best interest of other individuals.

Evaluating this situation requires us to weigh the benefits of disclosure against those of dissembling. Will hiding his diagnosis from Jeffrey protect him from depression or suicide, or will the anxiety caused by not knowing the meaning of his symptoms increase his emotional problems? Is suicide necessarily against Jeffrey's best interest? Is it best for a doctor to give a patient a placebo, which might offer some physical and emotional relief, or to let the patient know the truth, so the patient can make his or her own choices—from seeking unconventional treatments or a second opinion to choosing how to spend his last months? The final question, then, is can doctors know what is in their patients' best interest, and when if ever should they be given the authority to act on those judgments?

Eliot Freidson (1970a) has looked at the power dynamics underlying these different models. Doctors' power is greatest in two situations: when patients are completely incapacitated by coma, stroke, or the like, or when doctors have sufficiently greater cultural authority than their patients to argue convincingly that they can most accurately judge patients' best interests, whether that patient is a Jehovah's Witness who refuses a blood transfusion, a pregnant woman who refuses a cesarean section, or someone labeled mentally ill who opposes hospitalization. Doctors' power also increases when they work in group practice rather than in solo practice because doctors in group practice obtain most of their business through referrals from colleagues or MCO contracts rather than from satisfied patients and so have less fear of economic repercussions if they assert their power and alienate their patients. Finally, doctors' power is higher when interacting with patients who do not share the doctors' language, culture, and social status. In sum, doctors' power depends on their cultural authority, economic independence, cultural differences from patients, and assumed social superiority to patients. As this suggests, and given the demographic composition of contemporary medicine, doctors are most likely to adopt egalitarian interaction patterns with white, nonelderly, male, and middle- or upper-class patients (Street, 1991).

To explore *how* doctors maintain dominance during their meetings with patients, researchers have conducted detailed analyses of conversation patterns

Box 11.4   **"Hi, Lucille, This is Dr. Gold!"**
Lucille G. Natkins

I'm going in for a dilation and curettage (D&C) next week. But even as I worry about carcinomas and five-year survival rates, an incident from my last D&C keeps popping into my mind.

That operation occurred after I hadn't seen a gynecologist in years. On my internist's recommendation I saw a physician whom I'll call Dr. James Gold, diplomate, American Board of Obstetrics and Gynecology; fellow, American College of Surgeons; and associate attending physician at a large teaching hospital. It turned out that he was a contemporary, that he lived in my neighborhood, and that his children and mine were classmates. He'd gone to medical school with one of my friends and interned with another. No one would have worried about inviting us to the same dinner party.

One visit and several phone calls later—all conducted on a cordial "Dr. Gold" and "Mrs. Natkins" basis—surgery was scheduled and soon afterward I was wheeled into the operating room. As my vision blurred and my legs numbed, a voice cut through the anesthetic haze. "Hi, Lucille, this is Dr. Gold!" Stupor turned to rage. "You expletive, that's not the

way it goes! It goes 'Hi, Lucille, this is Jim' or 'Hi, Mrs. Natkins, this is Dr. Gold.'"

All soundless. I was out of it, zonked. The next thing I remember was a female voice saying, "Wake up, Lucille, the operation's over. Wake up, Lucille." Damn, I thought, not again.

The biopsy findings were negative. I was free to stop worrying about gynecological malignancies, but "Hi, Lucille" wouldn't leave me. There are more dignified positions in life than lying naked and horizontal, legs spread-eagle, while half a dozen strangers shove their fists into what was once (wisely) called "one's private parts." But that indignity was unavoidable. What, though, was the purpose of "Hi, Lucille, this is Dr. Gold" from someone who would have been Jim had we met socially, or "Wake up, Lucille" from someone who was ensuring my waking by slapping my face? What purpose other than to underscore my lack of dignity and helplessness?

"Hi, Lucille" was still rankling months later when my 80-year-old mother-in-law was hospitalized. Overwhelmed by crippling arthritis and a host of other problems, she asked the

between doctors and patients (Fisher, 1986; Katz, 1984; Waitzkin, 1991; West, 1984). Conversations between doctors and patients typically follow a pattern in which the doctor opens a topic with a question, the patient responds, and the doctor signals that the topic is closed (Mishler, 1990). The doctor can then raise the next topic or ask further questions for clarification and repeat the cycle. In either event, the doctor maintains control over the direction and length of the conversation. For example, a patient might come to a doctor complaining of various problems. The doctor will ask for further details about only some of those problems, typically ignoring how factors in patients' lives might cause health problems or how health problems might cause other problems in patients' lives. The doctor also can ask questions about problems the patient had not mentioned but the doctor

nurse, whose name pin read "T. Bass," to "please get my slippers from the bedroom." "Whatever are you talking about, Bertha," snapped T. Bass, who was, perhaps, all of 30 years old. "You're in the hospital, not your house." My mother-in-law stiffened and blanched. Reality therapy with a bludgeon.

I became a first-name freak, asking friends and colleagues who addressed them by first name without expecting reciprocity and, conversely, whom they addressed by first name while expecting to be called Mr. Price or Dr. Wand. No surprises in this survey. Inferiors are called by first name: children, menial workers, the elderly, and women.

I wrote to the hospital where my mother-in-law had been a patient, noting that the hospital system that was reducing an 80-year-old woman to a child was robbing her of the will and determination she needed to ensure her recovery. The administrator replied that he could not understand my charges of abuse. I wrote to a widely syndicated medical columnist, asking why his replies to women began "Dear Amy" and to men "Dear Mr. Hall." No answer. . . .

I chose a new gynecologist. But not by using physician referrals and checking medical directories as I would have before, when I thought I was sophisticated. "Is your gynecologist a nice person?" I asked friends. "Are you treated with dignity and consideration? Called by your first name or your last?" Another survey with few surprises. Not many women answered "yes," "yes," and "last name."

But some did. (And, yes, my new gynecologist is board-certified, as nearly everyone in a metropolitan area seems to be these days.) So far, so good, but next Friday both of us will have to pass our big tests in the operating room. Will I have malignant cells on my pelvic wall? Will he resist the temptation to say "Hi, Lucille" when I'm flat on my back and going down for the count?

Health and self-respect, I've learned, are both necessities.

Source: *Journal of the American Medical Association*, May 7, 1982. 247(17):2415. Copyright 1982, American Medical Association. Reprinted with permission.

expects to find, thereby defining certain problems but not others as relevant. In addition, doctors control conversations by asking close-ended rather than open-ended questions, making it difficult for patients to raise new topics.

Other techniques also enable doctors to control interactions with patients. Doctors interrupt patients far more often than the reverse, cutting off discussions and questions the doctors consider irrelevant or uncomfortable. They give general rather than specific answers to patients' questions, give information only when directly asked, or use euphemisms (such as "tumor" instead of "cancer") that leave patients confused about their situation. As a result, patients lack the information they need to challenge doctors' actions or make their own decisions. This in turn can create both stress and distrust when patients conclude that their doctors have withheld information.

CONCLUSIONS

Between 1850 and 1950, allopathic medicine attained and then enjoyed unprecedented autonomy and dominance, becoming the premiere example of a profession. In its battles for status with its many nineteenth-century rivals, allopathic medicine benefitted from the public's growing respect for and fascination with scientific knowledge and from the increase over time in the field's scientific foundations. It also benefitted from the public's assumptions that because allopathic doctors were disproportionately upperclass white men, they must be more competent than the minorities, women, and poorer persons who dominated some of the competing health care fields.

Since the 1950s, however, doctors' social status has declined and their control over working conditions, relationships with patients, and finances has diminished. Yet doctors continue to have far more autonomy and dominance than most other occupations, especially within the health care field. This continued professional dominance—and the continued internecine warfare between medicine and other health care occupations—affects all of us as consumers of health care because it sets the stage on which attempts to improve the health care system must occur.

Professional socialization, too, affects all of us as consumers. In its current form, this process is lengthy, arduous, and expensive, making it difficult if not impossible for many otherwise-qualified persons to become doctors and encouraging those who do become doctors to become emotionally hardened or financially driven. To these **unintended negative consequences** of medical training must be added the problems caused by a medical culture that emphasizes emotional detachment, clinical experience rather than scientific knowledge, intervention rather than relying on natural processes, and acute and rare illnesses rather than common and chronic illnesses.

As consumers of health care, we all benefit from the extensive training doctors receive. Those benefits, however, must be weighed against the costs we pay when our doctors also learn ways of interacting with patients and thinking about illness that can encourage overly aggressive, scientifically unjustified, or simply discourteous treatment. Only by directly confronting the nature of medical culture can we hope to change medical training and make future doctors better able to meet their patients' needs.

Currently, pressures to change the nature of medical culture and the doctor-patient relationship are coming from within as well as outside the medical field. Many doctors now believe that the rise in malpractice suits largely reflects patients' disenchantment with their relationships with doctors rather than problems in the quality of care. As a result, medical journals now often publish articles instructing doctors to reduce their malpractice risk by improving their relationships with patients (Annandale, 1989).

Deeply felt personal beliefs, and not just economic self-interest, have driven other doctors to work for changes in the system. Such beliefs have led to the founding of organizations such as Physicians for a National Health Plan and the American Holistic Medical Association. Similarly, the American College of Physicians, the professional organization for doctors in internal medicine, derives its strength partly from the growing number of doctors who favor its humanistic approach to medical care.

Finally, throughout the United States, medical students and professors are working to implement innovative programs for integrating more humanistic perspectives into the medical curriculum. At Harvard Medical School, for example, students now must take a three-year course specifically designed to improve relationships with patients and to humanize medical care (Tosteson et al., 1994). Beginning with role playing and discussing their personal experiences of illness, students are reminded what it is like to experience illness and health care. Subsequently, students learn how to interview patients, with the emphasis on listening to patients and understanding the psychosocial circumstances in which individuals experience illness. In this way, students can learn from the beginning of their training to see health care from patients' perspectives.

Similarly, "**cultural competence**" is now a commonly cited goal of medical education. Cultural competence refers to the ability of health care providers to understand at least basic elements of others' culture and thus to provide medical care in ways that better meet clients' emotional as well as physical needs. Cultural competence has been promoted as a means of increasing clients' willingness to seek medical care and satisfaction with care, and consequently as a means of improving health outcomes. Both the American Psychiatric Association and the American Academy of Family Physicians have officially endorsed including cultural competence in medical training, and numerous medical schools and residency programs now do so (Fadiman, 1997:270–272). In the long run, programs such as these may restructure medical culture and doctor-patient relationships.

SUGGESTED READINGS

Hafferty, Frederic W. 1991. *Into the Valley: Death and the Socialization of Medical Students.* New Haven: Yale University Press. Vividly describes how medical students develop their attitudes toward patient care. Far broader and more interesting than the title suggests.

Rothman, Ellen Lerner. 1999. *White Coat: Becoming a Doctor at Harvard Medical School.* New York: William Morrow. Rothman tells of her experiences in Harvard's revamped medical school program.

Starr, Paul. 1982. *The Social Transformation of American Medicine.* New York: Basic. Not easy reading, but *the* book for anyone seriously interested in the history of American medicine.

Conley, Frances K. 1998. *Walking Out On the Boys.* New York: Farrar Straus and Giroux. Conley, the first female full professor of neurosurgery in the United States, describes in this memoir the shocking discrimination still encountered by women in surgical training and practice.

GETTING INVOLVED

American Civil Liberties Union. 125 Broad St. NY 10004-2400. (212) 549-2500. www.aclu.org. Among other things, fights for the civil rights of patients.

American Medical Students Association. 1902 Association Drive, Reston, VA 20191. (800) 767-2266. www.amsa.org. Among other things, seeks to make medical education more humanistic. Open to premedical as well as medical students.

REVIEW QUESTIONS

What was the difference between *allopathic* and *homeopathic* doctors?

What was medical training like in 1850?

What could a doctor offer his patients in 1850? in 1900?

What does it mean to say that an occupation is a *profession*?

How did doctors achieve professional dominance?

What factors have reduced doctors' professional dominance?

How do doctors deal with medical errors? How does the treatment of medical errors reflect doctors' professional dominance?

What are the major medical norms and how do doctors learn them?

How do medical norms affect patient-doctor relationships?

What is cultural competence, and why is it important?

INTERNET EXERCISES

Go to the Web site for the University of California's Survey Documentation and Analysis Archive, first described in Chapter 3. Click on the GSS Cumulative Datafile, 1972–1996. Find the Select an Action section, then click the button for Frequencies or Crosstabulations. Next click on Start. A form with several blank spaces will appear on your screen. For row variable, type *conmedic.* For column variable type *class.* Click on the boxes to the left of Column Percentaging, Statistics, and Question Text. Then click the button to Run the Table. Repeat, using first sex and then health as column variables. Which groups have the least confidence in the people running the institution of medicine? Which variables have the most impact?

Go to the Web site for the Center for Responsive Politics (www.opensecrets.org) and see what you can learn about how medical organizations and other health care industry groups are working to affect elections and health-related laws in the United States. You might look for information on managed care legislation, tobacco control, or gun control, among other topics.

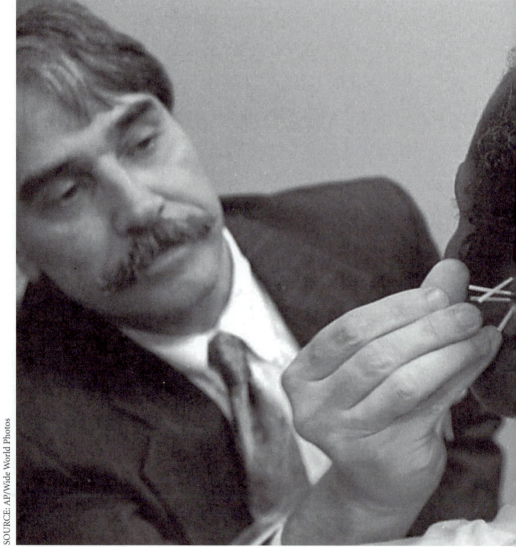

An American acupuncturist uses an ancient Chinese healing technique to treat drug addition.

CHAPTER 12

Other Mainstream and Alternative Health Care Providers

—

For more than a decade, Juliana van Olphen-Fehr ran an independent practice as a nurse-midwife delivering babies in women's homes. In the following story, she gives us a sense of what it is like to participate in a home birth:

> Late in the evening, Mona's contractions started getting quite intense. She paced around the room while we watched. She'd sit on the toilet frequently and Dave [her husband] rubbed her back when she was on the bed. . . . We tried to encourage Dave to go take a nap but he didn't want to leave Mona for a moment. He finally fell asleep in the bed while it was our turn to rub Mona's back. The night moved into early morning. The clock ticked away. We walked and talked.
>
> It's amazing how long it takes a baby to be born. As time passes slowly, labor gives one the opportunity to reflect on the process of birth. Each contraction comes and goes, [as] the uterus gets smaller and smaller [and] the baby is massaged down further and further into the pelvis. . . . Finally, the uterus, getting more powerful as it decreases in size, pushes the baby out of its first cradle, the pelvis, through the vagina, the passageway to life, into the outside world. The mother, feeling more and more pressure, joins the uterus in its expulsive efforts. She bears down gently and involuntarily at first but then more forcefully and purposefully as the baby approaches birth.
>
> In its natural environment, giving birth is like a musical masterpiece, building to its crescendo when the baby enters the world. Just as a symphony pulls its audience into its powerful rhythm, so does a laboring woman pull in her onlookers. All of those present at birth must be in synch just as all of the instruments in an orchestra must be in synch. This synchronization helps the mother keep her power to create her own masterpiece. . . .

Mona's labor built up to the point where she started to feel the urge to bear down. Her cervix was completely dilated and I felt the baby's head low in the vagina. She squatted while she pushed during the contractions and walked during the break between them. She found it most comfortable to lean on the banister in her hallway while she pushed. . . . Dave was still behind her, supporting her hips. I encouraged her to push while I got under her to monitor the baby's heartbeat.

Finally, the head appeared. Dave was behind Mona, sitting on the floor, I was beneath her in the front. Together we had our hands around the baby's head, supporting it as we coaxed her to push the baby out slowly. A beautiful little boy was born into Dave's and my hands. I held the baby as Dave eased Mona back onto his lap. His arms were around her as they both welcomed the baby into their arms. My birth assistant covered all three of them with blankets to keep the baby warm with their body heat. We turned the light low so the baby would open his eyes. In happy exhaustion, we sat back and through tears watched this family fall in love with each other (Van Olphen-Fehr, 1998:111–113).

Van Olphen-Fehr's story evokes for us both the joy that midwives can find in assisting at childbirth and some of the reasons health care consumers might choose a nontraditional option like home birth. Since this story took place, however, unaffordabe insurance premiums have forced virtually all nurse-midwives to abandon independent practice and to return to working directly under physician supervision. This situation illustrates the problems faced by nonmedical health care workers in trying to achieve professional status in a system characterized by **medical dominance.** This chapter first looks at the history and current status of four occupations now considered part of mainstream health care—nursing, nurse-midwifery, pharmacy, and osteopathy. As we will see, nursing in general, handicapped by its historically female tradition, has achieved only semiprofessional status, although nurse-midwives have gained a somewhat higher status by carving out a specialized niche for themselves within the health care system. Pharmacy, on the other hand, is now considered a profession, but has faced continued struggles to retain that status and its professional prerogatives, while osteopaths have attained professional status parallel to that of medicine. The chapter then considers the history and status of five occupations that, to a greater or lesser extent, remain outside of mainstream health care—chiropractic, lay midwivery, *curanderismo*, Christian Science practice, and traditional acupuncture. The history of chiropractic illustrates how, despite medical dominance, an alternative health care occupation can secure a role

Table 12.1   *Occupational Prestige, Rated by a Random Sample of Americans, 1989*

OCCUPATION	SCORE
Doctors (M.D. or D.O.)	86 (highest score possible)
Lawyers	75
Dentists	72
Pharmacists	68
RNs	66
Legislators	61
Chiropractors	60*
LPNs	60
Dental hygienists	52
Real estate sales	49
Waiters	28
Lay midwives	23*

*From 1970 survey; not available on 1989 survey. In general, scores are highly stable across time, so data on occupational prestige has not been collected since 1989. Note, though, that these scores were obtained before the resurgence of lay midwifery.

Source: National Opinion Research Center, *General Social Surveys, 1972–1991: Cumulative Codebook*, July 1991.

for itself by limiting its services to a narrow field. Finally, the histories of lay midwives, Christian Science practitioners, Mexican-American *curanderos*, and traditional acupuncturists show how occupations can remain marginal to the health care system, unable in the face of medical dominance to secure more than a small and precarious niche for themselves. Table 12.1 compares the occupational prestige of some of these fields.

MAINSTREAM HEALTH CARE PROVIDERS

Nursing: A Semiprofession

Although both the general public and sociologists (as indicated by number of research studies) often seem to equate health care workers with doctors, nurses form the true backbone of the health care system: hospital patients quickly learn that it is nurses who make the experience miserable or bearable and whose presence or absence often matters most. The history of nursing demonstrates the difficulties of achieving professional status for a "female" occupation.

The Rise of Nursing

Before the twentieth century, popular sentiment held that caring came naturally to women and, therefore, that families could always call on any female relative to care for any sick family member (Reverby, 1987). Hospitals, meanwhile, relied for custodial nursing care on the involuntary labor of lower-class women who were either recovering hospital patients or inmates of public **almshouses.** These beginnings in home and hospital created the central dilemma of nursing: nursing was considered a natural extension of women's character and duty rather than an occupation meriting either respect or rights (Reverby, 1987).

Nevertheless, increasingly during the nineteenth century, unmarried and widowed women sought work as nurses in both homes and hospitals. Few of these, however, had any training.

The need to formalize nursing training and practice did not become obvious until the Crimean War of the 1850s, when the Englishwoman Florence Nightingale demonstrated that trained nurses could substantially alleviate the horrors of war (Reverby, 1987). The acclaim Nightingale garnered for her war work enabled her subsequently to open new training programs and establish nursing as a respectable occupation.

Like most of her generation, Nightingale believed that men and women had inherently different characters and thus should occupy "separate spheres," playing different roles in society. To Nightingale, women's character, as well as their duty, both enabled and required them to care for others. She thus conceived of caring as nursing's central role. In addition, because her war work had convinced her of the benefits of strict discipline, she created a hierarchical structure in which nurses and nursing students would follow orders from their nursing supervisors. This structure, she hoped, would provide nurses with a power base within women's separate sphere parallel to that of doctors within their sphere. These principles became the foundation of British nursing. A few years later, when the U.S. Civil War made the benefits of professional nurses obvious to Americans, these principles were also adopted by American nursing.

By the early twentieth century, nursing schools had sprouted across the United States, as hospital administrators discovered that running a nursing school provided a ready pool of cheap labor. Within these hospital-based schools, education became, at best, secondary to patient care. A 1912 survey found that almost half of these schools had neither paid instructors nor libraries (Melosh, 1982:41). Students typically worked ten to twelve hours on the wards daily, with work assignments based on hospital staffing needs

SOURCE: *Harper's Weekly*, January 12, 1871. National Library of Medicine.

Nurses first won public respect through their work in the Civil War, the Franco-Prussian War, and other nineteenth century military conflicts.

rather than on educational goals. Formal lectures or training, if any, occurred only after other work was done.

This highly exploitative training system stemmed directly, if unintentionally, from the Nightingale model and its emphasis on caring and duty. As historian Susan Reverby notes (1987:75), "Since nursing theory emphasized training in discipline, order and practical skills, the ideological justification explained the abuse of student labor. And because the nursing work force was made up almost entirely of women, altruism, sacrifice, and submission were expected and encouraged."

Those women who, by the beginning of the twentieth century, sought to make nursing a profession by raising educational standards, establishing standards for licensure or registration, and improving the field's status found their hands tied by the nature of the field. According to Reverby (1987:122), to raise its status, nursing reformers

> had to exalt the womanly character and service ethic of nursing while insisting on the right of nurses to act in their own self-interest, yet not be "unladylike." They had to demand higher wages commensurate with their skills and position, but not appear "commercial." Denouncing the exploitation of nursing students as workers, they had to forge political alliances with hospital physicians and administrators who perpetrated this system of training. While lauding character and sacrifice, they had to measure it with educational criteria in order to formulate

registration laws and set admission standards. In doing so, they attacked the background, training, and ideology of the majority of working nurses. Such a series of contradictions were impossible to reconcile.

Political weaknesses also hamstrung nurses' attempts to increase their status. Like other white women, white nurses could not vote in most states until 1920, and most nonwhite nurses could not do so until considerably later. Moreover, nurses faced formidable opposition from doctors and hospitals that feared losing control over this cheap workforce. Nevertheless, by the 1920s, most states had adopted licensing laws for nursing schools and nurses. However, most laws were weak and poorly enforced (Melosh, 1982:40). As a result, the term "registered nurse" only became truly meaningful following World War II.

The Doctor-Nurse Game

The dilemmas nurses faced in gaining acceptance as a full profession are made vivid through what Leonard Stein (1967) dubbed the "**doctor-nurse game.**" According to Stein, "the object of the game is as follows: The nurse is to be bold, have initiative, and be responsible for making significant recommendations, while, at the same time, she must appear passive. This must be done in such a manner so as to make her recommendations appear to be initiated by the physician" (1967:699). Thus inexperienced doctors are expected to use subtle verbal cues to elicit treatment recommendations from more experienced nurses, and nurses are expected to just as subtly make their recommendations. Similarly, David Hughes (1988) observed that experienced emergency room nurses subtly tell inexperienced medical residents what to do by selecting which instruments to place on the table and in which sequence, and by telling patients what doctors will do next and why. In these ways, both doctors and nurses can maintain the illusion of medical control while recognizing the disparity between nurses' semiprofessional status and their expertise.

As Stein and his colleagues have noted in a more recent article, however, the doctor-nurse game has grown considerably less common over time. This is particularly true in areas such as emergency rooms and intensive care units, where the need for split-second decisions makes such subterfuge not only counterproductive but dangerous (Stein, Watts, and Howell, 1990). The shortage of trained nurses also has increased nurses' bargaining power, while (as described in the previous chapter) the status of doctors has declined, reducing doctors' ability to force nurses to play the "game." Simultaneously, nursing schools—now largely based in colleges rather than in hospitals—increasingly have encouraged novice nurses to eschew the submissiveness of the past and to view themselves as equal professional members of the health care team. Finally, the doctor-nurse game reflected broader assumptions about proper interactions between men and women.

As social ideas about women's roles have changed and as the number of both female doctors and male nurses has increased, the very premises of the doctor-nurse game have eroded.

Professionalizing Nursing

The Impact of Education

One major strategy used by nursing leaders to increase nurses' autonomy and improve their status and working conditions has been to increase educational requirements for the field (Melosh, 1982:67–76). Beginning in the 1960s, the American Nurses Association (ANA) promoted the development of two- and four-year college-based nursing programs and lobbied to make college education a requirement for nursing. The new college-based programs quickly proved popular, as changing social norms encouraged women to seek a college education in the hopes of improved employment opportunities. At the same time, however, the move toward higher education challenged the qualifications of those nurses—the majority—who had not attended college, especially because the college programs did not and still do not accept transfer credit from noncollege training programs. As a result, the drive toward professional status, or **professionalization,** inadvertently limited the ANA's power by alienating most practicing nurses from the organization. As a result, only a small fraction of nurses have ever belonged to it.

The increased emphasis on educational qualifications has reinforced nursing's hierarchical structure. At the bottom of the hierarchy are nursing assistants who, as described in Chapter 10, receive minimal training. Next are the **licensed practical nurses (LPNs),** who have approximately one year of classroom and clinical training and provide mostly custodial care to patients. The top tier comprises **registered nurses (RNs).** Eighteen percent of RN students are minorities, compared with 30 percent of the much lower-status LPN students (National League for Nursing, 1997a:58, 163).

Registered nurses themselves divide into four tiers. At the bottom of this hierarchy are **diploma nurses,** who receive their training through two- or three-year hospital-based diploma programs. Next are nurses who hold associate degrees in nursing from two-year community college programs and then nurses who hold Bachelor in Nursing degrees from four-year colleges or universities. Finally, at the top of the RN hierarchy are **advanced practice nurses,** such as nurse practitioners and nurse-midwives, who have postgraduate training in specialized fields. All advanced practice nurses enjoy considerably more autonomy, status, and financial rewards than do other nurses, including the right to prescribe some medications in most states (Lewin, 1993).

Nursing's leadership has achieved considerable success in its push to increase the educational qualifications of practicing nurses. Between 1977 and 1996, diploma nurses fell from 23.2 percent to only 6.0 percent of nursing graduates (National League for Nursing, 1997b:40). Associate degree programs, however, continue to attract more students than do bachelor degree programs, and grew from 46.7 percent of students in 1977 to 59.8 percent in 1996 (National League for Nursing, 1997b:40). Associate degree programs have remained popular because many hospitals believe those programs provide the best practical training and so prefer to hire graduates from those programs.

Although still only a small fraction of nurses, the greatest growth has occurred in graduate degree programs for advanced practice nurses. These programs first appeared during the 1960s, in response to projections of a coming shortage of doctors. As of 1996, 35,715 students were enrolled in masters degree programs and 2,994 in doctoral programs (National League for Nursing, 1997c:7, 12).

The increased educational qualifications of nursing have enabled it to achieve **semiprofessional** status, achieving some but not all of the hallmarks of a profession. Although most nurses consider themselves professionals and although nurses have more autonomy and status than in the past, they remain subordinate to doctors. Not only do doctors continue to determine much of nurses' working conditions, but doctors also are involved in setting educational and licensing standards for nursing. Moreover, despite the growth of nursing colleges and graduate degree programs, nursing has yet to develop public confidence that it has a truly independent knowledge base. However, the National Institute of Health now includes a separate institute specifically for funding nursing research, suggesting some change in public attitudes. Moreover, "in health policy circles, nurses are finding a stronger voice: a nurse now sits on the committee that accredits hospitals throughout the nation. Nurses run and accredit home-health services [and] about a dozen nurses are presidents or chief executives of hospitals" (Lewin, 1993).

Despite the increase in nurses' levels of education, caring seems to remain more central to nurses' work than to medical care, even at the highest levels of the nursing hierarchy. Using data collected during three years of observing nurse practitioners and family practice doctors, Sue Fisher (1995) concluded that nurse practitioners spend more than five times as long with each patient compared with doctors, using this additional time to gain a **holistic** sense of their patients' clinical problems and social situations. Whereas doctors typically rely on close-ended questions, tightly control which topics are discussed during patient visits, and seek to close discussions quickly, nurse practitioners rely heavily on open-ended questions, give patients more freedom to open topics, and do not push to close discussions. In addition, whereas many doctors routinely reinforce their dominance both verbally and nonverbally (by, for example, never addressing patients by name or implying that patients cannot accurately describe their own problems), nurse practitioners downplay differences in status between themselves and patients and assume that patients can accurately assess their own situations. On the other hand, like doctors, nurse practitioners retain final authority in patient/provider interactions—opening and closing discussions, asking most of the questions and thus determining which topics will be discussed, and, in the end, defining the nature of the problem.

The Impact of Changes in the Health Care System

Beginning in the late 1940s, hospitals assigned most direct patient care to LPNs and nursing assistants, and used RNs mostly as administrators. During the 1970s, however, **corporatization** and the resultant increased concern about controlling costs led hospitals to adopt "**primary nursing**" (Brannon, 1996). Under primary nursing, hospitals replaced many LPNs with RNs and reunified nursing tasks. Because RNs could perform a wider range of tasks than could LPNs and do so more efficiently, hospitals hoped to save costs (even though RNs earned more per hour than LPNs) by hiring fewer nursing personnel overall. Nursing leaders supported primary nursing because they believed it would increase job opportunities for RNs. They also hoped it would increase nurses' occupational status by allowing nurses to claim responsibility for a given set of patients and thus forcing doctors to recognize nurses' knowledge about the daily status and care of those patients.

Although primary nursing gave RNs greater control over patient care, their professional standing did not improve (Brannon, 1996). Instead, by making nurses more directly accountable for patient care to both medical and administrative supervisors, it reinforced nurses' relatively low position in the hospital hierarchy. In addition, by making RNs responsible for many of the labor-intensive, menial tasks formerly performed by LPNs, it deprofessionalized RNs' daily work. Moreover, in addition to these tasks, RNs remained responsible for the administrative and technical tasks that they had been doing all along. As a result, primary nursing intensified the workload of RNs (Aiken et al., 1996). On the other hand, the shift to primary nursing made it harder for hospitals to replace RNs with nonprofessional workers, thus fostering a nursing shortage during the 1980s and driving up wages. In turn, this shortage of nurses and their higher wages contributed to the decline in primary nursing during the 1990s, as hospitals scrambled to replace RNs with LPNs at the bedside (Brannon, 1996). At the same time, external cost control pressures (described in Chapter 10) have encouraged hospitals to shift services from inpatient care to less expensive outpatient clinics, where fewer RNs are needed, nursing salaries are lower, and nursing jobs are less prestigious. As a result, the shift of nursing out of hospitals has hurt RNs in their continuing struggle for professional status.

The rise of specialized nursing, on the other hand, has increased the professional status of at least some nurses. According to sociologist Andrew Abbott (1988), occupations rarely gain full professional dominance over directly competing occupations. Instead, occupations typically achieve professional status by carving out niches for themselves where there is less competition. Recent research suggests that nurses can gain increased status through doing what would otherwise be low status work if that work affords them both recognition of their specialized knowledge and public respect for taking on work perceived as dangerous and unpleasant (Aiken and Sloane, 1997). This appears to have happened serendipitously with the development of "dedicated" AIDS wards (devoted solely to caring for persons with AIDS) during the 1980s. A nationwide survey conducted in 1988 found that, compared with nurses on other wards, nurses on dedicated AIDS wards enjoyed greater professional status, control over their work environment, and professional autonomy, as well as better relations with physicians, less burnout, and less emotional exhaustion (Aiken and Sloane, 1997). On these wards, doctors were willing to cede some autonomy and responsibility to nurses because doctors were not particularly interested in providing the low-technology, palliative care persons with AIDS most often need. In addition, on these units, nurses gained specialized knowledge as great as, if different from, that of doctors. Equally important, whereas on general wards each doctor shares only a handful of patients with each nurse, on dedicated AIDS wards doctors and nurses routinely work together on the same patients, giving doctors more opportunities to witness nurses' expertise and thus making doctors more willing to treat nurses as colleagues. Finally, nurses' willingness to do dangerous and often unpleasant work caring for stigmatized patients enhanced their public image as dedicated professionals.

The Impact of Changing Gender Roles

Changing gender roles in the broader society has the potential either to help or hinder nursing's attempts to professionalize. Over the last three decades, as women have gained entry to other fields, the most academically qualified women students increasingly have chosen to enter medicine, pharmacy, or biological research instead of nursing (*New York Times,* 1999a; Williams, 1988). Thus, nursing no longer attracts the type of students it once could have counted on for its future leadership. In addition, nursing now attracts fewer white students and middle- or upper-class students. Given existing social prejudices, these changes are likely to reduce the status of nursing even if the quality of students remains constant.

Changing gender roles have not only encouraged women to seek careers other than nursing, but also have opened nursing to men. Men currently constitute 5.4 percent of all employed nurses and 12.1 percent of nursing students (Moses, 1997:33; National League for Nursing, 1997b:67). Because nursing is so strongly identified with femininity, working as a nurse presents men with a serious conflict between their gender identity and their work identity. Christine Williams (1989) found that men typically respond to this conflict by stressing the differences between what they do and traditional nursing—de-emphasizing nurturing while emphasizing their technical skills, administrative expertise, or use of physical strength.

Despite these difficulties, working as a nurse offers men substantial benefits. Williams (1989:95) points out that, "as in other female-dominated occupations, men are overrepresented in the most prestigious and best paying specialties" and in administrative positions. This occurs for two reasons. First, on average male nurses have more years of education than female nurses. Second, both male and female doctors more often respect, support, and socialize with male nurses than with female nurses, giving the men help in their careers and encouragement to enter more prestigious subfields. According to Williams (1992), whereas women in nontraditional fields (such as medicine) often encounter a "**glass ceiling**" caused by conscious discrimination and unconscious social expectations that limit their career progress, men in nursing, as in other predominantly female fields such as social work, encounter a "**glass escalator**" which moves them into administrative positions unless the men actively resist. It seems, then, that entering a traditionally female field such as nursing benefits individual male nurses. Whether the increasing entry of men into nursing will improve the overall status of the field, counteract the loss of academically superior and socially prestigious women students, and raise the status of the field overall, however, remains to be seen.

Nurse-Midwifery: The Limits of Specialization

The example of nurse-midwifery, one of the oldest forms of advanced practice nursing, illustrates both the benefits and the limitations of seeking professional status through carving out a specialized niche.

Throughout the nineteenth century, almost all American babies were delivered at home by lay midwives who lacked specialized training and worked within their own geographic or ethnic communities (Wertz and Wertz, 1989). By the 1920s, however, most Americans had come to believe that doctor-assisted childbirth was safer and, certainly, less painful. Yet few doctors were interested in providing care to poor or rural women. Responding to this need, in 1925 the Frontier Nursing Service opened the first school for nurse-midwives to serve Kentucky's rural poor. **Nurse-midwives** would be registered nurses who additionally received formal, nationally accredited training in midwifery. The students who trained in Kentucky learned not only to deliver babies but also to provide all needed prenatal and postnatal care. Seven years later, the Maternity Center Association began training nurse-midwives to serve New York City's urban poor.

These two organizations remained the only sources of nurse-midwives until the 1950s, when several universities, responding to widely publicized reports of an impending shortage of doctors, opened training programs. As of 1999, 50 colleges and universities offer accredited training programs in nurse-midwifery, mostly at the master's level. In addition, the American College of Nurse-Midwives recently began accrediting programs to train individuals who have no nursing background as "certified midwives." These programs will add basic education in health skills and general medical science to the usual graduate midwifery curriculum.

Like previous generations of nurse-midwives, current nurse-midwives are expected to work primarily for bureaucratic organizations in under-served poor and rural areas. However, unlike previous generations of nurse-midwives who had functioned largely independent of doctors and hospitals, these new nurse-midwives are expected to deliver babies solely in hospitals and to take responsibility solely for normal births, which doctors considered routine, uninteresting, and unremunerative.

From its beginnings, then, nurse-midwifery was designed to avoid threatening medical dominance. Nevertheless, during the 1970s and 1980s growing numbers of nurse-midwives began to pose a threat by opening private practices with only loose connections to the doctors who provided their back-up support. By 1990, 10 percent of nurse-midwives worked in private practice (Lehrman, 1992).

This threat to medical dominance, however, was short-lived, for changes in insurance coverage during the early 1990s made independent practice virtually impossible for nurse-midwives. The costs of a standard malpractice insurance policy rose from $35 in 1983 to as high as $13,500 in 1998 (Gordon, 1989; Rooks, 1997:86). Insurance became even more expensive for midwives who attended home births, as well as for the doctors who worked with them. Yet only about 10 percent of nurse-midwives (compared with 73 percent of obstetricians) have ever been sued for malpractice (American College of Obstetrician–Gynecologists, 1998; Lehrman, 1992). Moreover, studies have consistently found that nurse-midwives offer care to low-risk pregnant women at least as safe as if not safer than that offered by doctors, even after taking into account the small number of midwifery clients who in the end require medical care, apparently because nurse-midwives spend more time with their patients, both prenatally and during labor and delivery, and place greater emphasis on patient counseling, education, and emotional support (MacDorman and Singh, 1999; Rooks, 1997:295–343). As a result of the rise in insurance premiums, independent practice by nurse-midwives has virtually ended, and the percentage of nurse-midwives attending home births dropped from 14 percent in 1982 to less than 5 percent in 1994 (Rooks, 1997:192). Most nurse-midwives now work in hospitals or doctors' offices. In 1996, 96 percent of deliveries attended by nurse-midwives occurred in hospitals, 3 percent in freestanding birth centers, and only 1 percent in homes (American College of Nurse-Midwives, 1998).

Nurse-midwifery is legal in all 50 states, and nurse-midwives have legal authority to write prescriptions in 47 states (Reed, 1997). Thirty-one states

require private health insurers to reimburse nurse-midwives for their services, and all states reimburse midwives for serving **Medicaid** clients. However,
these regulations do not apply to employers who **self-insure,** setting aside a
pool of money to pay health care costs for their employees rather than offering health insurance as a benefit (American College of Nurse-Midwives,
1998); self-insurance now covers 70 percent of insured U.S. workers.

In sum, nurse-midwives have gained considerable autonomy and public
recognition, as well as an established place for themselves in the health care
system, through specialized training and providing care to a specific population. Their abilities to gain greater professional status and independence
from medical control, however, have been thoroughly blocked.

Pharmacy: The Push to Reprofessionalize

Unlike nursing, pharmacy meets the three criteria laid out in the previous
chapter that define a profession: the autonomy to set its own educational and
licensing standards and to police its members for incompetence or malfeasance; a body of specialized knowledge, learned through extended, systematic
training; and public faith that its work is grounded in a code of ethics. Like
medicine, however, pharmacy's history illustrates how corporatization can
limit an occupation's ability to retain crucial professional prerogatives. In addition, its history shows the continued difficulties health care occupations
face in maintaining professional status despite medical dominance.

Gaining Education, Losing Professional Prerogatives

Pharmacists' role has changed considerably during the last half century,
placing their professional status in jeopardy (Birenbaum, 1982). In the past,
pharmacists needed to use complex skills to store, compound, and accurately dispense the drugs that doctors prescribed. Now, however, pharmaceutical companies deliver drugs in forms suitable for dispensing, leaving
pharmacists with few tasks other than counting, selling, and occasionally
offering advice on drugs to consumers or health care providers. At the same
time, whereas before about 1970 more than half of pharmacists owned
their own businesses, with the associated responsibilities and rewards, now
most work as employees of drugstore chains, supermarket chains, or hospitals. Incomes for pharmacists remain reasonably high, with a median income of $59,276 (U.S. Bureau of Labor Statistics, 1998:197), but working
conditions can be poor, especially in chain stores, where twelve-hour shifts,
staffing shortages, and pressure to fill prescriptions quickly are common
(Stolberg, 1999b). Like doctors, then, pharmacists have experienced proletarianization: they are more economically vulnerable; have less decision-
making autonomy; and no longer set their own working conditions, own
their tools or workspaces, or maintain individual relationships with freely
chosen clients.

As pharmacists' role has shrunk, however, their education has expanded. Virtually all of the nation's pharmacy schools have replaced their older four- and five-year degree programs with six-year programs leading to Doctorates in Pharmacy. These new programs place less emphasis on technical aspects of drug manufacturing and more on the complex subject of drug effects and interactions (Broadhead and Facchinetti, 1985:427). These changes in education, combined with changes in pharmacists' role, have created an identity crisis, as pharmacists consider themselves professionals but increasingly find their professional autonomy constrained (Birenbaum, 1982; Broadhead and Facchinetti, 1985).

The Growth of Clinical Pharmacy

This identity crisis has stimulated interest among pharmacists in regaining their former level of professional status, or **reprofessionalizing.** To do so, pharmacists, beginning in the early 1970s, began touting research studies that suggested that many hospital patients become ill or die because of drug errors—summed up in the influential book *Pills, Profits, and Politics* (Silverman and Lee, 1974:262) as "wrong drug, wrong dose, wrong route of administration, wrong patient, or failure to give the prescribed drug." Because pharmacists considered themselves more knowledgeable than doctors about drug actions, reactions, and interactions, they argued that the best way to limit drug errors was to encourage **clinical pharmacy,** in which pharmacists would actively advise doctors on drug treatment, while less-skilled pharmacy technicians would take over the routine tasks of storing and dispensing drugs.

The push for clinical pharmacy garnered unintended support from changes in hospital procedures (Broadhead and Facchinetti, 1985). Most hospitals now have pharmacists dispense and deliver medications to each patient daily. This system gives pharmacists regular access to patient records, including all records regarding drug treatments, health status, and progress. As a result, pharmacists can evaluate the effects—both positive and negative—of doctors' drug prescriptions and learn to predict when prescriptions are likely to cause health problems.

Other support for clinical pharmacy has come from changes in the legal system. Whereas during the 1980s, hospitals discouraged pharmacists from documenting medication errors, out of fear that such documentation might *increase* hospitals' legal liability, during the 1990s, court decisions that held pharmacists legally responsible for monitoring medications led hospitals to encourage clinical pharmacy as a means of *reducing* hospitals' legal liability. Nevertheless, pharmacists' concern about preserving cordial roles with doctors, who remain the dominant professionals in the health care arena, has led them to use caution in critiquing doctors' medication decisions. As one pharmacist described

I would like to talk to the physician face-to-face. You're trying to correct the mistake in a nonthreatening way. You know, "I'm not trying to put you down for making this mistake, but it's something that I want you to reconsider." It's not that I'm afraid to confront a physician, bending over backwards because he's up there and I'm down here. It's just that I want to maintain a relationship and the way you interact is important (Broadhead and Facchinetti, 1985:432).

The Development of Pharmaceutical Care

The growth of clinical pharmacy had little impact on pharmacists who worked outside of hospitals and increased divisions between them and hospital pharmacists. The development of "pharmaceutical care," however, has given these two groups a unified program for cementing the professional status of pharmacists (personal communication, Dr. Jeanine Mount, August 1999). **Pharmaceutical care** refers to idea that pharmacy's central mission should be to advise consumers (rather than doctors, as in clinical pharmacy) regarding the proper use of medications, based on knowledge gathered through **controlled** studies based on **random** samples (Hepler and Strand, 1990). Its rapid adoption by virtually all pharmacy associations reflects both current concerns about cost control and the belief that pharmacy care will provide a tool to protect pharmacy's professional status.

The Impact of Managed Care

Ironically, **managed care** and **utilization review,** which have limited doctors' professional status, have increased the professional power and status of at least some pharmacists. In the last decade, many pharmacists working for health care businesses that use managed care (including insurance plans, hospitals, and nursing homes) have become actively involved in developing protocols for doctors to follow in prescribing drugs. Pharmacists also may participate in utilization review, monitoring doctors' use of prescription drugs. Similarly, some pharmacists now serve on committees responsible for developing **formularies**—official lists of drugs, published by health care businesses, that are considered the most cost-effective treatments for given conditions and that doctors working with these businesses are expected to prescribe. Doctors may prescribe other drugs with special authorization, but getting such authorization takes time and energy and requesting authorization too frequently can jeopardize a doctor's position within the business. As a result, doctors only rarely prescribe outside of formularies. For example, the antacid Tagamet dropped from being one of the five top-selling drugs in the country to near-oblivion as a result of pharmacists' success in replacing it on formularies with the safer, easier to use Zantac (*Fortune,* 1999). Similarly, pharmacists' decision to remove the weight-loss drug Redux from formularies because of their concerns about life-threatening side-effects forced the federal Food and Drug Administration to examine the drug more closely and, in the end, to ban its sale. As a result, Redux's manufacturer, Wyeth-Ayerst, lost millions of dollars in sales and additional millions in recall and legal fees.

The rise of managed care has also stimulated growth in **disease management** by pharmacists. Disease management is a form of pharmaceutical care, and refers to situations in which pharmacists are given responsibility for monitoring the use of prescription drugs by certain patients, typically those with chronic conditions that require constant attention to medication. Pharmacists engaged in disease management counsel patients, monitor the impact of medications on patients, and, in some circumstances, prescribe drugs themselves. Disease management has been adopted by managed care organizations as a way to control costs by shifting care from doctors to lower-paid pharmacists and by preventing medication errors. As of 1999, Mississippi reimburses pharmacists under Medicaid for disease management of certain groups of patients, Wisconsin and Missouri are considering adopting similar regulations, and 21 states have given pharmacists legal authority to initiate or modify drug treatment. Importantly, some of these laws allow direct reimbursement to pharmacists outside of MCOs, suggesting the possibility of growing professional autonomy for pharmacists in the future.

Osteopathy: A Parallel Profession

Osteopathy provides an example of a health care occupation that has achieved professional status almost equal to that of medicine. Osteopaths function as what Walter Wardwell (1979:239) has called "**parallel practitioners,**" performing basically the same roles as **allopathic doctors** while retaining professional autonomy and at least the remnants of a fundamentally different ideology about illness causation. The history of osteopathy demonstrates the potential costs of gaining professional status in the face of medical dominance.

Nineteenth Century Roots

Osteopathy was founded by Andrew Taylor Still, a self-taught allopathic doctor (Gevitz, 1988). The deaths in 1864 of three of his children from meningitis, coupled with his belief that the use of any drug was immoral, provoked Still to investigate alternatives to allopathic medicine. The system Still eventually developed drew on the popular contemporary concept of "magnetic healing" (Gevitz, 1988:126–127). **Magnetic healers** theorized that an invisible magnetic fluid flowed through the body and that illness occurred when that flow was obstructed, unbalanced, inadequate, or excessive. They believed that by moving their hands along patients' spinal cords, they could correct problems in the magnetic fluid and thus cure illness. Andrew Still adopted this theory essentially intact, although he attributed health and illness to problems in the flow of blood rather than the flow of magnetic fluid.

During the next few years, Still also studied the work of local bonesetters, whose work consisted primarily of setting broken and dislocated bones and joints and secondarily of treating joint problems through extending and manipulating limbs. Still's experiences convinced him that such manipulations could cure a wide variety of illnesses.

Combining the theories of magnetic healing and bonesetting, Still concluded that disease occurs when misplaced bones, especially of the spinal column, interfere with the circulation of blood. He named his new system of spinal manipulation "osteopathy" from the Greek words for "bone" and "sickness." After the germ theory of disease became widely accepted, Still incorporated it into his theory by arguing that spinal problems predispose individuals to infections and that correcting spinal problems can help the body fight infection. To date, no research has demonstrated clearly whether osteopathic treatment has any effect—whether positive or negative—on clients. (The same, of course, could be said for many of the drugs and procedures used by allopathic doctors, as the last chapter discussed.).

Professionalizing Osteopathy

In 1892, Still established the American School of Osteopathy and began accepting students for a four-month course of instruction. Still lengthened the course to twenty months after the governor of Missouri vetoed an osteopath licensing bill on the grounds that osteopaths lacked sufficient education; four years later, a similar bill passed easily. By 1920, osteopathic colleges required a period of instruction equal to that offered by medical colleges.

In 1897, Still helped found the American Osteopathic Association (AOA). As Gevitz (1988:132–133) describes,

> from its inception, the AOA actively worked to secure the conditions necessary for the movement to obtain professional recognition. It fought for independent boards of registration and examination to give the profession autonomy; it significantly lengthened the standard course of undergraduate training and supported ongoing research projects; and it championed a code of ethics while combating the growth of impostors and imitators.

The AOA proved highly successful. By 1901, and despite strong opposition from doctors and medical societies, fifteen states legally recognized osteopathy (Gevitz, 1988:132). By 1923, forty-six of the forty-eight states licensed osteopaths, although many states gave them only limited privileges and required them first to pass a basic sciences examination written and administered by allopath-controlled licensing boards.

Although threats from allopathic medicine have failed to eliminate osteopathy, changes from within raise questions about osteopathy's future as an independent field. By the 1920s, most osteopaths had concluded that to compete with allopathic doctors they would have to offer a similar range of patient services. As a result, osteopaths increasingly treated acute as well as chronic illness. Osteopathic colleges continued to teach spinal manipulation but added courses in surgery and obstetrics, often taught out of medical textbooks. By the end of the decade, in a major break with its founder, the AOA mandated that osteopathic colleges provide a course in "supplementary therapeutics," including drugs. Thus osteopathy began moving toward a merger with allopathic medicine.

Despite these changes, many allopathic doctors still disdained osteopaths. Although osteopathic education had improved, it had not kept up with the changes in allopathic education, leading many states to grant only restricted privileges to osteopaths. To combat this problem, the AOA adopted a series of reforms between 1935 and 1960, including requiring three years of college for admission to osteopathic colleges; improving the curriculum, facilities, and faculty at those colleges; and strengthening internship programs at osteopathic hospitals. As a result of these changes, by 1960 osteopaths had received unlimited privileges to practice in thirty-eight states (Gevitz, 1988:144).

The Waning of Osteopathic Identity

Despite these reforms, osteopaths still lacked the professional autonomy and status of allopathic doctors, who outnumbered them by at least twenty to one throughout the 1900s (Gevitz, 1988:146). This situation led osteopaths in California, the state where osteopathy was most entrenched, to strike a bargain in 1962 with their allopathic counterparts. Two thousand of the 2,300 California osteopaths agreed to dissolve their ties with the AOA, stop using their osteopathic degrees, and accept new medical degrees. The California osteopathic hospitals and colleges agreed to become allopathic institutions, and the state osteopathic organization agreed that the state would stop issuing osteopathic licenses.

Although at the time many osteopaths worried that this move would weaken osteopathy, the reverse proved true. A substantial number of both allopathic and osteopathic doctors opposed the merger, making any further mergers unlikely. In addition, the continuing professional problems of the former California osteopaths convinced osteopaths elsewhere that merging would not end their problems. Thus, interest in pursuing a broader merger never developed. Meanwhile, both federal and state legislators and regulators interpreted the AMA's support for the merger to mean that osteopathic and allopathic doctors were essentially equivalent. Partly as a result, by the 1970s, osteopaths had received unrestricted privileges in all fifty states and now have essentially the same relationship with insurance providers as do allopathic doctors. Currently, almost 39,000 osteopaths practice in the United States, more than twice the number in practice in 1976 (American Osteopathic Association, 1999, personal communication).

Osteopathy, then, no longer faces serious threats from the outside. Its existence remains threatened, however, by its success (Gevitz, 1988). Osteopaths now receive training and hospital privileges virtually identical to allopathic doctors and interact essentially as equals with other doctors. Although osteopaths occasionally use spinal manipulation, generally they use the same treatment modalities as allopaths. As a result, ties among osteopaths have waned while those to allopathic doctors have grown. At the same time, the virtual elimination of differences between allopathic and osteopathic treatment and theory has reduced osteopaths' sense of a strong separate identity.

On the other hand, the growth of the consumer health movement and the rise of interest in alternative medicine since the 1970s have given a new burst of life to osteopathy. Modern consumers are increasingly sympathetic to osteopaths' orientation toward patient care, which in general is more holistic and humanistic than that found among allopathic doctors. In addition, consumers increasingly have sought less interventionistic treatments, such as osteopathic manipulation, either instead of or in addition to allopathic treatment. As a result, whereas in 1968, only five osteopathic schools remained, graduating a total of 427 students, by 1998, 19 schools graduated 2,436 students (Barzansky et al., 1998:806). Similarly, whereas in 1970, osteopathic students represented 5 percent of all medical students, by 1996 they represented 12 percent.

In sum, the history of osteopathy demonstrates the difficulties a parallel profession faces in maintaining an independent identity once no longer forced to do so by discrimination from the dominant occupation, as well as the importance of retaining an ideological justification for its separate existence.

ALTERNATIVE HEALTH CARE PROVIDERS

The occupations described to this point all share to a greater or lesser extent allopathic medicine's understanding of how the body and illness work and all have found significant roles within the mainstream health care system. The occupations described in the remainder of this chapter are sufficiently divorced from mainstream American medicine—neither widely used nor taught in medical schools or other medical institutions—to be considered **alternative** or **complementary therapies,** even if they sometimes are covered by health insurance.

With a few exceptions (such as chiropractic, lay midwifery, and acupuncture), little is known about the effectiveness of alternative healing techniques, which include meditation, reflexology, faith healing, herbal therapies, and colonics. Because allopathic medicine has dominated the American health care system for so long, researching alternative therapies has been all but impossible. Scientific testing requires large investments of time and money, generally available only from the government, universities, or pharmaceutical companies. Until recently, researchers who wanted to study alternative techniques faced near-insurmountable barriers to obtaining funding, especially from pharmaceutical companies, which have no reason to fund research on herbs or techniques that they cannot patent. In addition, researchers who studied these techniques faced great difficulties in getting their results published in the prestigious medical publications that set the standards for health care practice.

In 1992, however, and in a major break with past policy, the U.S. Congress voted to establish within the National Institutes of Health (NIH) an Office for the Study of Unconventional Medical Practices (later renamed the Office of Alternative Medicine). The major impetus for this legislation came from former California Congressman Berkley Bedell, who had experimented with alternative therapies after his doctors diagnosed him with terminal cancer. His apparently successful experiences convinced him that such treatments warranted wider study and use. Bedell's success in getting this legislation passed reflects legislators' recognition of both the soaring costs of mainstream medical care and the growing public interest in alternative health care. In 1999, NIH budgeted $50 million for research into alternative healing, an increase from only $2 million in 1992, and upgraded the Office of Alternative Medicine into a full-fledged NIH center, the National Center for Complementary and Alternative Medicine.

Interest in alternative healing is growing not only among American consumers but also among allopathic doctors. As of 1999, 600 medical doctors and osteopaths belong to the American Holistic Medical Association, and 75 medical schools offer courses on alternative therapies. In addition, the University of Arizona offers a full program in "integrative medicine" combining allopathic and alternative techniques and philosophies. Even more impressive, a survey distributed to allopathic doctors in several communities in Washington State and New Mexico found that more than 60 percent had referred a patient to an alternative health care provider at least once during the preceding year (Borkan et al., 1994). However, referrals most often occurred at the patient's suggestion, when conventional treatment had failed, or when the physician believed that the patient's problem was emotional rather than physical.

In the remainder of this chapter, I describe five groups of alternative health care providers. The first two, chiropractors and lay midwives, at least sometimes use the language of science to justify their work. The three remaining groups, *curanderos,* Christian Science healers, and traditional acupuncturists, base their practices in traditional beliefs unrelated to the western scientific worldview.

Chiropractors: From Marginal to Limited Practitioners

Unlike osteopaths, **chiropractors** have fully retained their unique identity (Table 12.2). The history of chiropractic illustrates how an occupation

| Table 12.2 | **Limited and Marginal Health Care Occupations** | | |

		LIMITED RANGE OF CARE	
		Yes	No
MARGINAL SOCIAL POSITION	Yes	Lay midwives	Traditional healers
	No	Chiropractors	Allopathic doctors

made of **marginal practitioners,** who treat a wide range of physical ailments and illnesses but have low social status, can become, like podiatrists, optometrists, and dentists, **limited practitioners**—confining their work to a limited range of treatments and bodily parts and thereby gaining greater social acceptance (Wardwell, 1979:230).

Early History

The roots of chiropractic nearly mirror those of osteopathy. Chiropractic was founded in 1895 by Daniel David Palmer, who coined the term from the Greek words for "hand" and "practice." Like Still, Palmer studied magnetic healing and spinal manipulation and concluded that spinal manipulation could both prevent and cure illness. However, whereas Still argued that spinal problems foster disease by restricting blood flow, Palmer argued that spinal problems foster disease by restricting nerves.

In 1896, Palmer founded the first chiropractic school to teach his techniques of spinal manipulation. The field really began growing after his son, B. J. Palmer, took over the school in 1907. By 1916, about 7,000 chiropractors had opened practices; by 1930, that number had more than doubled, as schools opened around the country (Wardwell, 1988:159, 174).

Although from the beginning, some allopathic doctors studied chiropractic and taught at chiropractic schools, B. J. Palmer attempted to sharply separate chiropractic and allopathic medicine. Those who shared his philosophy and used only spinal manipulation became known as "straights." Most chiropractors, however, found Palmer's monocausal theory of illness too simplistic and limiting. As a result, Palmer's power waned as most chiropractors adopted a wide variety of therapeutic techniques. These "mixers" treated not only musculoskeletal problems but also other illnesses, as well as providing obstetrical and mental health care (Wardwell, 1988:162–165).

The Fight Against Medical Dominance

The American medical establishment greeted the emergence of chiropractic with the same hostility it had demonstrated toward osteopathy. To eliminate these competitors, the AMA and its regional organizations during the 1930s and 1940s filed lawsuits—many of them successful—against more than 15,000 chiropractors for practicing medicine without a license.

To further restrict chiropractic, the AMA pressed for legislation requiring prospective chiropractors to pass statewide basic science examinations written by allopathic-controlled boards. Ironically, this requirement strengthened rather than weakened chiropractic by forcing the field to raise its previously low educational standards. (As with early allopathic and osteopathic schools, early chiropractic schools accepted essentially all who could pay tuition and offered only a few months of training.) Standards improved most dramatically during the 1940s, when the National Chiropractic Association (NCA) established accrediting standards for schools and when tuition money from veterans studying chiropractic under the federal GI Bill provided the funds schools needed to meet those standards. Since 1968, all chiropractic schools have required two years of college for admission and most states require four years of chiropractic schooling for licensure.

Similarly, allopathic medicine's legal war against chiropractic in the end benefitted the latter field. Under heavy lobbying from the AMA, when Congress first authorized the **Medicare** program in 1965, it voted to refuse Medicare reimbursement to chiropractors, as well as to clinical psychologists, social workers, physical therapists, and a variety of other competitors with medicine. Outraged chiropractic patients responded with a massive public letter-writing campaign, which led Congress in 1972 to pass legislation extending Medicare coverage to chiropractic services, despite the lack of scientific research available at the time on its effects. This set the stage for state legislatures to require other insurance plans to reimburse for chiropractic care, at least in certain situations; by 1985, forty-two states had done so (Wardwell, 1988:179).

In 1974, the last of the fifty states passed legislation licensing chiropractors. Yet organized medicine continued to limit the ability of chiropractors to practice freely. In addition to fighting legislation designed to allow chiropractors to receive private insurance reimbursement, the AMA banned contact between chiropractors and allopaths, making it impossible for chiropractors and allopaths to refer to each other. In response, chiropractors and their supporters filed antitrust suits in the late 1970s against the AMA, various state medical associations, the American Hospital Association, and several other representatives of organized medicine (as well as the AOA) alleging that these organizations had restrained trade illegally. Chiropractors and their defenders eventually won or favorably settled out of court all the suits. As a result, overt opposition to chiropractic ended.

Current Status

These changes have allowed chiropractors to solidify their social position. Although, as has been true throughout its history, chiropractic remains most common in the Midwest and west and in smaller communities, use is widespread (U.S. Bureau of Labor Statistics, 1998:181–183). A 1997 national random survey found that 11.0 percent of English-speaking U.S. residents had visited a chiropractor in the last year, for an average of 9.8 visits (Eisenberg et al., 1998:1572). A separate survey of chiropractors' patient records found that chiropractic patients were typically between 30 and 50 years old and married, and slightly more likely to be female than male (Hurwitz et al., 1998).

In 1995, the sixteen schools of chiropractic graduated almost 3000 students, 29 percent of whom were female (Snyder, 1997). As of 1996, approximately 44,000 chiropractors worked in the United States, most (about 70 percent) in solo practice (U.S. Bureau of Labor Statistics, 1998:182). Median net income for chiropractors is $80,000—considerably below the $124,000 median for general and family practitioners but for a much shorter work week, averaging 42 hours (U.S. Bureau of Labor Statistics, 1998:182–183, 188). These figures alone suggest chiropractic's success.

That success, however, is bounded by chiropractors' status as limited practitioners. Insurers now often pay for chiropractic services—55.7 percent of those who use chiropractic services have full or partial coverage—but usually will do so only for treating specific conditions in specific ways (Eisenberg et al., 1998:1574; Shekelle, 1998). State licensure laws sometimes set similar limits, as does patient demand—despite chiropractic's desires to treat a broader range of problems, most patients go to chiropractors for treatment of acute lower back pain, and only one percent for anything other than musculoskeletal problems (Hurwitz et al., 1998).

Nevertheless, chiropractors continue to push for a wider role in health care. Many chiropractors believe spinal problems underlie all illness and spinal manipulation can cure most health problems, from asthma to cancer (*Consumer Reports*, 1994). In addition, many believe they can serve effectively as **primary care** providers and now advertise heavily that they offer care for the whole family throughout the life course.

Current research suggests that chiropractic care can help those with acute lower back pain but is unlikely to help others. A recent study in which patients with acute lower back pain were randomly assigned to receive either chiropractic care, physical therapy, or simply an educational booklet on managing back pain found that both chiropractic and physical therapy resulted in reduced symptoms compared with the educational booklet. However, the improvements were slight and no better than those achieved through the cheaper option of physical therapy. Moreover, the three therapies did not differ significantly in number of days of reduced activity or in rate of recurrence of back pain (Cherkin et al., 1998). Other studies suggest that spinal manipulation might help some patients with neck pain, but to date none has tested whether manipulation is *more* effective than other treatments or whether its risks (including delays in seeking medical care, strokes brought on by spinal manipulation, and radiation poisoning from the full-body x-rays used by some chiropractors) might outweigh any potential benefits (Shekelle, 1998). Finally, no reputable research has yet demonstrated any benefits from chiropractic for health problems other than neck and back injuries. Nor does it seem likely that future research will do so, as the basic principles of chiropractic simply do not mesh with current scientific understanding of human biology.

Lay Midwives: Limited but Still Marginal

The history of lay midwifery shows the difficulties members of an occupation face in gaining acceptance as limited practitioners when the occupation draws only from socially marginal groups—in this case, women, often from minority groups. Although until about 1910 lay midwives delivered the majority of American babies (Litoff, 1978:27), currently lay and nurse-midwives together deliver only 6.5 percent (National Center for Health Statistics, 1998a). However, that percentage, although small, has increased steadily since 1975, when the federal government began collecting statistics on midwife-assisted births. This section describes how these changes came about and how lay midwives have attempted to regain their lost position.

Midwives, Doctors, and the Struggle to Control Childbirth

Until well into the nineteenth century, Americans considered childbirth solely a woman's affair (Wertz and Wertz, 1989). Almost all women gave birth at home, attended by a lay midwife or by female friends or relatives. Although a few local governments during the colonial era licensed midwives, licensure laws did not survive past U.S. independence, so anyone who wanted to call herself a midwife could practice essentially without legal restrictions. Unlike nurse-midwives, who did not exist until the twentieth century, these **lay midwives** had no formal training but rather learned their skills through experience, and, sometimes through informal apprenticeships. Typically, they served only the women of their geographic or ethnic community. Doctors (all of whom were men) played almost no role in childbirth because Americans suspected the motives of any men who worked intimately with female bodies (Wertz and Wertz, 1989:97–98). Moreover, doctors had little to offer childbearing women beyond the ability to destroy and remove the fetus when prolonged labor threatened women with death. Midwives, meanwhile, could offer only patience, skilled hands, and a few herbal remedies.

During the late nineteenth century, Americans' willingness to have doctors attend childbirths gradually increased, as did doctors' interest in doing so. As described in the previous chapter, nineteenth century allopathic doctors faced substantial competition not only from each other but also from many other kinds of practitioners. As a result, doctors attempted to expand into various fields, from pulling teeth to embalming the dead to assisting in childbirth

(Starr, 1982:85). Doctors considered assisting in childbirth especially crucial because they believed that families who came to a doctor for childbirth would stay with him for other services (Wertz and Wertz, 1989:55).

As Americans' belief in science and medicine grew during the late nineteenth century, medical assistance in childbirth became more socially acceptable among the upper classes (Starr, 1982:59). Many women supported this change because it allowed them to obtain painkillers from doctors without feeling guilty for circumventing the biblical command to bring forth children in pain (Wertz and Wertz, 1989:110–113). In addition, because midwifery was not a respectable occupation for Victorian women, by the late nineteenth century middle- and upper-class women seeking a childbirth attendant had only two options: lower-class lay midwives or doctors of their own social class. As a result, having a doctor attend one's childbirth could both reflect and increase one's social standing (Leavitt, 1986:39; Wertz and Wertz, 1989). Ironically, however, doctors probably threatened women's health more than did midwives, for, although inexperienced or impatient midwives certainly could endanger women, doctors more often used surgical and manual interventions that could cause permanent injuries or deadly infections (Leavitt, 1983:281–292, 1986:43–58; Rooks, 1997).

The desire to obtain a monopoly on childbirth care had led doctors to voice opposition to midwives beginning in the mid-nineteenth century. These attacks escalated substantially in the early twentieth century (Sullivan and Weitz, 1988:9–14). Recent waves of immigrants had swelled the ranks of midwives and made them more visible and threatening to doctors, whose status, especially in obstetrics, remained low. Moreover, doctors now needed the business of poor women as well as wealthier women because the rise in scientific medical education had created a need for poor women patients who could serve as both research subjects and training material.

To expand their clientele, doctors attempted through speeches and publications to convince women that childbirth was inherently and unpredictably dangerous and therefore required medical assistance. In addition, doctors played on contemporary prejudices against immigrants, African Americans, and women to argue that midwives were ignorant, uneducable, and a threat to American values and that therefore midwifery should be outlawed. For example, writing in the *Southern Medical Journal*, Dr. Felix J. Underwood, the director of the Mississippi Bureau of Child Hygiene, described African-American midwives as "filthy and ignorant and not far removed from the jungles of Africa, with its atmosphere of weird superstition and voodooism" (1926:683), while the *Boston Medical and Surgical Journal* editorialized that midwives were "perhaps, an inevitable evil" in Europe but were "inconsistent with the methods and ideals of civilization and of medical science in this country" (1915:785).

Although these campaigns cost midwives many clients, they had little effect on the law. Many members of the public, and even many doctors (particularly those in public health), believed that trained midwives could provide satisfactory care, at least for poor and nonwhite women who might not be able to obtain doctors' services. Consequently, laws passed during this era tended to have quite lenient provisions. In the end, however, imposing lenient laws, rather than laws requiring upgraded midwifery training and skills, resulted in the deterioration of midwifery and its virtual elimination. The only exceptions were in immigrant and nonwhite communities in the rural south and southwest, where traditional midwives continued to conduct home births until at least the 1950s (Sullivan and Weitz, 1988:13–14).

The Resurgence of Lay Midwifery

By the second half of the twentieth century, childbirth had moved almost solely into hospital wards under medical care. Although childbearing women were grateful for the promise of pain relief and safety that doctors offered, all too often women nonetheless found the experience painful, humiliating, and alienating. Despite the absence of scientific support for such practices, doctors routinely shaved women's pubic area before delivery, strapped them on their backs to labor and delivery tables (the most painful and difficult position for delivering a baby), isolated them from their husbands during delivery and from their infants afterwards, and gave them drugs to speed up their labors or make them unconscious—all practices that scientific research would eventually find unnecessary or dangerous (Sullivan and Weitz, 1988).

Objections to such procedures sparked the growth of the natural childbirth movement during the 1960s and 1970s and forced numerous changes in obstetric practices. Most hospitals, for example, now offer natural childbirth classes. Critics, however, argue that the real purpose of these classes is to make women patients more compliant and convince them that they have had a natural childbirth as long as they remain conscious, even if their doctors use drugs, surgery, or forceps (Sullivan and Weitz, 1988:39).

By the late 1960s, many women had concluded that hospitals would never offer truly natural childbirth (Sullivan and Weitz, 1988:38–39). As a result, a tiny but growing number of women chose to give birth at home. For assistance, they turned to sympathetic doctors and to female friends and relatives, some of whom were nurses. Over time, women who gained experience in this fashion might find themselves identified within their communities as lay midwives. This new generation of lay midwives reflects the broader revolt against medicalized birth (Sullivan and Weitz, 1988:23–59).

Working as a lay midwife means long and uncertain hours with little pay. Most midwives, however, are motivated by ideological rather than economic concerns (Sullivan and Weitz, 1988:68–80). Although midwives recognize the need for obstetricians to manage the complications that occur in about 10 percent of births, they fear the physical and emotional dangers that arise when obstetricians employ interventionist practices, developed for the rare pathological case, during all births. Like nurse-midwives, lay midwives strongly believe in the general normalcy of pregnancy and childbirth and in the benefits of individualized, holistic maternity care in which midwife and client work as partners.

Lay midwives work almost solely in their homes or the homes of clients, and delivered 13,788 babies in 1996 (National Center for Health Statistics, 1998a:73). Research suggests that home births conducted by experienced lay midwives working with low-risk populations is at least as safe as doctor-attended hospital births, even after accounting for the small number of low-risk women who develop problems needing medical attention (Lewis, 1993; Sullivan and Weitz, 1988:112–132).

No national laws set the status of lay midwives. As of 1999, lay midwifery was definitely legal in twenty-nine states and illegal in seventeen, with the status in the remaining five states unclear (American College of Nurse-Midwives, 1999). In states where midwifery is illegal, midwives run the risk of prosecution for practicing medicine without a license and for child abuse, manslaughter, or homicide if a mother or baby suffers injury or death.

In states where lay midwifery is legal, midwives typically must abide by regulations restricting what sorts of clients they can take and what procedures they can use—for example, allowing them to take only clients under age 35 and forbidding them from suturing tears following deliveries. Licensed midwives typically must have a back-up doctor and must transfer their clients to medical care if the doctor so orders. Thus, licensure has given midwives some degree of freedom to practice in exchange for limited subordination to medicine (Sullivan and Weitz, 1988:97–111). At the same time, however, medical opposition remains strong and public support weak, although insurance companies do cover midwifery services in some states. As a result, lay midwives, even where licensed, cannot claim to have achieved social acceptance even as limited practitioners.

Curanderos

Curanderos are folk healers who function within Mexican and Mexican-American communities (Perrone et al., 1989; Roeder, 1988). In the United States, *curanderos* are used primarily by immigrants, as well as by some U.S.-born Mexican Americans, especially those who live in close-knit communities in southwestern towns and cities. In Denver, for example, doctors familiar with the Mexican-American community estimate that between 100 and 200 *curanderos* work out of their homes, advertising primarily by word of mouth (*New York Times,* 1999b). Some work for free, and some charge fees ranging from $5 to $100. A survey conducted in Denver found that 29 percent of a randomly selected sample of adult Hispanic patients at a low-_income clinic had visited a *curandero* at least once during their lives (*New York Times,* 1999b). Most did not use *curanderos* as a primary source of health care but instead went in addition to seeing a doctor, when medical care had failed, or when distance or poverty limited their access to medical care.

Theories and Treatments

Curanderos recognize both Western categories of disease, such as colds, and unique categories of illness, such as *susto* (Roeder, 1988). A common diagnosis, *susto* refers to an illness that occurs when fright "jars the soul from the body, in which case treatment consists of calling the soul back" (Roeder, 1988:324). *Curanderos* also sometimes trace illness to supernatural forces such as *mal de ojo,* or the evil eye.

Curanderos treat illness in a variety of ways, including herbal remedies, massage, prayer, and rituals designed to combat supernatural forces. They believe illness reflects all aspects of an individual's life—biology, environment, social setting, religion, and supernatural forces—and thus must be treated holistically. As a result, *curanderos* often spend considerable time listening to their clients. The successes *curanderos* sometimes achieve in treating their clients' illnesses thus derive not only from their knowledge of herbs and the healing powers of their clients' faith but also from the simple healing power of a sympathetic listener.

Becoming a Curandero

Individuals become *curanderos* through apprenticeships, typically with family members. Successful *curanderos* find that their practices evolve gradually from part-time work, paid primarily in goods and services, to more or less full-time, cash businesses.

The story of Gregorita Rodriguez, a *curandera* (female *curandero*) living in Santa Fe, New Mexico, who specializes in massage treatments, illustrates this process:

> Gregorita traces her own career as a *curandera* back to her grandmother, Juliana Montoya, who taught Gregorita's aunt, Valentina Romero, the art of *curanderismo*. When any of Gregorita's seventeen children became ill, she took them to her Aunt Valentina for treatment. *La curandera* taught Gregorita, encouraging her by asking, "Why don't you learn? Look, touch here." Using her children's bellies as a classroom, Gregorita felt the different abdominal disorders and learned how to manipulate the intestines to relieve the ailments. Another of her patients during this learning period was her husband. Responding to his complaints, Gregorita said, "Maybe I can do something for you." Mr. Rodriguez replied, "No, no, no! You are not going to boss me!" So, off he went to see Aunt Valentina, who was elsewhere delivering a baby. Finally, Gregorita got her chance. Her husband was desperate and allowed her to learn, all the time howling about how much she was hurting him. "Cranky," she described him, "especially when I felt a big ball in his stomach and had to work very hard. Slow, slow, I fixed him and he got better. When he went to my aunt, she said he was okay now. After that I treated

my husband and one of my sisters and then her family. That's the way it started" (Perrone et al., 1989:108–109).

After that, the neighbors began to come for treatment of their various ailments, and her reputation grew, but she was reluctant to compete with her aunt for business. In 1950, Aunt Valentina died and Gregorita came into her own, her credibility already well established.

Because she lacks any recognized training in health care, Gregorita cannot legally charge fees or bill insurance companies as a *curandera.* To circumvent these legal restrictions, she has become licensed as a massage therapist and bills her clients as such. As this suggests, even a folk healer who appears to function completely outside the bounds and control of the Western scientific world cannot avoid its authority altogether.

Christian Science Practitioners

Theories and Treatments

Christian Science is a Christian sect founded in New England in about 1875. Christian Scientists believe God creates only good, while evil, sickness, suffering, and death exist only because mortals believe in them. The practitioner's job, then, is to lead the sufferer, through prayer, study, and talk, to reject the "counterfeit reality" of the "material self" and to achieve the true reality of divine perfection.

According to Fox (1989:107), "Ideally, Christian Science treatment should be entirely and exclusively metaphysical. Practitioners are not even supposed to listen too attentively to patients' symptoms lest they be tempted to accept them as real; also, they idealize 'undifferentiated' treatment not directed toward a specific problem. There should be no counseling of patients on a human level, no appeal to psychological processes." Reality, however, rarely matches this ideal. Practitioners spend much of their time talking with clients about the emotional and moral problems underlying clients' "counterfeit" physical problems. Healing seems to rely heavily on practitioners' persuasive verbal skills (Fox, 1989).

Becoming a Practitioner

As with *curanderos,* becoming a practitioner is a gradual process (Fox, 1989). Most practitioners (almost all of whom are women) begin by healing family members and friends. During weekly religious services, satisfied patients may announce successful treatment by a particular practitioner. Over time, if a practitioner's personality and reputation seem suitable, other friends and acquaintances might turn to that practitioner for assistance. Eventually, individuals may apply to the central church office for listing in *The Christian Science Journal.* Approval comes after the practitioner submits letters of support from members of the congregation testifying to his or her effectiveness. After this, practitioners can open full-time offices. Currently, the *Journal* lists several thousand practitioners. The geographic distribution of practitioners across regions and between urban and rural communities reflects the distribution of the population as a whole. Care by practitioners is covered under Medicare, Medicaid, and many private health insurance plans (*Journal of the American Medical Association*, 1990).

Christian Scientists' opposition to medical care has precipitated a long history of legal battles in which doctors or states have sued for the right to force individuals to accept medical treatment. In general, courts have ruled that because Christian Scientists never seek medical care, doctors have no legal standing and cannot force care on adults. However, courts have ruled in favor of forcing care on children, arguing that the state has the right and duty to protect the health of children, and have found parents guilty of child abuse or involuntary manslaughter when children who received only spiritual treatment have died. (Box 12.1 discusses some of the ethical issues involved in the decision to use alternative therapies.)

Box 12.1　　**Ethical Debate: Choosing Alternative Options**

John and Mary Miller work as high school teachers in a medium-sized New England town. They have two children, both delivered by doctors in hospitals, and are now expecting their third child. Their experiences with those first two births convinced them that medically controlled childbirth is frightening, alienating, and potentially dangerous for both mother and infant.

A few weeks ago, the Millers learned that a woman in their church had given birth at home attended by a highly experienced, but illegal, lay midwife. The woman's glowing account of her labor and delivery—walking around her home to ease her contractions, listening to her stereo, having her husband bring her food and water when she requested it, feeling her body work to deliver her baby without the aid of drugs—has convinced John and Mary to hire this midwife to deliver their

child. Their parents, however, have expressed horror at their decision, fearing they could endanger the life and health of both Mary and the baby. Although the Millers remain committed to their decision, they worry about the legal consequences of using an illegal practitioner and wonder if they have made the right choice.

Do prospective parents have the right to choose unconventional childbirth attendants? More broadly, do individuals have the right to choose unconventional health care practices or treatments, for themselves or for their children, without interference from doctors and the courts?

As in the example from the previous chapter regarding telling the truth to someone who has cancer, the central issues in this case are autonomy and paternalism. Here the issue is not personal paternalism by doctors but state paternalism—the idea that the state has an obligation to protect the

Acupuncturists

Theories and Treatments

If anything, acupuncturists' ideas regarding health and illness bear even less relationship to the ideas of Western medicine than do those of *curanderos* and Christian Science practitioners. Acupuncture is one of the oldest forms of healing known. Its recorded history goes back 2,000 years, with strong prehistorical evidence going back to the Bronze Age.

Like all traditional Chinese medicine, acupuncture is based on the concept of *chi* (Fulder, 1984). This concept, which has no Western equivalent, refers to the vital life force, or energy. Health occurs when *chi* flows freely through the body, balanced between *yin* and *yang*, the opposing forces in nature. Because any combination of problems in the mind, body, spirit, social environment, or physical environment can restrict *chi*, treatment must be holistic.

---

welfare of its citizens, even when doing so means going against citizens' wishes.

Restricting individual autonomy is a serious matter, for it implies that an individual is not competent to decide what is in his or her own best interest. As the word implies, *paternalism* suggests that an individual is more like a child or even an animal than like an adult human. Requiring motorcyclists to wear helmets, for example, suggests motorcyclists are too ignorant or stupid to assess for themselves the advantages and disadvantages of helmets.

Should the need for paternalism outweigh the desire for autonomy in this case? One way to decide is to evaluate the evidence for and against home birth to see whether home birth is as dangerous and hospital birth as safe as most doctors claim. Another way is to consider more abstractly the relative value and appropriate roles of autonomy and state paternalism. An individual might, for example, conclude that home birth is unsafe but still believe that protecting individual autonomy is more important than protecting individuals from themselves.

In this situation, the ethical dilemma is muddied because two lives are at stake—both Mary's and her baby's. The issue, then, is not simply whether Mary has the right to decide for herself what sort of health care she wants, but also whether she has the right to make a decision that will affect her baby. To evaluate this situation, then, one also must decide, first, whether parents or the state can best and most appropriately judge children's interests and, second, in what circumstances state intervention is justified. Remember, though, that parents who choose home birth, like those who take their children to alternative healers, typically do so not because they don't care about their children's welfare, but because they care enough to flout social conventions when they believe it is in their children's best interests.

Following this theory, traditional Chinese healers consider both symptoms and diagnosis unimportant and focus instead on unblocking *chi*. Acupuncture is based on the theory that *chi* runs through the body to the different organs in channels known as meridians, which have no Western equivalents. To cure a problem in the colon, for example, acupuncturists apply needles to the index finger, which they believe connects to the colon via a meridian. In this way, they believe, they can stimulate an individual's *chi* and direct it to the parts of the body where it is needed. Acupuncturists decide on treatment through taking a complete history; palpating the patient's abdomen; measuring his or her blood pressure; and reading the twelve pulses that Chinese medicine recognizes.

Acupuncture is still used extensively in China, both alone and in conjunction with Western medicine, and is used increasingly in the West. To ascertain the impact of acupuncture, the U.S. National Institute of Health recently organized a consensus panel on the topic. (A consensus panel is a group of experts from diverse backgrounds brought together to reach joint conclusions on a topic.) The panel's 1998 report concluded that despite the paucity of high-quality research, there are sufficient data to conclude that acupuncture definitely alleviates nausea and some types of pain and definitely does not help in stopping smoking. The report also noted that acupuncture has fewer harmful side effects than modern medicine has and that many accepted western medical practices have no greater scientific evidence of efficacy. The **World Health Organization,** meanwhile, considers acupuncture effective for treating about fifty disorders, including the common cold, bronchial asthma, childhood myopia, and dysentery (Wolpe, 1985:420).

The Impact of Medical Dominance

Widespread American interest in acupuncture began during the 1970s, when the People's Republic of China first opened to U.S. travelers. Early travelers brought back near-miraculous tales of acupuncture anesthesia and treatment. Because American doctors had no scientific model that could account for acupuncture's effects, these tales threatened their position and worldview (Wolpe, 1985). As a result, various well-known doctors publicly denounced acupuncture, claiming it worked only as a placebo or only because Chinese stoicism or revolutionary zeal allowed them to ignore pain, even though acupuncture also had worked on animals and on Western travelers to China.

To remove this threat to their cultural authority, doctors worked to control the definition, study, and use of acupuncture (Wolpe, 1985). This proved relatively easy, for, unlike chiropractic or osteopathy, acupuncture at the time had few American supporters. As a result, in their writings and public pronouncements, doctors could strip acupuncture of its grounding in traditional Chinese medical philosophy and define it simply as the use of needles to produce anesthesia. Pressure from medical organizations led the National Institutes of Health to adopt a similar definition in funding research on acupuncture. At the same time, pressure from doctors led most states to adopt licensure laws allowing any doctors, regardless of training, to practice acupuncture but forbidding all others, no matter how well trained, from doing so except under medical supervision. Most traditional acupuncturists in the United States now work illegally within Asian communities. Some states, however, do allow nondoctors to perform acupuncture and some insurance companies will reimburse nondoctors for acupuncture treatments. Use of acupuncture appears to be growing: comparable national random surveys of English-speaking U.S. residents conducted in 1990 and 1997 found that the percentage reporting use of acupuncture rose from 0.4 percent to 1.0 percent (Eisenberg et al., 1998). Of these, 40.7 percent reported partial insurance coverage for treatment, but none reported full coverage. These figures suggest that acupuncture remains a marginal therapy and occupation, posing little threat to medical dominance.

CONCLUSIONS

As the discussions in this chapter have suggested, the health care arena is much broader than we usually recognize. Many alternatives to medical treatment exist far beyond those discussed herein. Most of these alternatives exist not so much in opposition to mainstream health care as in parallel, with those seeking care jumping back and forth across the tracks. For example, a woman might deliver her first child with a doctor, her second with a nurse-midwife, and her third with a lay midwife, and a man who experiences chronic back pain might see a chiropractor or acupuncturist either before, after, or in addition to seeing a medical doctor.

This chapter has highlighted the factors that help health care occupations gain professional autonomy in the face of medical dominance. Timing certainly seems to play a role: those occupations that emerged before medical dominance became cemented, such as osteopathy and chiropractic, have proved most successful. Social factors, too, consistently seem important: health care occupations with roots in and support from higher-status social groups have a better chance of winning professional autonomy than do those with lower-status roots and supporters.

Other occupations seem to retain some autonomy—if a marginal position in the health care arena—because they pose little threat to medical dominance. *Curanderos,* for example, attract a relatively small clientele of poor Mexicans and Mexican Americans who might not be able to pay for medical care or to communicate effectively with medical doctors anyway. As a result, doctors have little incentive to eliminate *curanderos'* practices. Acupuncturists, on the other hand, have attracted not only Asians and Asian Americans but also well-educated whites—including individuals with the skills and resources to publicize the virtues of acupuncture. Consequently, doctors have had a far greater vested interest in restricting acupuncturists' practices and in coopting acupuncture for their own purposes.

Not surprisingly, developing professional autonomy seems most difficult for those, like nurses, who work directly under medical control. In contrast, Christian Scientist practitioners, for example, have considerably more leeway to develop their practices without interference from medical doctors.

Finally and ironically, strict licensing laws, even when devised by doctors opposed to a field's growth, in the end can help occupations gain professional autonomy by forcing them to increase standards and thereby enabling them to gain additional status and freedom to practice.

To date, medical doctors have succeeded in retaining their professional autonomy and dominance partly because of their greater ability to provide scientific data supporting their theories and practices—or at least to convince the public that they have such data. It remains to be seen whether, with the increased federal support for research on alternatives and despite medical control of funding and publication mechanisms, those who favor alternative health care options will be able to use this research to increase scientific credibility and public support for their practices.

SUGGESTED READINGS

McGuire, Meredith B. 1988. *Ritual Healing in Suburban America*. New Brunswick, NJ: Rutgers University Press. A fascinating account of alternative therapies and why some middle-class Americans use them.

Reverby, Susan. 1987. *Ordered to Care: The Dilemma of American Nursing*. New York: Cambridge University Press. A history of American nursing, emphasizing the role of gender.

Sullivan, Deborah A., and Rose Weitz. 1988. *Labor Pains: Modern Midwives and Home Birth*. New Haven: Yale University Press. Describes the origins and social position of the home birth movement and modern lay midwives.

GETTING INVOLVED

Midwives Alliance of North America. 4805 Lawrenceville Hwy. Suite 116–279, Lilburn, GA 30047. (888) 923 6262. www.mana.org. Promotes communication between lay midwives and nurse-midwives and the legal rights of both groups.

Committee for Freedom of Choice in Medicine. 1180 Walnut Avenue, Chula Vista, CA 91911. (619) 429-8200. Supports consumers' freedom of choice to use any therapy that shows clear evidence of efficacy.

REVIEW QUESTIONS

How did the early history of nursing make it difficult for nurses to increase their status or improve their working conditions?

How have nurses attempted to professionalize? Why haven't these strategies succeeded?

How have changes in the health care system affected nurses' occupational status and position?

What factors have led to the development of clinical pharmacy, pharmacy care, and disease management? What factors have restrained their growth, or could do so in future?

How did osteopaths attempt to professionalize? What factors enabled them to succeed? What price has osteopathy paid for its success?

To what extent and in what ways have chiropractors succeeded in improving their occupational status?

How and why did doctors gain control over childbirth?

What factors led to the growth of nurse-midwifery? of lay midwifery? What is the difference between the two?

How do individuals become traditional healers? How does medical dominance affect their work and their lives?

INTERNET EXERCISES

The federal government's main Web site for consumer health is www.healthfinder.com. Browse the site, looking for links to Web pages related to fraud and quackery, accountability, and treatment errors. Do the site's organizers appear as concerned about fraud and similar problems among mainstream practitioners as among alternative practitioners? In what ways, if any, does the site's handling of alternative medicine differ from its handling of mainstream medicine?

Using the Internet, find policy statements related to home birth and midwifery from a variety of organizations (such as the World Health Organization, the Midwives Alliance of North America, the American College of Obstetricians-Gynecologists, and the American College of Nurse-Midwives). How do their positions differ? What evidence do they use to justify their positions?

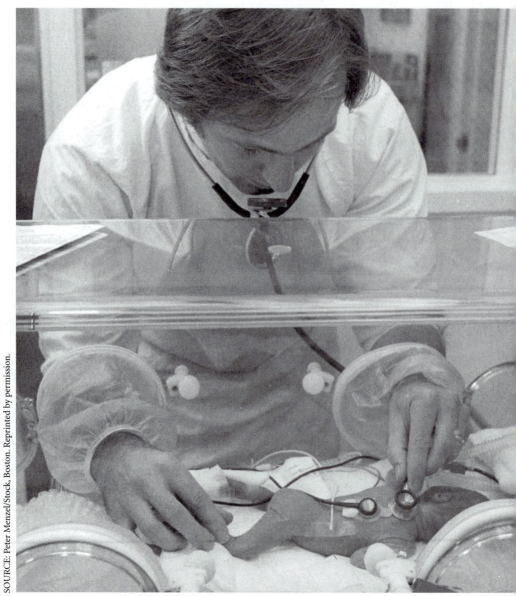

A baby being treated in the Stanford Hospital intensive care nursery, Palo Alto, California

CHAPTER 13

Issues in Bioethics

—

For centuries, doctors have formally recognized that ethical principles should underlie health care. The Hippocratic Oath, for example, written in about 400 B.C., instructed doctors to take only actions that would benefit their patients and to foreswear euthanasia, seducing patients, or divulging patients' secrets.

Despite this long tradition, however, only in the 1960s did **bioethics,** the study of all ethical issues involved in the biological sciences and health care, emerge as a distinct field. In this chapter, I describe the history and basic principles of bioethics, as well as how bioethics has become institutionalized in American health care and its impact on research and clinical practice.

To some students (and perhaps to some faculty as well), it might seem odd to include a chapter on bioethics in a sociology textbook. Yet at a fundamental level, the issues raised by bioethics are sociological issues, for many of the issues bioethicists ponder revolve, either explicitly or implicitly, around the impact of power differences between social groups (most importantly, physicians and patients). Even when exploring the same issues, however, bioethicists and sociologists do so through different lenses. Robert Zussman, a sociologist who has studied bioethics extensively, succinctly summarizes the difference:

> Put broadly, but with at least moderate accuracy, medical ethics may be thought of as the normative study of high principles for the purpose of guiding clinical decisions. In contrast, the sociology of medical ethics may be thought of as the empirical study of clinical decisions for the purpose of understanding the social structure of medicine. Clearly then, medical ethicists and sociologists of medical ethics travel much of the same terrain, but they do so traveling in different directions (Zussman, 1997:174).

— A HISTORY OF BIOETHICS

Since its beginning in 1848, the **AMA** has required its members to subscribe to its Code of Ethics. The Code, however, speaks more directly to medical etiquette—proper relations between doctors—than to anything we would now consider medical ethics or, more broadly, bioethics. Indeed, throughout the nineteenth century and well into the twentieth century, doctors' ideas regarding bioethics remained ill-defined and their commitment to bioethics minimal. Although doctors undoubtedly would have identified relieving human suffering as their primary goal, both in their research and in clinical practice doctors sometimes behaved in ways that would horrify modern doctors and bioethicists.

Dr. J. Marion Sims, considered the father of modern obstetrics, provides a vivid example of the limitations of nineteenth-century medical ethics (Barker-Benfield, 1976). Sims achieved fame during the 1840s for developing a surgical procedure to correct vesico-vaginal fistulae, tears in the wall between a woman's vagina and bladder usually caused by overaggressive medical intervention during childbirth. Women who suffered these fistulae could not control leakage of urine and often had to withdraw from social life altogether because of odor and the resulting social shame. To develop a surgical cure, Sims bought black women slaves who had fistulae and then operated on them as many as thirty times each, in an era before antibiotics and antisepsis and with only addictive drugs for anesthetics. When Sims announced his new surgical technique, the medical world and the public greeted him with acclaim. No one questioned his research ethics.

Almost a century later, Nazi doctors working in German concentration camps also used socially disvalued populations for equally barbaric—and even less justifiable—experiments. The world's response to these experiments would mark the beginnings of modern bioethics.

The Nazi Doctors and the Nuremberg Code

In 1933, the German people voted the Nazis, under Adolf Hitler's leadership, into power. At that time, Germany's medical schools and researchers were known and respected worldwide and its system of health care was considered one of the best and most comprehensive (Redlich, 1978).

Shortly after coming to power, the Nazi government passed the Law for the Prevention of Congenitally Ill Progeny (Lifton, 1986). This law required the sterilization of anyone considered likely to give birth to children with diseases that doctors considered genetic, including mental retardation, schizophrenia, manic-depression, epilepsy, blindness, deafness, or alcoholism. Under this law, government-employed doctors sterilized between 200,000 and 300,000 persons. Two years later, in 1935, the government passed the Law to Protect Genetic Health, prohibiting the marriage of persons with certain diseases.

Both these laws reflected a belief in **eugenics,** the theory that the population should be "improved" through selective breeding and birth control. The eugenics movement had many followers throughout the Western world. By 1920, twenty-five U.S. states had passed laws allowing sterilization of those believed (usually incorrectly) to carry genes for mental retardation or criminality. Several states also passed laws forbidding interracial marriage and marriage by persons with illnesses considered genetic (Lifton, 1986).

As the power of the Nazis grew in Germany, and as public response to their actions both within and outside Germany proved mild, the Nazis adopted ever bolder eugenic actions (Lifton, 1986; Redlich, 1978). Beginning in 1939, the Nazis began systematically killing patients in state mental hospitals. Doctors played a central role in this program, selecting patients for death and supervising their poisoning with lethal drugs or carbon monoxide gas. Doctors and nurses also watched silently while many more patients starved to death. In total, between 80,000 and 100,000 adults and 5,000 children died (Lifton, 1986). Shortly after, the Nazi government began systematically killing Jews, Gypsies, and others whom they considered racially inferior. By the end of World War II, the Nazis had murdered between five million and ten million people in their concentration camps.

At least 350 doctors played major roles in this genocidal policy (Lifton, 1986; Redlich, 1978). As prisoners entered the concentration camps, medical officers of the Nazi SS corps decided which to kill immediately and which to use for forced labor. When shooting those marked for death proved too expensive, doctors developed more efficient means of mass murder using carbon monoxide gassing. Medical corpsmen, supervised by doctors, conducted the murders. Those whom doctors selected for forced labor, meanwhile, usually died in a matter of weeks from starvation, overwork, or the epidemic diseases that ravaged the camps. In addition, doctors working in the concentration camps, including university professors and highly respected senior medical researchers, performed hundreds of patently unethical experiments on prisoners, such as studying how quickly individuals would die once exposed to freezing cold or given experimental poisons and whether injecting dye into prisoners' eyes would change their eye color. Doctors also used prisoners to gain surgical experience by, for example, removing healthy ovaries or kidneys or creating wounds on which to practice surgical treatments.

Following the Nazi defeat, the Allied victors prosecuted 23 of these doctors for committing "medical crimes against humanity," eventually sentencing seven to death and nine to prison (Lifton, 1986). The decisions in these cases contained the basis for what is now known as the **Nuremberg Code,** a set of internationally recognized principles regarding the ethics of human experimentation (Box 13.1). The Code requires researchers to have a medically justifiable purpose, do all within their power to protect their subjects from harm, and insure that their subjects give **informed consent,** that is, voluntarily agree to participate in the research with a full understanding of the potential risks and benefits.

The 1960s: The Rise of Bioethics

Because the trials had received relatively little publicity in the United States, and because Americans—both doctors and the general public—tended to view Nazi doctors as Nazis, not as doctors, few drew connections between Nazi practices and American medical practices (Rothman, 1991). As a result, discussion of bioethics remained largely dormant in the years following the Nuremberg Trials. During the 1960s, however, as health care costs rose exponentially, ethical questions regarding access to health care became topics of popular discussion. New technologies, too, such as the development of organ transplants and of life support systems for comatose persons, raised issues not only of equity and access, but also of how to balance the benefits of new technologies against their dangers. From these issues would emerge a heightened interest in bioethics.

These issues first came to a head with the development of kidney dialysis, a technology that could keep alive persons whose kidneys had failed

---

Box 13.1 | *Principles of the Nuremberg Code*

1. The voluntary consent of the human subject is absolutely essential.

2. The experiment should be such as to yield fruitful results for the good of society, unprocurable by other methods or means of study, and not random and unnecessary in nature.

3. The experiment should be so designed and based on the results of animal experimentation and a knowledge of the natural history of the disease or other problem under study that the anticipated results will justify the performance of the experiment.

4. The experiment should be so conducted as to avoid all unnecessary physical and mental suffering and injury.

5. No experiment should be conducted where there is an *a priori* reason to believe that death or disabling injury will occur. . . .

6. The degree of risk to be taken should never exceed that determined by the hu-

manitarian importance of the problem to be solved by the experiment.

7. Proper preparations should be made and adequate facilities provided to protect the experimental subjects against even remote possibilities of injury, disability, or death.

8. The experiment should be conducted only by scientifically qualified persons. The highest degree of skill and care should be required through all stages of the experiment of those who conduct or engage in the experiment.

9. During the course of the experiment, the human subject should be at liberty to bring the experiment to an end. . . .

10. During the course of the experiment, the scientist in charge must be prepared to terminate the experiment at any stage if he has probable cause to believe . . . that a continuation of the experiment is likely to result in injury, disability, or death to the experimental subject.

(Fox and Swazey, 1974). Demand for dialysis far outstripped supply, forcing selection committees made up of doctors and, in some cases, lay people to decide who would receive this life-saving treatment and who would die. Forced to choose from among the many who, on medical grounds, were equally likely to benefit from the treatment, these committees frequently based their choices on social criteria, such as sex, age, apparent emotional stability, social class, and marital status. When news of these committees' work reached the public, the resulting outcry led to new federal regulations designed to allocate kidney dialysis more fairly.

Although the dialysis issue sparked public concern about medical *practice*, medical *research* still remained outside the bounds of public discussion. In 1966, however, one article changed this. Writing in the *New England Journal of Medicine,* respected medical professor Henry Beecher (1966) described 22 research studies, published in top journals in the recent past, that had used ethically questionable methods. In one study, for example, soldiers sick with streptococcal infections received experimental treatments instead of penicillin, causing twenty-five soldiers to develop rheumatic fever. In another, doctors working without parental consent catheterized and x-rayed the bladders of healthy newborns to see how they worked.

To determine the frequency of such studies, Beecher looked at one hundred consecutive research studies published in a prestigious medical journal. In twelve of the one hundred studies, researchers either had not told subjects of the risks involved in the experiments or had not even told them they were in an experiment. Yet no journal reviewer, editor, or reader had questioned the ethics of these studies.

Beecher's article sent ripples of concern not only through the medical world but also through the general public, as news of the article spread through the mass media. This public concern translated into pressure on Congress and, in turn, pressure on the U.S. Public Health Service (PHS), the major funder of medical research. To demonstrate to Congress that they could deal with the problem on their own and to keep public concern from turning into budget cuts, the PHS in 1966 published guidelines for protecting human subjects in medical research (Rothman, 1991).

The responses to Beecher's article and the dialysis issue demonstrate the increased role that the mass media and the general public had begun to play in health care decision-making. Meanwhile, the growth of the civil rights and women's rights movements stimulated discussion both about patients' rights generally and about birth control and abortion specifically. The patients' rights movement would also draw energy from the publication in 1969 of Dr. Elizabeth Kübler-Ross' book *On Death and Dying,* which called attention to the dehumanizing aspects of modern medical treatment of the dying.

According to historian David Rothman (1991), the concept of patients' rights also found fertile ground during the 1960s because of changes in the relationships between doctors and their patients. Before World War II, Americans typically received their health care at home or in a nearby office from general practitioners they had known for years. Doctors and their clients lived in the same neighborhoods and often shared the same ethnic and social class background. By the 1960s, however, as medical practice shifted from general to specialty care, from home and office to hospital, and from talking and direct physical interventions to impersonal technological interventions, the ties binding doctors and clients had weakened. In these circumstances, trust between doctors and clients diminished and public demands for control over medical work grew. Similarly, medical research had shifted from small-scale, rare events in which doctors typically conducted experiments first on themselves and then on their families and neighbors to large-scale business enterprises with only weak links between doctors and subjects.

By the late 1960s, writers could look at developments around the country and proclaim the birth of the bioethics movement (Fox, 1974; Jonsen et al., 1978; Rothman, 1991). Over the next few years, several important organizations devoted to bioethics were founded, including the Hastings Center for Bioethics, the Society for Health and Human Values, and the Center for Bioethics at Georgetown University, and bioethics secured at least a small place in medical education.

The 1970s: Willowbrook, Tuskegee, and Karen Quinlan

The Willowbrook Hepatitis Study

During the 1970s, three cases further stimulated popular, legal, and medical interest in bioethics. The first of these, the Willowbrook hepatitis experiments, reached public attention in 1971. Willowbrook State School was an institution for mentally retarded children run by the state of New York. Conditions in Willowbrook were horrendous, with children routinely left naked, hungry, and lying in urine and excrement. As a result, hepatitis, a highly contagious, debilitating, and sometimes deadly disease, ran rampant among the children and, to a lesser extent, the hospital staff.

In 1956, to document the natural history of hepatitis and to test vaccinations and treatments, two professors of pediatrics from New York University School of Medicine began purposely infecting children with the disease. In addition, to test the effectiveness of different dosages of gamma globulin, which the researchers knew offered some protection against hepatitis, they injected some children with gamma globulin but left others unvaccinated for comparison. The children's parents had consented to this research, but had received only vague descriptions of its nature and potential risks.

The researchers offered several justifications for their work. First, they argued, the benefits of the research outweighed any potential risks. Second, they had infected the children only with a relatively mild strain of the virus and therefore had decreased the odds that the children would become infected with the far less common but considerably more dangerous strain that also existed in the school. Third, the children who participated in the experiments lived in better conditions than did the others in the institution and therefore were protected against the many other infections common there. Fourth, the researchers argued that the children would probably become infected with hepatitis anyway, given the abysmal conditions in the institution. Finally, the researchers felt they should not be held accountable because the parents had given permission. Using these arguments, the researchers had obtained approval for their experiments from the state of New York, the Willowbrook State School, and New York University. Over a fifteen-year period, they published a series of articles based on their research, without any reviewers, editors, or readers raising ethical objections.

In 1970, however, Methodist theologian Paul Ramsey (1970) exposed the ethical flaws of these experiments in his influential book, *The Patient as Person*. Shortly thereafter, in the spring of 1971, an exchange of letters and editorials debating the ethics of these experiments appeared in the prestigious British medical journal, *The Lancet*. Ramsey and others wrote in *The Lancet* that parents had not given truly *voluntary* consent because they could get their children admitted to Willowbrook only by allowing them to participate in the hepatitis experiments. In addition, parents had not given truly *informed* consent because researchers had not told them that gamma globulin could provide long-term immunity to hepatitis. Writers to *The Lancet* also questioned why the researchers experimented on children, who could not give informed consent, rather than on the hospital staff. Finally, these writers questioned why the researchers—who, after all, were pediatricians—had chosen to take advantage of this "opportunity" to study hepatitis rather than trying to wipe out the epidemic. This debate over the Willowbrook studies was taken up by the New York media and, in the ensuing public outcry, the research ground to a halt.

The Tuskegee Syphilis Study

A year later, in 1972, the Tuskegee Syphilis Study made headlines (Jones, 1993). Begun by the PHS in 1932, the study, which was still underway, hoped to document the natural progression of untreated syphilis in African-American men. At the time it began, medical scientists understood the devastating course of syphilis in whites (which, in its later stages, can cause neurological damage and heart disease), but, reflecting the racist logic of the times, suspected its progression took a different and milder form in African Americans.

For this study, researchers identified 399 desperately poor and mostly illiterate African-American men with untreated late-stage syphilis living in the Tuskegee, Alabama, area. The men were neither told they had syphilis nor offered treatment. Instead, researchers informed them that they had "bad blood," a term used locally to cover a wide variety of health ailments. The researchers then told the men that if they participated in this study of bad blood, they would receive free and regular (if infrequent) health care, transportation to medical clinics, free meals on examination days, and payment of burial expenses—enormous inducements given the men's extreme poverty.

At the time the study began, treating syphilis was difficult, lengthy, and costly. The development of penicillin in the early 1940s, however, gave doctors a simple and effective treatment. Yet, throughout the course of the study, researchers not only did not offer penicillin to their subjects but also kept them from receiving it elsewhere. During World War II, researchers worked with local draft boards to prevent their subjects from getting drafted into the military, where the subjects might have received treatment. When federally funded venereal disease treatment clinics opened locally, researchers enlisted the support of clinic doctors to keep research subjects from receiving treatment. Similarly, they enlisted the cooperation of the all-white County Medical Society to ensure that no local doctor gave penicillin to their subjects for any other reason.

The Tuskegee Syphilis Study, which treated African-American men as less-than-human guinea pigs, was not the work of a few isolated crackpots. Rather, it was run by the PHS with additional funding from the widely respected Milbank Fund. The study received significant cooperation from the state and county medical associations and even from doctors and nurses affiliated with the local Tuskegee Institute, a world-renowned college for African Americans. Over the years, more than a dozen articles based on the study appeared in top medical journals, without anyone ever questioning the study's ethics. Yet the study patently flouted the Nuremberg Code and, after 1966, the PHS's own research ethics guidelines. Not until 1972 did the study end, following a newspaper exposé and the resulting public outcry. By that time, at least twenty-eight and possibly as many as one hundred research subjects had died as a result of syphilis, and an unknown number had succumbed to syphilis-related heart problems (Jones, 1993). In addition, the study indirectly caused untold additional deaths by convincing many in the African American community to distrust public health workers. That legacy has lasted to the present day, contributing to suspicions among African Americans that the federal government created **HIV** to control population growth in their community (Jones, 1993; Thomas and Quinn, 1991).

Karen Quinlan and the Right to Die

Several years later, in 1975, public attention would focus on Karen Quinlan, whose case raised issues not of medical experimentation but of medical treatment. At the age of 21, after ingesting a combination of drugs at a party, Quinlan fell into a coma. Initially, her parents encouraged her doctors to make all efforts to keep her alive and return her to health. Her parents soon learned, however, that she had suffered extensive brain damage and would never regain any mental or physical functioning. At this point, and as her body gradually and permanently curled into a fetal position, they asked that she be removed from all life-support machines and allowed to die. When the doctors refused, the parents took their fight to the courts. After almost a year of legal battles, Quinlan's parents won the right to remove her from the mechanical respirator that was keeping her alive. Against her parents' wishes, however, the hospital staff slowly weaned Quinlan off the respirator so her body would learn to breathe without it. In addition, the staff continued to give her regular intravenous feedings and antibiotics. As a result, Quinlan survived for another ten years.

The Quinlan case gained enormous public attention and sympathy for the right to die and highlighted the problems involved in having too much, rather than too little, access to medical care and technology. In addition, the Quinlan case signaled the entry of lawyers and the legal system into health care decision-making.

The 1980s: Reproductive Technology

The 1980s and 1990s saw both further development of medical technology and further questions regarding whether medicine might have gone too far in its ability to control human life and death.

One area that sparked many of these questions is **reproductive technology,** or medical developments that allow doctors to control the process of human conception and fetal development. Reproductive technology first came to the public's attention in 1978, with the birth of Louise Brown, the world's first "test-tube baby." Louise's mother was unable to conceive a baby because her fallopian tubes, through which eggs must descend to reach sperm and be fertilized, were blocked. Using a technique known as *in vitro fertilization,* her doctors removed an egg from her body, fertilized it with her husband's sperm in a test-tube, and then implanted it in her uterus to develop. Nine months later, Louise Brown was born.

Louise Brown's birth raised questions about how far doctors should go in interfering in the normal human processes of reproduction. Subsequent cases raised even trickier questions. For example, courts have had to decide whether fetuses should be placed for adoption when the biological parents have died, and whether custody of fetuses following divorce should go to the parent who wants the fetuses implanted or the one who wants them destroyed. More recently, doctors and others have debated whether couples should be allowed to hire women to carry their fetuses to term for them and whether postmenopausal women should be allowed to have a baby using another woman's egg.

More broadly, these cases have raised basic questions regarding the morality of intervening so directly in the process of human reproduction, including whether individuals are harmed or helped by having access to such technologies. Those who favor the new reproductive technologies argue that the technologies give couples greater control over their destinies. Those who oppose the new technologies, on the other hand, argue that these technologies seduce couples into spending enormous amounts of time and money in a usually futile effort to have children biologically their own, rather than finding other ways to make meaningful lives for themselves, and that these technologies encourage the idea that children are purchasable commodities (Rothman, 1989).

The 1990s: Setting Priorities and Enhancing Human Traits

Setting Priorities

Whereas earlier debates on funding health care focused on deciding which *individuals* should get specific scarce resources such as kidney dialysis, during the 1990s debates focused on setting priorities to help decide which *procedures* should be funded. This debate came to the fore in 1989 with passage of legislation establishing the Oregon Health Plan (OHP), which promised to provide free care to all Oregonians who had incomes below the federal poverty level (Leichter, 1997). To extend coverage to individuals not eligible for **Medicaid,** the OHP currently provides a somewhat limited package of services. To decide each year which services to offer, OHP first prioritizes all the potential health care services it might offer. It then prospectively contracts with **managed care** organizations to purchase services for its members, beginning at the top of its priority list and working its way down until it reaches its budget limit. Thus, if it runs out of funding, some services are cut, but no individuals are dropped from the program.

The OHP legislation marked the first time that a governmental body in the United States explicitly rationed health care—deciding in advance that some procedures simply cost too much to provide to some populations. The explicit use of rationing resulted in an outcry across the country, both from those who considered it discrimination against persons with disabilities and those who believed it was unethical to ration care only for the poor. As a result, it took the state almost five years to win federal approval to pilot the program, and ethical questions continue to plague the system.

Yet rationing has always existed in the U.S. health care system (Callahan, 1998). For example, few eyebrows had been raised when California decided to ration care a few years earlier by tightening eligibility for Medicaid. In that case, however, rationing was implicit—people did not get health care because they were dropped from Medicaid and couldn't afford to purchase care on their own, not because others decided certain services shouldn't be offered. Yet in the absence of some system for prioritizing services, health care dollars routinely are spent on services that offer little benefit or that offer great benefit only to a few (see, for example, Callahan, 1998; Weitz, 1999; Zussman, 1992), while much larger groups go without basic services.

The Oregon system at least rationalizes rationing, deciding, for example, that it makes more sense to fund vaccinations for thousands of children than kidney transplants for a handful.

Enhancing Human Traits

The 1990s also saw growing interest in the ethical issues embedded in medical interventions designed to enhance human traits. No clear definition of such **enhancements** exist, but the term is used to refer to techniques generally believed to improve human traits beyond a level considered normal rather than to treat conditions considered deviant or defective. This is a necessarily subjective definition, as individual judgments regarding what is normal vary greatly. Nevertheless, we would probably all acknowledge a qualitative difference between providing cosmetic surgery to a person with a severely burned face versus providing it to a professional model who desires more prominent cheekbones. Similarly, there is a qualitative difference between using psychotropic drugs to avoid schizophrenic episodes and using them to get extra energy and improve one's test scores—a process psychiatrist Peter Kramer (1993) refers to as "cosmetic psychopharmacology."

Ethical questions regarding enhancements have emerged along with their use. For example, is it ethically justifiable for individuals to improve their offspring through genetic pre-selection or fetal surgery and, if so, will those who do not use these technologies become a "genetic underclass"? Should health insurance cover cosmetic (as opposed to reconstructive) surgery? Should it cover drugs such as Viagra, which helps men achieve erections and can improve quality of life perhaps beyond the norm for a given age? And should psychotropic drugs be prescribed to individuals who do not have diagnosable mental illnesses but who want to be more sociable, alert, or assertive (Whitehouse et al., 1997)? These questions raise issues not only about the ethics of covering costs for enhancements while others still lack basic services but also more broadly about whether enhancements could provide unethical advantages. Given that the Olympics forbids athletes from taking drugs that give some an advantage over others, why are actresses allowed to get breast implants or business people allowed to take Prozac to enhance their alertness? Conversely, is it ethical to restrain the options of those who would provide or purchase such services? Questions such as these are increasingly common, as evidenced by the special supplement the *Hastings Center Report*, an influential bioethics journal, published on this topic in January–February 1998.

INSTITUTIONALIZING BIOETHICS

Increased concern about bioethcs stimulated the development of formal mechanisms designed to ensure that health care and health research will be conducted ethically. In this section, I look at three of those mechanisms:

hospital ethics committees, research ethics committees, and professional ethics committees.

Hospital Ethics Committees

The origins of hospital ethics committees can be traced to the 1950s. Like the Seattle Kidney Center, many other hospitals used committees to select patients for kidney dialysis. Similarly, hospitals routinely used committees to decide which women merited abortions on medical grounds. At the time, the legal status of abortion was unclear, and the moral status of abortion was just starting to become a public issue (Luker, 1984). Because psychiatric problems were considered justifiable medical grounds for abortion, wealthy women easily could find doctors who would testify to committees that abortion was psychiatrically needed. Poor women, on the other hand, typically could obtain abortions only if their lives were physically endangered. In reality, therefore, these committees made their decisions more on social than on medical grounds and primarily existed to protect doctors who performed abortions from legal or social sanction (Luker, 1984).

Other hospital ethics committees arose in the aftermath of the 1982 "Baby Doe" case, in which parents of a newborn who was mentally retarded and had a defective digestive system decided that they did not want the defect corrected by surgery. The doctors complied with their decision, and the baby died six days later. After news of the case broke, the federal government implemented regulations forbidding hospitals that received federal funds from withholding medical or surgical treatment from disabled infants. The regulations also urged but did not require hospitals to establish Infant Care Review Committees to prospectively evaluate decisions regarding withholding treatment from disabled infants. Many hospitals continued to use these committees even after the Supreme Court threw out the regulations in 1986. Since then, the use and purpose of hospital ethics committees has grown enormously, with the percentage of large hospitals (i.e., 200 or more beds) having ethics committees rising from 1 percent in 1982 to more than 60 percent by 1988 (Cohen, 1988).

More recently, however, hospitals have come to recognize the inherent difficulties in making decisions expeditiously by committee and so have shifted from relying on ethics *committees* to relying more on ethics *consultants*—individuals trained in bioethics and hired specifically to consult with hospital personnel regarding ethical issues. Ethics committees, meanwhile, have shifted from focusing on individual cases to consulting, advising, and providing information regarding broad ethical concerns, such as how to implement requests not to resuscitate terminally ill patients and whether hospital staff have an obligation to provide care to those with HIV disease (Fost and Cranford, 1985).

Research Ethics Committees

Although universities and hospitals had begun establishing committees to review research ethics in the 1960s, such committees did not become common until the 1970s. In the aftermath of the Tuskegee scandal, Congress in 1974 created the National Commission for the Protection of Human Subjects of Biomedical and Behavioral Research. The Commission's reports laid the groundwork for current guidelines regarding research ethics. That same year, the National Research Act mandated the development of **Institutional Review Boards** (**IRBs**) charged with reviewing all federally funded research projects involving human subjects. Such boards now exist at all universities and other research institutions.

Professional Ethics Committees

Many professional organizations now also have ethics committees that establish guidelines for professional practice. The American Fertility Society, for example, has published a statement of principles regarding the moral status of human embryos created in the laboratory, and the ethics committee of the American College of Obstetrics and Gynecology has published guidelines regarding the ethics of selectively aborting fetuses when a woman who has used fertility drugs becomes pregnant with multiple fetuses.

THE IMPACT OF BIOETHICS

The growth of the bioethics movement and the institutionalizing of bioethics in U.S. hospitals and universities have made ethical issues more visible than ever before. Articles on bioethics, virtually nonexistent before the 1960s, now appear routinely in medical journals, while in both the clinical and research worlds, ethics committees have proliferated.

These developments have led some observers to conclude that the bioethics movement has fundamentally altered the nature of medical work. According to historian David Rothman (1991:2):

> By the mid-1970s, both the style and the substance of medical decision-making had changed. The authority that an individual physician had once exercised covertly was now subject to debate and review by colleagues and laypeople. Let the physician design a research protocol to deliver an experimental treatment, and in the room, by federal mandate, was an institutional review board composed of other physicians, lawyers, and community representatives to make certain that the potential benefits to the subject-patient outweighed the risks. Let the physician attempt to allocate a scarce resource, like a donor heart, and in the

room were federal and state legislators and administrators to help set standards of equity and justice. Let the physician decide to withdraw or terminate life-sustaining treatment from an incompetent patient, and in the room were state judges to rule, in advance, on the legality of these actions.

Other observers, however, contend that the impact of the bioethics movement has been more muted (e.g., Annas, 1991). These critics argue that hospital, research, and professional ethics committees, like the earlier hospital abortion committees, exist primarily to offer legal protection and social support to researchers and clinicians, not to protect patients or research subjects. Further, they argue, although clinicians have become more concerned with *documenting* their allegiance to ethics guidelines, they have not become any more concerned with *following* those guidelines. Finally, sociologist Daniel F. Chambliss (1996) argues that bioethics' emphasis on helping individual health care providers make more ethical decisions simply does not apply to nonmedical health care workers like nurses, who often understand clearly what they should do ethically but lack the power to do so. For example, nurses often have a much better understanding than doctors of how much a patient is suffering and thus more often believe treatment should be discontinued unless it will improve quality as well as length of life. Yet nurses rarely can act on that belief because they lack the necessary legal standing, economic independence, and social status.

The following sections evaluate the impact of bioethics on health care research, medical education, and clinical practice.

The Impact on Research

According to ethicist George Annas, the bioethics movement, as institutionalized in research ethics boards and committees, has affected medical research only slightly. In his words, the

> primary mission [of research ethics committees] is to protect the institution by providing an alternative forum to litigation or unwanted publicity. . . . [For this reason] its membership is almost exclusively made up of researchers (not potential subjects) from the particular institution. These committees have changed the face of research in the U.S. by requiring investigators to justify their research on humans to a peer review group prior to recruiting subjects. But this does not mean that they have made research universally more "ethical." In at least a few spectacular instances, these committees have provided ethical and legal cover that enabled experiments to be performed that otherwise would not have been because of their potentially devastating impact on human subjects (1991:19).

As an example, Annas cites the case of "Baby Fae" (not her real name), who died in 1984 soon after doctors replaced her defective heart with a baboon's heart. Although all available evidence indicated that cross-species transplants could not succeed, the doctors who performed the surgery had received approval from their hospital's IRB. A subsequent review found that Baby Fae's parents had not given truly informed consent because the doctors had not suggested seeking a human transplant, had disparaged available surgical treatments, and had unreasonably encouraged the parents to believe that a baboon transplant could succeed.

More recently, serious questions about IRBs' ability to restrain unethical practices have arisen in the context of clinical drug trials for HIV disease in **developing nations.** In the last few years, doctors have learned that the risk of an HIV-infected pregnant woman infecting her baby before or during birth can be reduced by more than 50 percent if the woman takes several doses of the drug AZT (zidovidine) prenatally. Unfortunately, the costs of this drug regimen make it prohibitive for many women in the United States and for almost all in developing nations. As a result, during the 1990s scientists conducted 18 clinical trials in Africa, all sponsored by western organizations and all with IRB approval, to test whether less expensive regimens (either lower dosages or different drugs) might provide the same results at lower cost. For these studies, women were divided into an experimental group that received the less expensive regimen and a control group. According to both U.S. and international research ethics guidelines, the control group should have received the usual standard of care in the countries *sponsoring* the research—in this case, the standard drug regimen. Instead, in 15 of the 18 studies, the researchers argued that the standard treatment regimen for poor women in Africa was no treatment, and so their control groups received none. These studies were all halted in 1997, following an editorial condemning them in the *New England Journal of Medicine* and wide coverage in the news media (Lurie and Wolfe, 1997). Debate over these studies continues (see, for example, Crouch and Arras, 1998), as supporters argue that the drug trials—which eventually produced much less expensive means of preventing prenatal HIV transmission—were justified, while opponents argue that it was especially wrong to conduct ethically questionable research on vulnerable individuals who could not benefit from the research (since their babies would already have been born by the time the study results were known).

Lack of resources and conflicts of interest also limit the effectiveness of IRBs. IRB members are unpaid volunteers, who typically must review between 300 and 2,000 proposed experiments yearly and who, in many cases, have vested interests in approving research proposals so their institutions can obtain research funding (Hilts, 1999). Meanwhile, final responsibility for overseeing IRBs falls to the federal Office of Protection from Research Risks, which has only three full-time employees. These conditions make thorough review of human subjects research impossible.

Finally, even when IRBs work as designed, their authorizing statutes restrict them from addressing the broader issues of whether the benefits potentially available through research outweigh the potential for harm to society and whether the money allotted for a given research project could produce more beneficial effects if spent elsewhere (Williams, 1984). Yet these are often the most important questions to ask.

Nevertheless, and despite the limitations of IRBs and research ethics committees, the rise of bioethics largely has curbed the most egregious abuses of human subjects. According to David Rothman,

> The experiments that Henry Beecher described could not now occur; even the most ambitious or confident investigator would not today put forward such protocols. Indeed, the transformation in research practices is most dramatic in the area that was once most problematic: research on incompetent and institutionalized subjects. The young, the elderly, the mentally disabled, and the incarcerated are not fair game for the investigator. Researchers no longer get to choose the martyrs for mankind (1991:251).

Indeed, the balance has shifted to such an extent that we now sometimes read news stories not of researchers pressuring individuals to become research subjects but, rather, of desperately ill individuals pressuring researchers to accept them as research subjects for experimental treatments.

The Impact on Medical Education

One obvious result of the bioethics movement has been the incorporation of ethics training into medical education, with courses now common at U.S. medical schools. As critics have noted, however, those courses are too often divorced from real life, aimed at teaching students ethical principles and legal norms through classroom lectures rather than at teaching students how to negotiate the everyday ethical dilemmas they face. To achieve this latter goal, the University of Pennsylvania Medical School includes in its ethics course sessions in which students discuss ethical dilemmas they have encountered during their clinical training, such as pressures placed on them to perform medical procedures on unwilling patients (Christakis and Feudtner, 1993). Discussing situations like these can help students devise strategies for responding more ethically in future.

Other observers, however, have noted that a course like this also has its limits, for it assumes that students who are already undergoing socialization to medical culture still can identify ethically problematic aspects of that culture (Hafferty and Franks, 1994). Moreover, this strategy does not challenge the ways in which ethics are discounted in the "hidden curriculum" of medical practice and culture. For example, a structure that expects students both to provide care *and* to learn techniques on unsuspecting patients inherently teaches students to view patients at least partly as objects rather than as subjects. From this perspective, only through "the integration of ethical principles into the everyday work of both science and medicine" can we expect new doctors to adopt more ethical approaches to care (Hafferty and Franks, 1994:868).

The Impact on Clinical Practice

At a fundamental level, the bioethics movement challenges doctors' clinical autonomy, for it "substitutes principles and general rules for the case-by-case analysis that has long characterized medical practice . . . and attempts to reformulate medical problems as moral, rather than technical, issues" (Zussman, 1992:10–11).

According to George Annas (1991), professional ethics committees emerged to counter this challenge. Annas argues that the true purpose of these committees is not to foster more ethical behavior but to protect professional autonomy by providing clinicians with legal protection against accusations of unethical behavior. For example, published guidelines from the Ethics Committee of the American Fertility Society refer to human embryos created in the laboratory merely as "preembryos," even though they do not differ biologically from other embryos, and leave it up to each clinic to establish policies for their use. Similarly, published guidelines from the American College of Obstetrician-Gynecologists on whether to selectively abort fetuses when several embryos become implanted simultaneously in a woman's uterus state only that doctors and patients should make their decisions jointly. Such guidelines seem designed more to provide legal cover to clinicians than to encourage more ethical practices (Annas, 1991).

Relatively few studies have looked at the impact of bioethics on actual clinical practices. One series of studies looked at the impact of New York's 1987 law establishing formal policies for writing "do not resuscitate" orders (orders forbidding health care workers from intervening if the lungs or heart of a terminally ill patient stop functioning). These studies found that after the law's passage, doctors significantly altered how they documented their actions but not how they acted (Zussman, 1992:162). Similarly, studies have found that hospitals sharply limit access of patients, family, and nonmedical staff to ethics consultations. As a result, consultations primarily function to provide additional institutional support to doctors confronted by families or patients they consider disruptive, such as those who challenge doctors' decisions regarding how aggressively to treat a given condition (Kelly et al., 1997; Orr and Moon, 1993). These findings have led researchers to conclude that the true purpose of ethics consultations is to reinforce doctors' power.

On the other hand, significant change has occurred in doctors' decisions regarding whether to disclose diagnoses to their patients. Whereas in 1961, 88 percent of doctors reported that they generally would not tell a patient that he or she had cancer, by 1979, virtually all—98 percent—reported that they would do so (Novack et al., 1979). These statistics do not, however, necessarily mean that doctors now support client autonomy over professional authority, but could instead suggest that doctors have found this a way to limit their responsibility for making unpleasant decisions (Zussman, 1992:85–90).

The most extensive study of the impact of bioethics on clinical practice appears in *Intensive Care: Medical Ethics and the Medical Profession*, by sociologist Robert Zussman. Zussman spent more than two years observing and interviewing in the intensive care units of two hospitals. His research suggests both the impact and the limitations of the bioethics movement.

Although cases such as Karen Quinlan's and Baby Doe's might suggest that doctors often want to use aggressive treatment despite the objections of patients and families, Zussman found that on intensive care wards the reverse is usually the case. Knowing that most of their patients will die, doctors on these wards often hesitate before beginning aggressive treatment, which might only escalate costs, increase their work as well as their patients' suffering, and prolong the dying process. Patients, however, and, more important, their families (for, in most cases, the patients are incapable of communicating) often face a sudden and unexpected medical crisis. Unable to believe the situation hopeless, they demand that health care workers "do everything." In these situations, the doctors Zussman studied expressed allegiance to the principle that families have the right to make decisions regarding treatment. In practice, however, doctors found ways to assert their discretion, if no longer the authority they had in years past.

Doctors asserted their discretion in several ways. First, doctors made decisions without asking the family on the assumption that the family would agree with their decisions. Second, doctors sometimes ignored a family's stated decisions, arguing that forcing a family to make life or death decisions that would later cause them guilt or grief was cruel. Third, doctors might both ascertain and respect the family's wishes, but only after first shaping those wishes through selectively providing information. This information included defining the patient as terminally ill or not—a highly significant designation, for ethical guidelines permit health care workers to withhold or terminate treatment only for terminally ill patients. Fourth, when doctors failed to shape a family's wishes, the doctors could discount those wishes on the grounds that the family was too emotionally distraught to decide rationally.

Finally, and perhaps most important, doctors continued to assert their discretion by defining the decision to withhold or terminate treatment as a technical rather than an ethical problem. The following example from Zussman's research demonstrates this process:

The Countryside ICU [Intensive Care Unit] staff was considering whether or not to write a Do Not Resuscitate order for Mr. Lake, a 73-year-old man who had been admitted to the unit with acute renal [kidney] failure, a gastrointestinal bleed, pneumonia, and sepsis [infection]. Ken [the medical director of the ICU] asked what they should do "if the family wanted a full court press." One of the residents started to say what he thought were the "interesting ethical issues." But Ken cut him off, arguing that the decision depended entirely on prognostics: "There are no ethical issues.... I'm not an ethicist. I'm a doctor." When the resident attempted to distinguish different circumstances preceding codes [decisions not to resuscitate], Ken broke in again: "A code is a code. It's a medical decision, not an ethical decision (Zussman, 1992:150; ellipses in original).

Once doctors succeeded in defining treatment decisions as purely technical issues, they could define the family's stated wishes as uneducated and irrelevant. Doctors could end discussion regarding treatment decisions by declaring it simply a technical fact that any treatment would be futile. Similarly, doctors might acknowledge families' general wishes regarding how aggressively treatment should proceed, but then define each specific intervention as a technical decision best left to doctors. Because most treatment decisions involve not dramatically pulling a plug but rather a series of small, minute-to-minute actions, leaving doctors in control of these "technical" matters gives doctors power far outweighing families' general statements regarding whether to pursue aggressive treatment.

Summing up his findings, Zussman writes (1992:159–160),

> The picture I have drawn corresponds neither to an image of unbridled professional discretion nor to one of patients' rights triumphant. As many observers of contemporary medicine have argued, the discretion of physicians in clinical decisions (like the discretion of professionals in other fields) depends on their ability to make successful claims to the exclusive command of technical knowledge. Yet, while . . . physicians . . . make such claims, they do not always succeed either in convincing themselves that they are legitimate or in converting them to influence over patients and their families, for the claims of physicians are met by the counterclaims of patients and, more important, families. . . . The institutionalization of patients' rights, in law and in hospital policy, . . . empower[s] families when they do insist on doing everything. In such a situation, physicians may continue to exercise considerable influence and enjoy considerable discretion. By no means have they been reduced to the role of technicians and nothing more. But at the same time, they must, at the very least, take the wishes of patients and families into account.

In sum, Zussman's findings suggest that the bioethics movement has had a real, if limited, impact on the day-to-day world of health care.

CONCLUSIONS

This chapter has explored the history of the bioethics movement and its impact on health care research and practice. As we have seen, bioethics and sociology have much in common. At a very basic, if typically unacknowledged level, bioethics, like sociology, is about power. The abuses of the Nazi doctors, for example, not only illuminate the horrors possible when ethical principles are ignored but also show how social and occupational groups can obtain power over others and the potentially deadly consequences when this happens. Conversely, sociology, in similarly unacknowledged ways, is at a basic level an ethical enterprise. Underlying abstract, technical sociological discussions about the nature of society often lurk hidden assumptions about what society *should* be like and how society should be changed. These assumptions often draw on philosophies regarding justice, autonomy, human worth, and other basic ethical issues. Yet, in the same way that bioethicists often ignore the sociological implications of their work, sociologists often ignore the ethical implications of the questions they ask, the research they conduct, and the findings their research generates. It seems, then, that bioethicists and sociologists can provide each other with broader perspectives that can only enrich our understanding of both fields—encouraging bioethicists to see not only individual cases but broader social and political issues and encouraging sociologists to see the world and their work in it as an ethical as well as a political and intellectual enterprise.

SUGGESTED READINGS

Hastings Center Report. An eminently readable and always fascinating monthly journal on bioethics, published by the Hastings Center.

Jones, James. 1993. *Bad Blood: The Tuskegee Syphilis Experiment.* Rev. ed. New York: Free Press. A fascinating—and horrifying—account of the Tuskegee Syphilis Experiment.

Lifton, Robert J. 1986. *The Nazi Doctors: Medical Killing and the Psychology of Genocide.* New York: Basic Books. A frightening account of the role doctors played in the Nazi genocide.

Zussman, Robert. 1992. *Intensive Care: Medical Ethics and the Medical Profession.* Chicago: University of Chicago Press. An engrossing sociological analysis of the impact of modern bioethics.

GETTING INVOLVED

The Hastings Center. Rt. 9D, Garrison, NY 10524-5555. (914) 424-4040. www.thehastingscenter.org. A nonprofit organization, the Center is committed to research, lobbying, and public education on bioethics. Publishes the excellent *Hastings Center Report.*

REVIEW QUESTIONS

What is the Nuremberg Code, and how and why did it come into existence?

What factors led to the emergence of the bioethics movement in the late 1960s?

Why do researchers now consider the Tuskegee Syphilis Study and the Willowbrook experiments to have been unethical?

What are the ethical problems involved in the new reproductive technology? in enhancements?

What impact has bioethics had on health care and on health research?

INTERNET EXERCISES

Using InfoTrac©, look for articles on cosmetic surgery from a variety of sources. (You can access InfoTrac at http://www.infotrac-college.com/wadsworth, if your professor ordered it when ordering this textbook.) Do these articles suggest that there are any ethical or social issues inherent in cosmetic surgery, such as whether it is morally right or wrong, or whether social forces rather than objective aesthetic concerns press individuals to have this surgery? If yes, what ethical or social issues do they identify? If no, how do you explain why they do not recognize any ethical or social issues?

The ELSI program is a part of the Human Genome Project (which is itself a part of the National Institute of Health) designed to investigate the ethical, legal, and social implications ("ELSI") of human genetics research. Find the ELSI Web site, and learn about the types of research that have been sponsored by this program.

GLOSSARY

accommodation: Technique individuals use to smooth interactions with those they consider potential sources of trouble and to smooth interactions between those persons and others.

achieved statuses: Earned qualifications and positions such as ranking in the top third of one's class.

acquired immune deficiency syndrome (AIDS): The end stage of HIV disease. See *HIV disease.*

active voluntary euthanasia: When individuals help sick persons to kill themselves.

activity-passivity: Model of doctor-patient interaction in which doctor is active and patient is passive, such as during emergency surgery.

actuarial risk rating: A system in which insurers try to maximize their financial gain by identifying and insuring only those populations that have low health risks.

acute illness: Any illness that strikes suddenly and disappears rapidly (within a month or so). Examples include chicken pox, colds, and influenza.

ADA: See *Americans with Disabilities Act.*

addiction: The physical state in which an individual who has used a drug regularly will experience withdrawal if he or she ceases using the drug. See *withdrawal.*

advanced practice nurses: Individuals who, after becoming registered nurses, additionally receive specialized postgraduate training. Includes nurse-midwives and nurse-practitioners. See *registered nurses.*

AIDS: Acquired Immune Deficiency Syndrome. See *HIV disease.*

aligning actions: Actions or, more specifically, interpretations of actions designed to make behavior appear reasonable and normal in a given cultural context.

allopathic doctors: Nineteenth-century forerunners of contemporary medical doctors. Also known as "regular" doctors.

almshouse: An institution, also known as a poorhouse, in which all public wards, including orphans, criminals, the disabled, and the insane, received custodial care.

alternative therapies: Treatments rarely taught in medical schools and rarely used in hospitals.

AMA: See *American Medical Association.*

ambulatory care: Outpatient care.

American Medical Association (AMA): The main professional association for medical doctors.

Americans with Disabilities Act (ADA): Federal law, passed in 1990, that outlaws discrimination against those with disabilities in employment, public services (including transit), and public accommodations (such as restaurants, hotels, and stores). It requires that existing public transit systems and public accommodations be made accessible, along with all new public buildings and major renovations of existing buildings.

ascribed statuses: Innate characteristics such as ethnicity or gender.

assistant doctors: Chinese health care workers who receive three years of postsecondary training, similar to that of doctors, in both western and traditional Chinese medicine.

assisted living facilities: Institutions, typically consisting of small, private apartments, that offer two levels of services: basic medical and nursing services and help with basic tasks of daily living for disabled residents and comfortable living situations and social activities for healthy residents.

attendings: Doctors who have completed their training and who supervise residents. These doctors can be employed by a hospital or can work in private practice. See *residents.*

avoidance: Actively working to remain ignorant in order to maintain one's emotional balance and images of the future.

balance bill: To bill patients for the difference between the amount their insurance will pay for a given procedure and the amount the doctor would normally charge for that procedure.

barefoot doctors: Former term for agricultural workers who received about three months of training in health care and provided basic health services to members of their agricultural production teams. Now known as "village doctors."

bioethics: The study of all ethical issues involved in the biological sciences and health care.

blaming the victim: Process through which individuals are blamed for causing the problems from which they suffer.

Blue Cross: A group of private nonprofit companies offering insurance that reimburses individuals primarily for the costs of hospital care, not including doctors' bills. Blue Cross insurance is often offered and bought in conjunction with Blue Shield insurance. See *Blue Shield.*

Blue Shield: A group of private non-profit companies offering insurance that reimburses individuals primarily for the costs of receiving care from doctors, especially care received in hospitals. Blue Shield insurance is often offered and bought in conjunction with Blue Cross insurance. See *Blue Cross.*

board and care homes: Residential facilities that provide assistance in daily living but not nursing or medical care.

capitation: A system in which doctors are paid a set annual fee for each patient in their practice, regardless of how many times they see their patients or what services the doctors provide for their patients.

CDC: See *Centers for Disease Control and Prevention.*

Centers for Disease Control and Prevention (CDC): Federal agency responsible for tracking the spread of diseases in the United States.

challenging: Rejecting the social norms that attach stigma to a behavior or condition, including illness or disability. See *stigma.*

chiropractors: Health care practitioners who specialize in spinal manipulation, trace illness and disability to misalignments of the spine, and believe spinal manipulation can cure a wide range of acute and chronic health problems.

chronic illness: Illness that develops in an individual gradually or is present from birth and that will probably continue at least for several months and possibly until the person dies. Examples include muscular dystrophy, asthma, and diabetes.

claims harassment: When insurance companies establish bureaucratic structures that make it virtually impossible for consumers to file claims or to fight the insurer if a claim is denied.

clinical pharmacy: A subfield of pharmacy in which pharmacists participate actively in decisions regarding drug treatment.

CMHC: See *Community Mental Health Centers.*

cognitive norms: Socially accepted rules regarding proper ways of thinking. For example, someone should not think that he is Napoleon or that his radio is sending him secret messages from outer space.

commercial insurance: Insurance offered by companies that function on a for-profit basis.

commodification: Process of turning people into products that can be bought or sold.

Community Mental Health Centers (CMHCs): Centers established by the federal government to provide outpatient mental health care within the community to prevent the development of chronic mental illness and the need for inpatient care.

community rating: A system for calculating insurance premiums in which each individual pays a premium based on the average risk level of his or her community as a whole.

complementary therapies: Treatments rarely taught in medical schools and rarely used in hospitals.

compliance: Whether individuals do as instructed by health care workers.

conflict perspective: View that society is held together by power and coercion, with dominant groups imposing their will on subordinate groups.

control: A process through which researchers statistically eliminate the potential influence of extraneous factors. For example, because social class and race often go together, researchers who want to investigate the impact of social class have to be sure that they are not really seeing the impact of race. To study the impact of social class on mental illness, therefore, researchers would have to look separately at the relationship between social class and mental illness among whites and then at the relationship among blacks to control for any effect of race.

convergence hypothesis: The thesis that health care systems become increasingly similar over time as a result of similar scientific, technological, economic, and epidemiological pressures.

cooptation: A process through which an individual, organization, or movement exchanges some or all of its initial philosophy and goals in exchange for social acceptance and financial support.

copayment: Fee paid by persons who have certain forms of health insurance each time they see a care provider. Fees can range from nominal sums to 20 percent of all costs.

core nations: According to world systems theory, those nations with a highly diversified, industrialized economy, providing a relatively high standard of living for most citizens.

corporatization: The growing role of investor-owned corporations in the health care field.

cost shifting: Raising payments charged some individuals for services rendered to make up for losses incurred when services are provided to other individuals who cannot or will not pay for services.

covering: Attempting to deflect attention from deviance, including illnesses or disabilities.

cultural competence: The ability of health care providers to understand at least basic elements of others' culture and, thus, to provide medical care in ways that better meet clients' emotional as well as physical needs.

curanderos: Folk healers who function within Mexican and Mexican-American communities.

deductible: Dollar amount of health care expenses an individual with some forms of health insurance must pay annually before his or her insurance plan will begin covering the remaining costs of health care.

defensive medicine: Tests and procedures that doctors perform primarily to protect themselves against lawsuits rather than to protect their patients' health.

deinstitutionalize: To remove individuals, such as mentally retarded and mentally ill persons, from large institutions and return them to the community.

demedicalization: The process through which a condition or behavior becomes defined as a natural condition or process rather than an illness.

depersonalization: A sense that one no longer is, or is considered, fully human.

depoliticize: To define a situation in a way that hides or minimizes the political nature of that situation.

deprofessionalized: Referring to the lessening or loss of professional status of an occupational group.

developing nations: Nations characterized by a relatively low gross national product per capita. These countries typically have high rates of illiteracy, infant mortality, and other related problems, and have economies that rely heavily on a few industries or products.

deviance: Behavior that violates a particular culture's norms or expectations for proper behavior, and therefore results in negative social sanctions. See *negative social sanctions.*

deviance disavowal: The process through which individuals attempt to prove that, despite their apparent deviance, they are no different from other people.

diagnosis related groups (DRGs): System established by the federal government that sets, for all Medicaid patients and for each possible diagnosis, an average length of hospital stay and cost of inpatient treatment. Under the DRG system, hospitals are paid the established cost for each patient with a given diagnosis, regardless of actual cost of treatment.

Diagnostic and Statistical Manual (DSM): Manual published by the American Psychiatric Association and used by mental health workers to assign diagnoses to clients. Generally, this manual must be used if mental health workers want to obtain reimbursement for their services from insurance providers.

diploma nurses: Nurses who hold diplomas from hospital-based schools, rather than holding associates or bachelors degrees from colleges or universities.

disability: Restrictions or lack of ability to perform activities resulting from physical limitations or from the interplay between those limitations, social responses, and the built or social environment.

Disability Rights Movement: A political movement that fights to obtain equal rights in education, transportation, the workplace, and all other aspects of society for persons with disabilities.

disclosing: Making one's deviance more widely known by telling others about it or making it more visible.

discrimination: Differential and unequal treatment grounded in prejudice. See *prejudice.*

disease: A biological problem within an organism.

disease management: A form of pharmaceutical care in which pharmacists are responsible for counseling certain patients (typically those with chronic conditions) on their prescription drugs, monitoring the impact of the drugs, and, sometimes, prescribing drugs. See *pharmaceutical care.*

doctor-assisted euthanasia: When doctors help patients to kill themselves.

doctor-nurse game: "Game" in which the nurse is expected to make recommendations for medical treatment in such a way that the recommendations appear to have come from the doctor.

DRG: See *Diagnosis Related Groups.*

dysfunctional: That which threatens to undermine social stability.

emergency room abuse: Term used by hospitals to refer to patients who have neither health insurance nor money to pay for care and who therefore turn to hospital outpatient clinics and emergency rooms for treatment of chronic as well as acute health problems.

emotion work: Attempts by individuals to suppress socially inappropriate feelings and develop more appropriate feelings or to find ways of explaining their feelings so that the feelings appear reasonable and socially acceptable.

endemic: Referring to diseases that appear at a more or less stable rate over time within a given population.

enhancements: Techniques deemed to improve human traits beyond a level generally considered normal rather than to treat conditions considered deviant or defective. This distinction is artificial, but occasionally useful.

entrepreneurial system: A system based on capitalism and free enterprise.

environmental racism: The disproportionate burden of environmental pollution experienced by racial and ethnic minorities.

epidemic: Either a sudden increase in the rate of a disease or the first appearance of a new disease.

epidemiological transition: The shift from a society burdened by infectious and parasitic diseases and in which life expectancy is low to one characterized by chronic and degenerative diseases and high life expectancy.

epidemiology: The study of the distribution of disease within a population.

etiology: Cause.

eugenics: The theory that the population should be "improved" through selective breeding and birth control.

evidence-based medicine: Medical therapies whose efficacy has been confirmed by large, randomized, controlled clinical studies.

family leave programs: Programs that allow individuals to take time off from work without risking their jobs to care for family members. Some programs offer paid leave, others only unpaid leave.

fee-for-service: The practice of paying doctors for each health care service they provide, rather than paying them a salary.

fee-for-service insurance: Insurance that reimburses patients for all or part of the costs of the health care services they have purchased. This contrasts with health maintenance organizations, in which patients pay one charge in advance in exchange for any health care they might require during a given period.

feeling norms: Socially defined expectations regarding the range, intensity, and duration of appropriate feelings and regarding how one should express those feelings in a given situation.

feeling work: Efforts made by individuals to avoid being labeled mentally ill by making their emotions match social expectations. Individuals can (1) change or reinterpret the situation causing them to have unacceptable feelings, (2) change their emotions physiologically, through drugs, meditation, biofeedback, or other methods, (3) change their behavior, acting as if they are feeling more appropriate emotions or (4) reinterpret their feelings, telling themselves, for example, that they are only tired rather than worried.

feminization of aging: The steady increase in the proportion of Americans over age 65 who are female.

fetal rights: The growing body of legal, medical, and public opinion that holds that fetuses have rights separate from and sometimes contrary to those of their mothers.

financially progressive: Describes any system in which poorer persons pay a smaller proportion of their income for a given good or service than wealthier persons.

Flexner Report: The report on the status of American medical education produced in 1910 by Abraham Flexner for the Carnegie Foundation. This report identified serious deficiencies in medical education and helped to produce substantial improvements in that system.

for-profit, private hospitals: Hospitals run with the primary goal of producing a profit each year for shareholders.

formulary: Official list of drugs that doctors in a managed care organization can prescribe without special authorization. See *managed care.*

functionalism: View of society as a harmonious whole held together by socialization, mutual consent, and mutual interests.

gatekeeper: A primary care doctor in managed care plans such as HMOs who serves as the intermediary between patients and specialists. Patients must get a referral from their gatekeeper doctor before they see a specialist or must pay out of pocket to see the specialist. See *primary care doctors* and *health maintenance organizations.*

genetic paradigm: A way of looking at the world that emphasizes genetic causes.

glass ceiling: The invisible but real forces of discrimination and prejudice that keep members of a group from rising above a certain level in an organization or profession.

glass escalator: The invisible but real social forces that give members of a group an extra assist in rising in an organization or profession.

globalization: The process through which ideas, resources, and persons increasingly operate within a worldwide rather than local framework. For example, the globalization of tourism means that U.S. tourists now consider Africa a plausible destination.

government hospitals: Hospitals established by state and federal governments to provide services to those groups that would not otherwise receive care.

Great Confinement: The shift, from the 1830s on in both Europe and the United States, toward confining mentally ill persons in large public institutions instead of in almshouses, small private "madhouses," or at home with relatives.

group rates: Insurance rates set by an insurance company for all members of a large group, regardless of their individual health status or risk factors. Typically much lower than individual rates.

guidance-cooperation: Model of doctor-patient interaction in which the doctor guides and the patient cooperates, such as when a patient follows a doctor's advice regarding treating an injury.

health belief model: Model that predicts that individuals will following medical advice when they (1) believe they are susceptible to a particular health problem, (2) believe the health problem they risk is a serious one, (3) believe compliance will significantly reduce their risk, and 4) do not perceive any significant barriers to compliance.

health maintenance organizations (HMOs): Organizations that provide health care based on pre-paid group insurance. Patients pay a fixed yearly fee in exchange for a full range of health care services, including hospital care as well as doctor's services.

heroic medicine: System of treatment used by allopathic doctors before about 1860 that emphasized curing illnesses by purging the body through bloodletting, causing extreme vomiting, or using repeated laxatives and diuretics. See *allopathic doctors.*

HIV (human immunodeficiency virus): The virus that causes HIV disease.

HIV disease: A disease in humans caused by HIV and producing a gradual breakdown in the body's immune system, typically leading to death about twelve years after infection.

HMO: See *Health Maintenance Organization.*

holistic treatment: Treatment that assumes that all aspects of an individual's life and body are interconnected—that, for example, to treat an individual with cancer effectively, health care workers must look at all organs of the body, not only the one that currently has a tumor, as well as at the individual's psychological and social functioning.

home health aides: Workers, typically untrained, who provide essentially custodial care within individuals' homes.

homeopathic doctors: Popular nineteenth-century health care workers who treated illnesses with extremely dilute solutions of drugs that, at full strength, produced similar symptoms to a given illness.

horizontal integration: Situation in which a large corporation owns several institutions that provide the same type of service, such as several nursing homes.

hospices: Institutions designed to meet the needs of the dying.

house staff: Interns and residents.

Human Immunodeficiency Virus: See *HIV.*

illness: The social experience of having a disease.

illness behavior: The process of responding to symptoms and deciding whether to seek diagnosis and treatment.

immersion, illness as: Situation in which illness becomes so demanding that one must structure one's life around it.

incidence: Number of new cases of an illness or health problem occurring within a given population during a given time period (for example, the number of children born with Down Syndrome in the United States during 1993).

Independent Living Movement: The political movement, begun by Americans with disabilities, to provide services to people with disabilities so that they can live in the community rather than having to live in institutions.

Indian Health Service: The federally funded program that provides free health care to all registered members of Indian tribes who live on or near their reservations.

individual practice associations (IPAs): HMOs that contract with individual medical practitioners, working out of private offices, to see HMO patients in addition to their private patients. See *health maintenance organizations.*

individualism: A set of cultural beliefs and practices that encourages the autonomy, equality and dignity of individuals and that downplays the importance of connections to social groups.

industrialized nations: Nations characterized by a relatively high gross national product per capita. These countries typically have diversified economies and low rates of illiteracy, infant mortality, and other related problems.

informed consent: Voluntary agreement to participate in medical research or to receive a medical procedure or treatment, with a full understanding of the potential risks and benefits.

inpatient: Hospital patient who is formally admitted and kept overnight.

Institutional Review Boards (IRBs): Federally mandated committees charged with reviewing the ethics of research projects involving human subjects. No research can be conducted using federal funds unless it first receives IRB approval.

internal colonialism: The treatment of minority groups within a country in ways that resemble the treatment of native peoples by foreign colonizers.

interruption, illness as: Situation in which illness is experienced as only a small and temporary part of one's life.

intrusion, illness as: Situation in which illness demands time, accommodation and attention and forces one to live from day to day.

IPA: See *individual practice association.*

IRB: See *Institutional Review Board.*

irregular practitioners: Nineteenth-century health care practitioners other than allopathic doctors, such as homeopaths, midwives, botanic doctors, bonesetters, and patent medicine makers.

lay midwives: Midwives who do not have formal training in midwifery and who typically learn through experience or apprenticeship.

learned helplessness theory: Theory developed initially by Martin Seligman stating that depression develops when individuals learn to feel they are helpless to control their lives.

Licensed Practical Nurses (LPNs): Individuals, not registered nurses, who assist nurses primarily with the custodial care of patients. They usually have completed approximately one year of classroom and clinical training.

life events: Any changes that force readjustments in individuals' lives, including marriage or divorce, starting or leaving school, and gaining or losing a job.

life expectancy: The average number of years that individuals of a given group born in a given year are expected to live.

lifeworld: The everyday needs of people and ways in which they interact and live their lives.

limited practitioners: Occupational groups, such as chiropractors, that have gained social acceptance by confining their work to a limited range of treatments and certain parts of the body.

long-term care insurance: Insurance designed specifically to pay the costs of nursing home care, board and care homes, in-home nursing care, and other long-term health-related needs.

LPN: See *Licensed Practical Nurse.*

magic bullets: Drugs that prevent or cure illness by attacking one specific etiological factor.

magnetic healers: Nineteenth-century health workers who believed that an invisible magnetic fluid flowed through the body and that illness occurred when that flow was obstructed, unbalanced, inadequate, or excessive. Their treatments consisted of moving their hands along patients' spinal cords to "free" blocked magnetic fluid.

managed care: A system that controls health care spending by monitoring closely how health care providers treat patients and where and when patients receive their health care.

manufacturers of illness: Those groups, such as alcohol and tobacco manufacturers, that promote illness-causing behaviors and social conditions.

marginal practitioners: Occupational groups such as faith healers that treat a wide range of physical ailments and have low social status.

master status: A status viewed by others as so important that it overwhelms all other information about that individual. For example, if we know someone as the local scoutmaster, know he is a Republican and likes to play chess, and then learn he is gay, we might start thinking about him and interacting with him solely on the basis of his sexual orientation, essentially forgetting or ignoring the other information we have about him.

Medicaid: Joint federal/state health insurance program that pays the costs of health care for people with incomes below a certain (very low) amount. Most Medicaid recipients are poor mothers and their children. Medicaid can cover the costs of both preventive and therapeutic medical care and both inpatient and outpatient hospital care, but details of coverage vary considerably from state to state, with some providing considerably more services than others.

medical dominance: Professional dominance by doctors. See *professional dominance.*

medical model of illness: The way in which doctors conceptualize illness. This model consists of five doctrines: that disease is deviation from normal, specific and universal, caused by unique biological forces, analogous to the breakdown of a machine, and defined and treated medically through a neutral scientific process.

medical model of mental illness: A model of mental illness assuming that (1) objectively measurable conditions define mental illness, (2) mental illness stems largely or solely from something within individual psychology or biology, (3) mental illness will worsen if left untreated but might improve or disappear if treated promptly by a medical authority, and (4) treating someone who might be healthy is safer than not treating someone who might be ill.

medical norms: Expectations doctors hold regarding how they should act, think, and feel.

medical savings accounts (MSAs): Tax-free, dedicated savings accounts, similar to Individual Retirement Accounts, in which individuals can put aside money to use for purchasing health care privately. See *Medicare.*

medicalization: Process through which a condition or behavior becomes defined as a medical problem requiring a medical solution.

medically indigent: Persons who earn too much to receive government-provided health care but too little to pay medical bills or purchase health insurance.

Medicare: Federal insurance, based on the Social Security system, that offers hospital insurance and medical insurance to those over age 65 and to permanently disabled persons.

medigap policies: Insurance policies available for purchase by persons who receive Medicare to pay for prescription drugs and other medical services not available through Medicare. See *Medicare.*

miasma: According to pre-twentieth century doctors, disease-causing air "corrupted" by foul odors or fumes.

microcredit: Programs that offer loans of $100 or less to residents of developing nations.

minority group: Any group that, because of its physical or cultural characteristics, is considered inferior and subjected to differential and unequal treatment.

model programs: Programs for treating persons with serious mental illnesses that aim to avoid stigma, dehumanization, and hierarchical patient/staff relationships and that offer a range of social and economic services as well as psychiatric care.

models of doctor-patient interactions: See definitions for three models: *activity-passivity, guidance-cooperation,* and *mutual participation.*

moral status: A status that identifies in society's eyes whether one is good or bad, worthy or unworthy.

moral treatment: A nineteenth-century practice aimed at curing persons with mental illness by treating them with kindness and giving them opportunities for both work and play.

morbidity: Symptoms, illnesses, injuries, or impairments.

mortality: Deaths.

mortification: A process, occurring in total institutions, through which one's prior self-image is partially or totally destroyed and replaced by a personality suited for life in the institution. See *total institutions.*

mutual participation: Model of doctor-patient interaction in which doctor and patient are equal participants, with both assumed to have useful information regarding how a particular health problem should be dealt with. For example, doctors and patients might work together to establish the balance between diet and insulin for a person with diabetes.

national health insurance: A system in which all citizens of a country receive their health coverage from a single governmental insurance plan.

National health Service (NHS): A system in which the government directly pays all costs of health care for its citizens.

naturalistic theories of illness: Theories tracing illness to heat, cold, wind, damp, or other natural events that upset the body's equilibrium.

negative social sanctions: Punishments meted out to those considered deviant by society. Negative social sanctions can range from ridicule and isolation to imprisonment and execution.

neonatal infant mortality: Deaths of infants during the first 27 days after birth.

network model HMOs: HMOs that contract with multiple group medical practices to provide services to HMO patients. See *health maintenance organizations.*

new social movements: Groups of individuals who reject modern society's emphasis on science and rationality, value human interaction, and hope to create a more humane society primarily through living their lives in ways that reflect their ideals rather than through organized political activity.

normalize: Make something seem like the normal course of events. In the context of medical error, this refers to emphasizing how medical errors can happen to anyone. In the context of mental illness, this refers to explaining to oneself and others how unusual behavior is not really a sign of mental illness.

norms: Social expectations for appropriate behavior.

Nuremberg Code: A set of internationally recognized principles regarding the ethics of human experimentation which emerged during the post-World War II Nuremberg trials for medical crimes against humanity. The Code stipulates that researchers must have a medically justifiable purpose, do all within their power to protect their subjects from harm, and ensure that their subjects give voluntary, informed consent.

nurse-midwives: Registered nurses who receive additional formal, nationally accredited training in midwifery.

nursing assistants: Individuals, often untrained, who provide basic custodial care for patients, most often in nursing homes and hospitals. See *nursing homes.*

nursing homes: Facilities that primarily provide nursing and custodial care to many individuals over a long period of time. Skilled nursing homes also provide some medical care.

outpatient: Hospital patient who is neither formally admitted nor kept overnight.

pandemic: A world-wide epidemic. See *epidemic.*

parallel practitioners: Those occupational groups, such as osteopaths, that perform basically the same roles as allopathic doctors while retaining occupational autonomy. See *allopathic doctors.*

pass: To hide one's deviance (such as illnesses or disabilities) from others.

passive euthanasia: When health care workers allow patients to die through inaction.

patient dumping: When voluntary or for-profit hospitals surreptitiously transfer to public hospitals patients who cannot pay for their care.

performance norms: Socially accepted rules for how one should perform one's roles. For example, we expect mothers to keep their children clean and paid workers to arrive on time each day.

peripheral nations: According to world systems theory, those nations where modernization and industrialization have developed slowly if at all and the standard of living is generally low.

personalistic theories of illness: Theories that hold that illness occurs when a god, witch, spirit, or other supernatural power deservedly or maliciously lashes out at an individual.

pharmaceutical care: The idea that pharmacists' central mission should be to advise consumers regarding the proper use of medications, based on pharmacists' knowledge of randomized controlled studies.

placebo: Anything offered as a cure that has no known biological effect. Approximately 30 percent of the time, placebos will produce cures through their psychological effects.

positive social sanctions: Rewards of any sort, from good grades to public esteem.

postneonatal infant mortality: Deaths of infants between the 28th day after birth and eleven months after birth.

practice protocols: Guidelines that establish norms of care for particular medical conditions under particular circumstances based on careful review of clinical research.

prejudice: Unwarranted suspicion or dislike of individuals because they belong to a particular group.

prevalence: Total number of cases of an illness or health problem within a given population at a particular point in time (for example, the number of persons living in the United States who have Down Syndrome). This includes both those first diagnosed that year and those diagnosed in previous years but still alive.

primary care doctors: Those doctors in family or general practice, internal medicine, and pediatrics who are typically the first doctors individuals see when they need medical care.

primary nursing: A system in which tasks previously distributed among RNs, LPNs, and other auxiliary nurses are reunified and performed by an RN, who now has primary responsibility for direct patient care.

primary practitioners: See *primary care doctors.*

primary prevention: Strategies designed to keep people from becoming ill, including vaccinating, using seatbelts, encouraging exercise, and pasteurizing milk.

profession: An occupation that (1) has the autonomy to set its own educational and licensing standards and to police its members for incompetence or malfeasance, (2) has its own technical, specialized knowledge, learned through extended, systematic training, and (3) has the public's confidence that it follows a code of ethics and works more from a sense of service than a desire for profit.

professional dominance: A profession's freedom from control by other occupations or groups and ability to control other occupations working in the same sphere. Only priests, for example, can decide whether someone can become a priest, and priests control the training and work responsibilities of lay religious workers in their churches.

professional socialization: The process of learning the skills, knowledge, and values of an occupation.

professionalization: Process through which an occupation achieves professional status.

progressive financing system: Any system in which low-income individuals pay a lower dollar amount and a lower proportion of their income than do high-income individuals for the same services.

proletarianization: Process through which the status of members of an occupation declines from professionals to workers.

prospective reimbursement: With regard to doctors, a system in which they are paid in advance a set fee per patient regardless of how many times they see that patient or what procedures they perform. With regard to hospitals, a system in which the government pays hospitals a set amount for each Medicare or Medicaid patient based on the average cost of treating someone with that patient's diagnosis.

random samples: Samples selected in such a way that each member of a population has an equal chance of being selected. When a sample is randomly selected, we can be fairly certain that the selected individuals will represent statistically the population as a whole.

rate spiral: A situation in which insurers raise prices, encouraging relatively healthy persons to risk going without coverage while relatively ill persons continue to purchase insurance out of necessity. As a result, the costs of providing coverage increase for insurers, leading them to raise prices still further and leading even more relatively healthy persons to drop their coverage.

rates: Proportions of populations that experience certain circumstances.

RBRVS: See *resource-based relative value scale.*

Red Cross health workers: See *street doctors.*

reductionism: Philosophy stressing reductionistic treatment. See *reductionistic treatment.*

reductionistic treatment: Treatment based on the assumption that each part can be treated separately from the whole, in the same way that one can replace an air filter in a car without worrying whether the problem with the air filter has caused or stemmed from problems in the car's electrical system.

Registered Nurses (RNs): Individuals who have received at least two years of nursing training and passed national licensure requirements. As used in normal conversation, "nurse" generally means "registered nurse."

regressive financing system: Any system in which low-income individuals pay a higher proportion of their income than high-income individuals pay.

regular doctors: Nineteenth-century forerunners of contemporary medical doctors. Also known as allopathic doctors. See *allopathic doctors.*

reliability: The likelihood that different people using the same measure will reach the same conclusions.

remedicalization: The process through which mental illness is increasingly regarded by doctors and others as rooted in biology and amenable only to biological treatments.

reprofessionalizing: Regaining former professional status.

reproductive technololgy: Medical developments that offer control over human conception and fetal development.

residents: Individuals who have graduated medical school and received their M.D. degrees, but who are now engaging in further on-the-job training needed before they can enter independent practice.

resource-based relative value scale (RBRVS): A complex formula designed to curb the costs of Medicare by limiting reimbursement to doctors to the estimated actual costs of services in a particular geographic area.

respite care: Any system designed to give family caregivers a break from their responsibilities.

restructuring of medicine: Concept that medicine as a profession has maintained its dominance by reorganizing specialties not by clinical territory but by functional sector.

retrospective reimbursement: A system in which insured individuals first receive care from health care providers and pay their bills and then their insurance provider reimburses them for all or part of these costs.

right to die: The right to make decisions concerning one's own death.

RNs: See *Registered Nurses.*

role strain: Problems individuals experience within their major social roles as workers, parents, students, and so on.

secondary prevention: Strategies designed to reduce the prevalence of disease through early detection and prompt intervention, such as screening for diabetes and pap smears.

self-fulfilling prophesy: A situation in which individuals become what they are expected to be. For example, when it is assumed that no girls can throw a ball properly, girls might never be taught to do so, might never think it worth trying on their own, and, consequently, would not be able to do so.

self-insuring: Putting aside a pool of money from which to pay all health care expenses for one's employees rather than contracting with an insurance firm.

semiperipheral nations: According to world systems theory, those nations whose economy, standard of living, and relative power place them at a level midway between the core and peripheral nations.

semiprofessional: Referring to those occupations that have achieved some but not all the hallmarks of a profession.

sick role: The set of four social expectations in western society regarding how society should view sick people and how sick people should behave. First, the sick person is considered to have a legitimate reason for not fulfilling his or her normal social role. Second, sickness is considered beyond individual control, something for which the individual is not held responsible. Third, the sick person must recognize that sickness is undesirable and work to get well. Fourth, the sick person should seek and follow medical advice.

sickness funds: Insurance programs organized by social groups in German cities, occupations, and industries to provide insurance for their residents or members. Also known as social insurance.

single-payer system: A health care system in which a single government health insurance organization covers all residents of a nation.

snowballing: A process through which the perceived effect on life of each problematic behavior and emotion increases as the total number of problems increases. This process increases the odds that individuals will define the person experiencing the problems—whether self or other-as mentally ill.

social class: The combination of an individual's education, income, and occupational status or prestige; some researchers use only one of these indicators to measure social class while others combine two or more indicators.

social construction: Ideas created by a social group, as opposed to something that is objectively or naturally given.

social control: Means used by a social group to ensure that individuals conform to the norms of that group. Social control can be formal (such as execution or commitment to a mental hospital) or informal (such as ridicule or shunning). See *norms.*

social control agents: Those individuals or groups of individuals who have the authority to enforce social norms, including parents, teachers, religious leaders, and doctors. See *norms.*

social drift theory: A theory holding that lower-class persons have higher rates of illness because middle-class persons who become ill drift over time into the lower class.

social epidemiology: The study of the distribution of disease within a population according to social factors (such as social class, use of alcohol, or unemployment) as opposed to biological factors (such as blood pressure or genetics).

social insurance: See *sickness funds.*

Social Security: Federally funded program that, since 1935, has provided financial assistance to any disabled adults who have held a job for a specified period of time, as well as to the elderly, the blind, and disabled children.

social stress theory: A theory that holds that lower-class persons have higher rates of mental illness because of the stresses of lower class life.

sociological perspective: A perspective regarding human life and society that focuses on identifying social patterns and grappling with social problems rather than on analyzing individual behavior and finding solutions for personal troubles.

sociology in medicine: An approach to the sociological study of health, illness, and health care that focuses on research questions of interest to doctors.

sociology of medicine: An approach that emphasizes using the area of health, illness, and health care to answer research questions of interest to sociologists in general. This approach often requires researchers to raise questions that could challenge medical views of the world and existing power relationships within the health care world.

sponsorship: Process through which successful professionals in a given field actively help new members establish their careers.

stereotypes: Oversimplistic assumptions regarding the nature of group members, such as assuming that black people are unintelligent.

stigma: Any personal attribute that would be deeply discrediting should it become known.

street doctors: Chinese health care workers with little formal training who work in urban outpatient clinics under the supervision of a doctor. Sometimes known as "Red Cross health workers," these workers offer primary and basic emergency care as well as health education, immunization, and assistance with birth control.

superego: According to Freud, that portion of the personality that represents internalized ideas about right and wrong. Similar to the idea of a conscience.

symbolic interactionism: A theoretical perspective that argues that identity develops as part of an ongoing process of social interaction. Through this process, individuals learn to see themselves through the eyes of others, adopt the values of their community, and measure their self-worth against those values.

technological imperative: Belief that technology is always good, so any existing technological interventions should be used.

tertiary prevention: Strategies designed to minimize deterioration and complications among those who already have a disease.

third-party reimbursement: Payment from insurance programs for services rendered.

total institutions: Institutions in which all aspects of life are controlled by a central authority and in which large numbers of like-situated persons are dealt with *en masse.* Examples include mental hospitals, prisons, and the military.

unintended negative consequences: Unplanned, harmful effects of actions that had been expected to produce only benefits.

universal coverage: Health care systems that provide access to health care for all legal residents of a nation.

utilization review: A system in which insurance companies require doctors to get approval before ordering certain tests, performing surgery, hospitalizing a patient, or keeping a patient hospitalized more than a given number of days.

validity: The likelihood that a given measure accurately reflects reality and measures what researchers believe it measures.

vertical integration: Situation in which a large corporation owns several institutions that provide different types of service within the health care field, such as both nursing homes and drug manufacturing companies.

veterans hospitals: Hospitals established by the U.S. federal government to serve the health needs of those who have served in the Armed Forces.

vigilance: Seeking knowledge so that one can feel able to respond appropriately and therefore feel more in control.

village doctors: Chinese agricultural workers who receive a few months of training in health care and provide basic health services to members of their agricultural production team. More rigorously trained than the "barefoot doctors" they replaced. See *barefoot doctors.*

voluntary hospitals: Hospitals that are financially based in voluntarism, or charity, rather than a profit motive. Same as nonprofit institutions.

WHO: See *World Health Organization.*

withdrawal: The predictable combination of distressing physical symptoms experienced by a person who stops using a drug to which he or she is addicted.

World Health Organization (WHO): United Nations organization charged with documenting health problems and improving world health.

world systems theory: Theory stating that the wealthier nations of the world have achieved and maintained their present economic position through exploiting the resources of the poorer nations.

REFERENCES

Abbott, Andrew. 1988. *The System of Professions.* Chicago: University of Chicago Press.

Abel, Emily K. 1986. "The hospice movement: Institutionalizing innovation." *International Journal of Health Services* 16:71–85.

_____. 1990. "Family care of the frail elderly." Pp. 65–91 in *Circles of Care: Work and Identity in Women's Lives,* edited by Emily K. Abel and Margaret K. Nelson. Albany: State University of New York Press.

_____. In press. *Hearts of Wisdom: American Women Caring for Kin, 1850–1940.* Cambridge, MA: Harvard University Press.

Abel, Emily K., and Margaret K. Nelson. 1990. "Circles of care: An introductory essay." Pp. 4–34 in *Circles of Care: Work and Identity in Women's Lives,* edited by Emily K. Abel and Margaret K. Nelson. Albany: State University of New York Press.

Abraham, Laurie Kaye. 1993. *Mama Might be Better Off Dead: The Failure of Health Care in Urban America.* Chicago: University of Chicago Press.

Aiken, Linda H., and Douglas M. Sloane. 1997. "Effects of specialization and client differentiation on the status of nurses: The case of AIDS." *Journal of Health and Social Behavior* 38:203–222.

Aiken, Linda H., Julie Sochalski, and Gerard F. Anderson. 1996. "Downsizing the hospital nursing workforce." *Health Affairs* 15:88–92.

Albrecht, Gary L. 1992. *The Disability Business: Rehabilitation in America.* Newbury Park, CA: Sage.

Albrecht, Gary, Vivian Walker, and Judith Levy. 1982. "Social distance from the stigmatized: A test of two theories." *Social Science and Medicine* 16:1319–1327.

Alpert, Sheri. 1993. "Smart cards, smarter policy: Medical records, privacy, and health care reform." *Hastings Center Report* 23(6):13–23.

Altman, Lawrence K. 1994. "Infectious diseases on the rebound in the U.S., a report says." *New York Times* May 10:B7.

American Academy of Pediatrics. 1997. "Considerations related to the use of recombinant human growth hormone in children (RE9701)." *Pediatrics* 99:122–129.

American Association of Medical Colleges. 1999. *Medical School Admissions Requirements, 2000–2001.* Washington, DC: American Association of Medical Colleges.

American Bar Association, Commission on Mental and Physical Disability Law. 1998. "Study finds employers win most ADA Title I judicial and administrative complaints." *Mental and Physical Disability Law Reporter* 22:403–407.

American College of Nurse-Midwives. 1998. *Basic Facts about Certified Nurse-midwives.* Washington, DC: American College of Nurse-Midwives.

American College of Nurse-Midwives. 1999. *Direct Entry Midwifery: A Summary of State Laws and Regulations.* Washington, DC: American College of Nurse-Midwives.

American College of Obstetrician-Gynecologists. 1998. "Malpractice costs for ob-gyns continue to rise; Risks affecting physicians earlier in their career." Press release. March 30. http://www.acog.org/from_home/publications/press_releases/nr3-30-98.htm.

American College of Physicians. 1998. "The physician workforce and financing of graduate medical education." *Annals of Internal Medicine* 128:142–148.

American Hospital Association. 1994a. *Guide to the Health Care Field.* Chicago: American Hospital Association.

———. 1994b. *Hospital Statistics.* Chicago: American Hospital Association.

———. 1998. *Hospital Statistics.* Chicago: American Hospital Association.

American Society of Plastic and Reconstructive Surgeons. 1989 (July 1). *Comments on the Proposed Classification of Inflatable Breast Prosthesis and Silicone Gel Filled Breast Prosthesis.* Arlington Heights, IL: American Society of Plastic and Reconstructive Surgeons.

Andreasen, Nancy C. 1984. *The Broken Brain: The Biological Revolution in Psychiatry.* New York: Harper & Row.

Andrulis, Dennis P., Katherine L. Acuff, Kevin B. Weiss, and Ron J. Anderson. 1996. "Public hospitals and health care reform: Choices and challenges." *American Journal of Public Health* 86:162–165.

Annandale, Ellen C. 1989. "The malpractice crisis and the doctor-patient relationship." *Sociology of Health and Illness* 11:1–23.

Annas, George J. 1991. "Ethics committees: From ethical comfort to ethical cover." *Hastings Center Report* 21(May–June):18–21.

———. 1992. "The prostitute, the playboy, and the poet: Rationing schemes for organ transplantation." Pp. 549–553 in *Intervention and Reflection: Basic Issues in Medical Ethics,* fourth edition, edited by Ronald Munson. Belmont, CA: Wadsworth.

Anspach, Renee R. 1990. "The language of case presentation." Pp. 319–330 in *The Sociology of Health and Illness: Critical Perspectives,* edited by Peter Conrad and Rochelle Kern. New York: St. Martin's.

Appelbaum, Paul S. 1994. *Almost a Revolution: Mental Health Law and the Limits of Change.* New York: Oxford University Press.

Armstrong, Elizabeth M. 1998. "Diagnosing moral disorder: The discovery and evolution of fetal alcohol syndrome." *Social Science and Medicine* 47:2025–2042.

Armstrong, Pat, and Hugh Armstrong. 1996. *Wasting Away: The Undermining of Canadian Health Care.* New York: Oxford University Press.

———. 1998. *Universal Health Care: What the United States Can Learn from the Canadian Experience.* New York: New Press.

Arno, Peter S., Karen Bonuck, and Robert Padgug. 1995. "The economic impact of high-tech home care." Pp. 220–234 in *Bringing the Hospital Home: Ethical and Social Implications of High-tech Home Care,* edited by John D. Arras. Baltimore: Johns Hopkins University Press.

Aronowitz, Robert A. 1998. *Making Sense of Illness: Science, Society, and Disease.* New York: Cambridge University Press.

Arras, John D., and Nancy Neveloff Dubler. 1995. "Ethical and social implications of high-tech home care." Pp. 1–34 in *Bringing the Hospital Home: Ethical and Social Implications of High-tech Home Care,* edited by John D. Arras. Baltimore: Johns Hopkins University Press.

Association of Asian Pacific Community Health Organizations. 1997. *Taking Action: Improving Access to Health Care for Asian Pacific Islanders.* Oakland, CA: Association of Asian Pacific Community Health Organizations.

Astin, John A. 1998. "Why patients use alternative medicine: Results of a national study." *Journal of the American Medical Association* 279:1548–1553.

Avison, William R., and R. Jay Turner. 1988. "Stressful life events and depressive symptoms: Disaggregating the effects of acute stressors and chronic strains." *Journal of Health and Social Behavior* 29:253–264.

Bacquet, C.R., J.W. Horm, T. Gibbs, and P. Greenwald. 1991. "Socioeconomic factors and cancer incidence among blacks and whites." *Journal of the National Cancer Institute* 83:551–557.

Barker-Benfield, Graham J. 1976. *The Horrors of the Half-Known Life: Male Attitudes Toward Women and Sexuality in Nineteenth Century America.* New York: Harper.

Barrow, Susan M., Daniel B. Herman, Pilar Cordova, and Elmer L. Struening. 1999. "Mortality among homeless shelter residents in New York City." *American Journal of Public Health* 89:529–534.

Barstow, Anne Llewellyn. 1994. *Witchcraze: A New History of the European Witch Hunts.* San Francisco: Pandora.

Barzansky, Barbara, Harry S. Jonas, and Sylvia I. Etzel. 1998. "Educational programs in U.S. medical schools, 1997–1998." *Journal of the American Medical Association* 280:803–805.

Basch, Paul F. 1990. *Textbook of International Health.* New York: Oxford University Press.

Beauchamp, Tom, and James Childress. 1989. *Principles of Bioethics,* third edition. New York: Oxford University Press.

Becker, Howard S., Blanche Geer, Everett C. Hughes, and Anselm Strauss. 1961. *Boys in White: Student Culture in Medical School.* Chicago: University of Chicago Press.

Becker, Marshall H. (ed.). 1974. *The Health Belief Model and Personal Health Behavior.* San Francisco: Society for Public Health Education.

_____. 1993. "A medical sociologist looks at health promotion." *Journal of Health and Social Behavior* 34:1–6.

Beecher, Henry K. 1966. "Ethics and clinical research." *New England Journal of Medicine* 274:1354–1360.

Beeson, Paul B. 1980. "Changes in medical therapy during the past half century." *Medicine* 59:79–99.

Berner, Robert. 1999. "Health care: Pharmacists are starting to move in on doctors' turf." *Wall Street Journal* Jan. 28:B1+.

Bernstein, Barbara, and Robert Kane. 1981. "Physicians' attitudes toward female patients." *Medical Care* 19:600–608.

Birenbaum, Arnold. 1982. "Reprofessionalization of pharmacy." *Social Science and Medicine* 16:871–878.

Billings, Paul R., Mel A. Kohn, Margaret de Cuevas, Jonathan Beckwith, Joseph S. Alper, and Marvin R. Natowicz. 1992. "Discrimination as a consequence of genetic testing." *American Journal of Human Genetics* 50:476–482.

Bland, Karina. 1999. "Tucson facility offers some women a new beginning." *Arizona Republic,* September 19:A1+.

Blauner, Robert. 1972. *Racial Oppression in America*. New York: Harper & Row.

Blaxter, Mildred. 1983. "The causes of disease: Women talking." *Social Science and Medicine* 17:59–69.

Blendon, Robert J., Karen Donelan, Robert Leitman, Arnold Epstein, Joel C. Cantor, Alan B. Cohen, Ian Morrison, Thomas Moloney, Christian Koeck, and Samuel W. Levitt. 1993. "Physicians' perspectives on caring for patients in the United States, Canada, and West Germany." *New England Journal of Medicine* 328: 1011–1016.

Block, Susan D., Nancy Clark-Chiarelli, Antoinette S. Peters, and Judith D. Singer. 1996. "Academia's chilly climate for primary care." *Journal of the American Medical Association* 276:677–682.

Blum, Robert W., Brian Harmon, Linda Harris, Lois Bergeisen, and Michael D. Resnick. 1992. "American Indian-Alaska Native Youth Health." *Journal of the American Medical Association* 267:1634–1644.

Bodenheimer, Thomas. 1999. "The American health care system—physicians and the changing medical marketplace." *New England Journal of Medicine* 340(7):584–588.

Borkan, Jeffrey, Jon O. Neher, Ofra Anson, and Bret Smoker. 1994. "Referrals for alternative therapies." *Journal of Family Practice* 39:545–550.

Bosk, Charles L. 1979. *Forgive and Remember: Managing Medical Failure*. Chicago: University of Chicago Press.

Boston Medical and Surgical Journal. 1915. "The midwife versus the pregnancy clinic." 173:784–785.

Boston Women's Health Book Collective. 1998. *Our Bodies, Ourselves for the New Century*. New York: Simon & Schuster.

Bourre, Marie I. 1987. "The stumbling disease: A case study of stigma among Azorean-Portuguese." *Social Science and Medicine* 24:209–217.

Bradsher, Keith. 1995. "Gap in wealth in U.S. called widest in west." *New York Times* April 17:A1+.

Brandt, Allan M., and Paul Rozin. 1997. *Morality and Health*. New York: Routledge.

Brannon, Robert L. 1996. "Restructuring hospital nursing: Reversing the trend toward a professional work force." *International Journal of Health Services* 26:643–654.

Braverman, Paula A., Susan Egerter, Trude Bennett, and Jonathan Showstack. 1991. "Differences in hospital resource allocation among sick newborns according to insurance coverage." *Journal of the American Medical Association* 266:3300–3308.

Brickner, Philip W., Linda Keen Scharer, Barbara A. Conanan, Marianne Savarese, and Brian C. Scanlan. 1990. *Under the Safety Net: The Health and Social Welfare of the Homeless in the United States*. New York: Norton.

Brill, H., and Robert E. Patton. 1962. "Clinical statistical analysis of population changes in New York State mental hospitals since the introduction of psychotropic drugs." *American Journal of Psychiatry* 119:20–35.

Broadhead, Robert S., and Neil J. Facchinetti. 1985. "Drug iatrogenesis and clinical pharmacy: The mutual fate of a social problem and a professional movement." *Social Problems* 32:425–436.

Brooks, Nancy A., and Ronald R. Matson. 1987. "Managing multiple sclerosis." *Research in the Sociology of Health Care* 6:73–106.

Brothwell, Don R. 1993. "Yaws." Pp. 1096–1100 in *Cambridge World History of Human Disease,* edited by Kenneth F. Kiple. New York: Cambridge University Press.

Brown, Phil. 1985. *The Transfer of Care: Psychiatric Deinstitutionalization and Its Aftermath.* Boston: Routledge & Kegan Paul.

_____. 1990. "The name game: Toward a sociology of diagnosis." *Journal of Mind and Behavior* 11:385–406.

Brumberg, Joan Jacobs. 1997. *The Body Project: An Intimate History of American Girls.* New York: Random House.

Brundin, Jennifer. 1993. "How the U.S. press covers the Canadian health care system." *International Journal of Health Services* 23:275–277.

Bullard, Robert D. 1983. "Solid waste sites and the black Houston community." *Sociological Inquiry* 53:273–288.

_____. 1993. *Confronting Environmental Racism: Voices from the Grassroots.* Boston: South End.

_____. (ed.). 1997. *Unequal Protection: Environmental Justice and Communities of Color.* San Francisco: Sierra Club Books.

Bunker, John P., Howard S. Frazier, and Frederick Mosteller. 1994. "Improving health: Measuring effects of medical care. *Milbank Quarterly* 72:225–258.

Burbridge, Lynn C. 1993. "The labor market for home care workers." *Gerontologist* 33:41–46.

Burger, Jerry M. 1981. "Motivational biases in the attribution of responsibility for an accident: A meta-analysis of the defensive-attribution hypothesis." *Psychological Bulletin* 90:496–512.

Burns, Thomas J., Andrew I. Batavia, and Gerben DeJong. 1993. "The health insurance work disincentive for persons with disabilities." *Research in the Sociology of Health Care* 11:57–68.

Burstin, Helen R., Stuart R. Lipsitz, Troyen A. Brennan. 1992. "Socioeconomic status and risk for substandard medical care." *Journal of the American Medical Association* 268:2383–2387.

Bury, Michael. 1982. "Chronic illness as biographical disruption." *Sociology of Health and Illness* 4:167–182.

_____. 1991. "Sociology of chronic illness." *Sociology of Health and Illness* 13:451–468.

Butler, Sandra, and Barbara Rosenblum. 1991. *Cancer in Two Voices.* San Francisco: Spinsters.

Butterfield, Fox. 1999. "Prisons brim with mentally ill, study finds." *New York Times* July 12:A10.

Caldwell, John C. 1993. "Health transition: The cultural, social, and behavioral determinants of health in the third world." *Social Science and Medicine* 36:125–135.

Caldwell, John C., Pat Caldwell, and Israel O. Orubuloye. 1992. "The family and sexual networking in Sub-saharan Africa: Historical regional differences and present day implications." *Population Studies* 46:385–392.

Caldwell, John C., Pat Caldwell, and Pat Quiggin. 1989. "The social context of AIDS in sub-Saharan Africa." *Population Development Review* 15:185–234.

Caldwell, John C., Israel O. Orubuloye, and Pat Caldwell. 1991. "The destabilization of the traditional Yoruba sexual system." *Population Development Review* 17:229–262.

Callahan, Daniel. 1998. *False Hopes: Why America's Quest for Perfect Health Is a Recipe for Failure.* New York: Simon & Schuster.

Camacho, David E. (ed.). 1998. *Environmental Injustices, Political Struggles: Race, Class, and the Environment.* Durham, NC: Duke University Press.

Campbell, Jennifer A. 1999. "Health insurance coverage, 1998." *Current Population Reports* Series P60–208.

Cancian, Francesca M., and Stacey J. Oliker. 2000. *Caring and Gender.* Thousand Oaks, CA: Pine Forge.

Casper, Monica J. 1998. *The Making of the Unborn Patient: A Social Anatomy of Fetal Surgery.* New Brunswick, NJ: Rutgers University Press.

Castle, Nicholas G., and Vincent Mor. 1998. "Physical restraints in nursing homes: A review of the literature since the Nursing Home Reform Act of 1987." *Medical Care Research and Review* 55:139–170.

Center for the Evaluative Clinical Sciences, Dartmouth Medical School. 1996. *The Dartmouth Atlas of Health Care.* Chicago: American Hospital Association.

Centers for Disease Control and Prevention. 1993. "Standards for pediatric immunization practices." *Morbidity and Mortality Weekly Reports* 42(RR-5):1–13.

_____. 1997. "Vaccination coverage by race/ethnicity and poverty level among children aged 19–35 months—United States, 1996." *Journal of the American Medical Association* 278:1655–1656.

_____. 1998a. *HIV/AIDS Surveillance Report* 10(2):2–43.

_____. 1998b. *National Diabetes Fact Sheet: National Estimates and General Information on Diabetes in the United States,* revised edition. Atlanta: U.S. Department of Health and Human Services, Centers for Disease Control and Prevention.

Cerne, Frank. 1993. "Homeward bound: Hospitals see solid future for home health care." *Hospitals* 67(4):52–54.

Chambliss, Daniel F. 1996. *Beyond Caring: Hospitals, Nurses, and the Social Organization of Ethics.* Chicago: University of Chicago Press.

Charmaz, Kathy. 1991. *Good Days, Bad Days: The Self in Chronic Illness and Time.* New Brunswick, NJ: Rutgers University Press.

Chase-Dunn, Christopher. 1989. *Global Formation: Structure of the World Economy.* London: Basil Blackwell.

Chasnoff, Ira J., Harvey J. Landress, and Mark E. Barrett. 1990. "The prevalence of illicit drug use during pregnancy and discrepancies in mandatory reporting in Pinellas County, Florida." *New England Journal of Medicine* 322:1202–1206.

Chavis, Benjamin F. 1993. "Foreword." Pp. 3–5 in *Confronting Environmental Racism: Voices from the Grassroots,* edited by Robert D. Bullard. Boston: South End.

Chavkin, Wendy, Vicki Breitbart, Deborah Elman, and Paul H. Wise. 1998. "National survey of the states: Policies and practices regarding drug-using pregnant women." *American Journal of Public Health* 88:117–119.

Cherkin, Daniel C., Richard A. Deyo, Michele Battié, Janet Street, and William Barlow. 1998. "A comparison of physical therapy, chiropractic manipulation, and provision of an educational booklet for the treatment of patients with low back pain." *New England Journal of Medicine* 339:1021–1029.

Christakis, Dmitri A., and Christopher Feudtner. 1993. "Ethics in a short white coat: The ethical dilemmas that medical students confront." *Academic Medicine* 68:249–254.

_____. 1997. "Temporary matters: The ethical consequences of transient social relationships in medical training." *Journal of American Medical Association* 278:739–743.

Chrostowski, Keith. 1999. "Gun firms increasingly on defense." *Arizona Republic* May 9:A24.

Claiborne, William. 1999. "Indian health chief crusades for his people: In family tradition, Trujillo seeks better care for natives." *Washington Post* March 29:A17+.

Cobey, James C., Annette Flanagin, and William H. Foege. 1993. "Effective humanitarian aid: Our only hope for intervention in civil war." *Journal of the American Medical Association* 270:632–634.

Cohen, Cynthia B., (ed.). 1988. "Ethics committees." *Hastings Center Report* 18(1): 11–14.

Cohen, Jordan J. 1998. "Why doctors don't always go where they're needed." *Academic Medicine* 73:1277.

Colino, Stacey. 1998. "24 hours to a healthier you." *Glamour* 96(7):152+.

Committee on Ways and Means, U.S. House of Representatives. 1997. *Medicare and Health Care Chartbook.* Washington, DC: U.S. Government Printing Office.

Commonwealth Fund Commission on Women's Health. 1999. *Health Care Access and Coverage for Women: Changing Times, Changing Issues?* New York: Commonwealth Fund.

Conrad, Peter. 1985. "The meaning of medications: Another look at compliance." *Social Science and Medicine* 20:29–37.

_____. 1987. "The experience of illness: Recent and new directions." *Research in the Sociology of Health Care* 6:1–32.

_____. 1997. "Public eye and private genes: Historical frames, news construction, and social problems." *Social Problems* 44:139–154.

Conrad, Peter, and Joseph W. Schneider. 1992. *Deviance and Medicalization: From Badness to Sickness.* Philadelphia: Temple University Press.

Consumer Reports. 1990. "The crisis in health insurance." 55:533–544.

_____. 1992. "Health care in crisis: The search for solutions." 57:579–592.

_____. 1994. "Chiropractors." 59:383–390.

_____. 1996. "Drug advertising: Is this good medicine?" 61(6):62–63.

_____. 1997. "How will you pay for your old age?" 62(10):35–38.

_____. 1998. "Medicare: New choices, new worries." 63(9):27–39.

Consumers Union. 1999. *Blueprint for Fair Share Health Care.* Washington, DC: Consumers Union.

Coombs, Robert H., Sangeeta Chopra, Debra R. Schenk, and Elaine Yutan. 1993. "Medical slang and its functions." *Social Science and Medicine* 36:987–998.

Corbin, Juliet M., and Anselm Strauss. 1987. "Accompaniments of chronic illness: Changes in body, self, biography, and biographical time." *Research in the Sociology of Health Care* 6:249–282.

_____. 1988. *Unending Work and Care: Managing Chronic Illness at Home.* San Francisco: Jossey-Bass.

Corea, Gena. 1985. *The Hidden Malpractice: How American Medicine Mistreats Women.* New York: Harper.

Costello, Anthony, and Harshpal S. Sachdev. 1998. "Protecting breast feeding from breast milk substitutes." *British Medical Journal* 316:1103–1104.

Costello, Timothy W., and Joseph T. Costello. 1992. *Abnormal Psychology.* New York: Harper Perennial.

Council on Ethical and Judicial Affairs, American Medical Association. 1990. "Black-white disparities in health care." *Journal of the American Medical Association* 263:2344–2346.

_____. 1991. "Gender disparities in clinical decision making." *Journal of the American Medical Association* 266:559–562.

Council on Scientific Affairs, American Medical Association. 1990. "Home care in the 1990s." *Journal of the American Medical Association* 263:1241–1244.

_____. 1992. "Violence against women." *Journal of the American Medical Association* 267:3184–3189.

Cousineau, Michael R. 1997. "Health status of and access to health services by residents of urban encampments in Los Angeles." *Journal of Health Care for the Poor and Underserved* 8:70–82.

Crawford, Robert. 1979. "Individual responsibility and health politics." Pp. 247–268 in *Health Care in America: Essays in Social History,* edited by Susan Reverby and David Rosner. Philadelphia: Temple University Press.

Crosby, Alfred J. 1986. *Ecological Imperialism: The Biological Expansion of Europe, 900–1900.* New York: Cambridge University Press.

Crossette, Barbara. 1996. "Agency sees risk in drug to temper child behavior." *New York Times* February 29:A7.

Crouch, Robert A., and John D. Arras. 1998. "AZT trials and tribulations." *Hastings Center Report* 28(6):26–34.

Curtis, Kim. 1999. "Bias against employees with mental ills lingers." *Arizona Republic* May 20: D1.

Daniels, Cynthia R. 1993. *At Women's Expense: State Power and the Politics of Fetal Rights.* Cambridge, MA: Harvard University Press.

D'Arcy, Carl, and Joan Brockman. 1976. "Changing public recognition of psychiatric symptoms? Blackfoot revisited." *Journal of Health and Social Behavior* 17:302–310.

Davis, Fred. 1961. "Deviance disavowal: Management of strained interaction by the visibly handicapped." *Social Problems* 9:120–132.

Deam, Jenny. 1999. "Patients' orders: Doctors getting more and more requests for prescription drugs." *Denver Rocky Mountain News* July 27:3D.

DeCew, Judith Wagner. 1994. "Drug testing: Balancing privacy and public safety." *Hastings Center Report* 24(2):17–23.

DeJong, Gerben, Andrew I. Batavia, and Robert Griss. 1989. "America's neglected health minority: Working-age persons with disabilities." *Milbank Quarterly* 67(suppl.2):311–351.

De Lew, Nancy, George Greenberg, and Kraig Kinchen. 1992. "A layman's guide to the U.S. health care system." *Health Care Financing Review* 14:151–165.

Dettwyler, Katherine A. 1995. "Beauty and the breast." Pp. 167–213 in *Breastfeeding: Biocultural Perspectives,* edited by Patricia Stuart-Macadam and Katherine A. Dettwyler. New York: Aldine De Gruyter.

Dey, Achintya N. 1997. "Characteristics of elderly nursing home residents: Data from the 1995 National Nursing Home Survey." *Advance Data* 289.

Diamond, Timothy. 1992. *Making Gray Gold: Narratives of Nursing Home Care.* Chicago: University of Chicago Press.

Diller, Lawrence H. 1998. *Running on Ritalin: A Physician Reflects on Children, Society, and Performance in a Pill.* New York: Bantam.

DiMaggio, Paul J., and Walter W. Powell. 1983. "The iron cage revisited: Institutional isomorphism and collective rationalizing in organizational fields." *American Sociological Review* 48:147–160.

Dixon-Mueller, Ruth. 1990. "Abortion policy and women's health in developing countries." *International Journal of Health Services* 20:297–314.

Dobash, Russell P., and Rebecca Emerson Dobash. 1998. *Rethinking Violence against Women.* Newbury Park, CA: Sage.

Dolenc, Danielle A., and Charles J. Dougherty. 1985. "DRGs: The counterrevolution in financing health care." *Hastings Center Report* 15(3):19–29.

Donelan, Karen, Robert J. Blendon, John Benson, Robert Leitman, and Humphrey Taylor. 1996. "All payer, single payer, managed care, no payer: Patients' perspectives in three nations." *Health Affairs* 15:254–265.

Donnelly, John, and Dave Montgomery. 1999. "TB from ex-Soviet states resists most drugs." *Arizona Republic* March 21:A27+.

Dreze, Jean, and Amartya Sen. 1989. *Hunger and Public Action.* Oxford, England: Clarendon.

Dubos, Rene. 1961. *Mirage of Health.* New York: Anchor.

Dunn, Daniel, William Hsiao, Thomas Ketchan, and Peter Brown. 1988. "A method for estimating the preservice and postservice work of physicians' services." *Journal of the American Medical Association* 260:2371–2378.

Durham, Mary L. 1998. "Mental health and managed care." *Annual Review of Public Health* 19:493–505.

Eaton, William W. 1980. "A formal theory of selection for schizophrenia." *American Journal of Sociology* 86:149–158.

Eckert, J. Kevin, and Stephanie M. Lyon. 1992. "Board and care homes: From the margins to the mainstream in the 1990s." Pp. 97–114 in *In-Home Care for Older People: Health and Supportive Services,* edited by Marcia G. Ory and Alfred P. Duncker. Newbury Park, CA: Sage.

Economic Report of the President. 1993. Washington, DC: U.S. Government Printing Office.

Eisenberg, David M., Roger B. Davis, Susan L. Ettner, Scott Appel, Sonja Wilkey, Maria Van Rompay, and Ronald C. Kessler. 1998. "Trends in alternative medicine use in the United States, 1990–1997: Results of a follow-up national survey." *Journal of the American Medical Association* 280:1569–1575.

Emanuel, Ezekiel J., Elisabeth R. Daniels, Diane L. Fairclough, and Brian R. Clarridge. 1998. "The practice of euthanasia and physician-assisted suicide in the United States: Adherence to proposed safeguards and the effects on physicians." *Journal of the American Medical Association* 280:507–513.

Engelhardt, H. Tristram Jr. 1986. *Foundations of Bioethics.* New York: Oxford University Press.

Ensel, Walter M., and Nan Lin. 1991. "The life stress paradigm and psychological distress." *Journal of Health and Social Behavior* 32:321–341.

Epstein, Steven. 1996. *Impure Science: AIDS, Activism, and the Politics of Knowledge.* Berkeley: University of California Press.

Equal Employment Opportunity Commission. 1999. *Americans with Disabilities Act of 1990 (ADA) Charges, FY 1992 - FY 1998.* www.eeoc.gov/stats/ada.html.

Fadiman, Anne. 1997. *The Spirit Catches You and You Fall Down: A Hmong Child, Her American Doctors, and the Collision of Two Cultures.* New York: Farrar, Straus and Giroux.

Fairfield, Kathleen M., David M. Eisenberg, Roger B. Davis, Howard Libman, and Russell S. Phillips. 1998. "Patterns of use, expenditures and perceived efficacy of complementary and alternative therapies in HIV-infected patients." *Archives of Internal Medicine* 158:2257–2264.

Farmer, Paul. 1999. *Infections and Inequalities: The Modern Plagues.* Berkeley: University of California Press.

Feagin, Joe R., and Melvin P. Sikes. 1994. *Living with Racism: The Black Middle-class Experience.* Boston: Beacon.

Federal Interagency Forum on Child and Family Statistics. 1999. *America's Children 1999.* Washington, DC: U.S. Government Printing Office.

Federal Trade Commission. 1999. *Self-Regulation in the Alcohol Industry.* Washington, DC: Government Printing Office.

Feinstein, Jonathan S. 1993. "The relationships between socioeconomic status and health: A review of the literature." *Milbank Quarterly* 71: 279–322.

Ferguson, Tom (ed.). 1980. *Medical Self-Care: Access to Health Tools.* New York: Summit.

Feshbach, Morris. 1999. "Dead souls." *Atlantic Monthly* 283(1):26–27.

Feshbach, Morris, and Alfred Friendly. 1992. *Ecocide in the USSR: Health and Nature Under Siege.* New York: Basic .

Figert, Anne E. 1996. *Women and the Ownership of PMS: The Structuring of a Psychiatric Disorder.* New York: Aldine De Gruyter.

Fine, Michelle, and Adrienne Asch. 1988a. "Disability beyond stigma: Social interaction, discrimination, and activism." *Journal of Social Issues* 44:3–21.

_____. 1988b. "Introduction: Beyond pedestals." Pp. 1–38 in *Women with Disabilities: Essays in Psychology, Culture, and Politics,* edited by Michelle Fine and Adrienne Asch. Philadelphia: Temple University Press.

Fingarette, Herbert. 1988. *Heavy Drinking: The Myth of Alcoholism as a Disease.* Berkeley: University of California Press.

Finn Paradis, Lenora, and Scott B. Cummings. 1986. "The evolution of hospice in America toward organizational homogeneity." *Journal of Health and Social Behavior* 27:370–386.

Fisher, Barbara, Mel Hovell, C. Richard Hofstetter, and Richard Hough. 1995. "Risks associated with long-term homelessness among women: Battery, rape, and HIV infection." *International Journal of Health Services* 25:351–369.

Fisher, Ian. 1998. "Families provide medical care, tubes and all." *New York Times* June 7:1+.

Fisher, Sue. 1986. *In the Patient's Best Interest: Women and the Politics of Medical Decisions.* New Brunswick, NJ: Rutgers University Press.

_____. 1995. *Nursing Wounds: Nurse Practitioners, Doctors, Women Patients and the Negotiation of Meaning.* New Brunswick, NJ: Rutgers University Press.

Foner, Nancy. 1994. *The Caregiving Dilemma: Work in an American Nursing Home.* Berkeley, CA: University of California Press.

Fortune Magazine. 1999. "Why drug companies fear and loathe pharmacists: The *real* power in health care." March 29:46–47.

Fost, Norman, and Ronald E. Cranford. 1985. "Hospital ethics committees: Procedural aspects." *Journal of the American Medical Association* 253:2687–2692.

Foster, George. 1976. "Disease etiologies in non-western medical systems." *American Anthropologist* 78:773–782.

Fox, Margery. 1989. "The socioreligous role of the Christian Science practitioner." Pp. 98–114 in *Women as Healers: Cross-Cultural Perspectives,* edited by Carol Shepherd McClain. New Brunswick, NJ: Rutgers University Press.

Fox, Renee C. 1974. "Ethical and existential developments in contemporaneous American medicine: Their implications for culture and society." *Milbank Memorial Fund Quarterly* 52:445–483.

_____. 1977. "The medicalization and demedicalization of American society." *Daedalus* 106:9–22.

Fox, Renee C., and Judith Swazey. 1974. *The Courage to Fail.* Chicago: University of Chicago Press.

Frank, Richard G., and Thomas G. McGuire. 1998. "The economics of behavioral health carve-outs." Pp. 41–50 in *Managed Behavioral Health Care: Current Realities and Future Potential,* edited by David Mechanic. San Francisco: Jossey-Bass.

Franks, Peter, Carolyn M. Clancy, and Martha R. Gold. 1993. "Health insurance and mortality." *Journal of the American Medical Association* 270:737–741.

Freidson, Eliot. 1970a. *Professional Dominance: The Social Structure of Medical Care.* New York: Atherton.

_____. 1970b. *Profession of Medicine: A Study of the Sociology of Applied Knowledge.* New York: Dodd, Mead.

_____. 1975. *Doctoring Together: A Study of Professional Social Control.* New York: Elsevier.

_____. 1984. "The changing nature of professional control." *Annual Review of Sociology* 10:1–20.

_____. 1985. "The reorganization of the medical profession." *Medical Care Review* 42:11–35.

_____. 1986. "The medical profession in transition." Pp. 63–79 in *Applications of Social Science to Clinical Medicine and Health Policy,* edited by Linda Aiken and David Mechanic. New Brunswick, NJ: Rutgers University Press.

_____. 1994. *Professionalism Reborn.* Chicago: University of Chicago Press.

Freud, Sigmund. 1925. [1971]. *The Standard Edition of the Complete Psychological Works of Sigmund Freud.* Volume 19. London: Hogarth.

Fulder, Stephen. 1984. *Handbook of Complementary Medicine.* London: Hodder and Stoughton.

Gallagher, Sally, and David Mechanic. 1996. "Living with the mentally ill: Effects on the health and functioning of other household members." *Social Science and Medicine* 42:1691–1701.

Garrett, Laurie. 1994. *The Coming Plague: Newly Emerging Diseases in a World Out of Balance.* New York: Farrar, Straus & Giroux.

Gaston, Robert S., Ian Ayres, Laura G. Dooley, and Arnold G. Dietheim. 1993. "Racial equity in renal transplantation: The disparate impact of HLA-based allocation." *Journal of the American Medical Association* 270:1352–1356.

Geiger, H. Jack, and Robert M. Cook-Deegan. 1993. "The role of physicians in conflicts and humanitarian crises: Case studies from the field missions of Physicians for Human Rights, 1988 to 1993." *Journal of the American Medical Association* 270:616–620.

Gerber, J. 1990. "Enforced self-regulation in the infant formula industry." *Social Justice* 17:98–112.

Gething, Lindsay. 1992. "Judgments by health professionals of personal characteristics of people with a visible physical disability." *Social Science and Medicine* 34:809–815.

Gevitz, Norman. 1988. "Osteopathic medicine: From deviance to difference." Pp. 124–156 in *Other Healers: Unorthodox Medicine in America,* edited by Norman Gevitz. Baltimore: Johns Hopkins University Press.

Gilman, Sander L. 1999. *Making the Body Beautiful: A Cultural History of Aesthetic Surgery.* Princeton, NJ: Princeton University Press.

Glazer, Nona Y. 1993. *Women's Paid and Unpaid Labor: The Work Transfer in Health Care and Retailing.* Philadelphia: Temple University Press.

Goffman, Erving. 1961. *Asylums.* Garden City, NY: Doubleday.

_____. 1963. *Stigma: Notes on the Management of Spoiled Identity.* Englewood Cliffs, NJ: Prentice-Hall.

Goleman, Daniel. 1995. "Making room on the couch for culture," *New York Times* December 5:C1+.

Gonzales, Martin L. 1999. *Socioeconomic Characteristics of Medical Practice 1999.* Chicago: American Medical Association.

Good, Mary-Jo DelVecchio. 1995. *American Medicine: The Quest for Competence.* Berkeley: University of California Press.

Gordon, Rena. 1989. "The effects of malpractice insurance on certified nurse-midwives." *Journal of Nurse-Midwifery* 35:99–106.

Gore, Susan, and Thomas W. Mangione. 1983. "Social roles, sex roles, and psychological distress: Additive and interactive models of sex differences." *Journal of Health and Social Behavior* 24:300–312.

Gostin, Lawrence O., Chai Feldblum, and David W. Webber. 1999. "Disability discrimination in America: HIV/AIDS and other health conditions." *Journal of the American Medical Association* 281:745–752.

Gostin, Lawrence O., Zita Lazzarini, T. Stephen Jones, and Kathleen Flaherty. 1997. "Prevention of HIV/AIDS and other blood-borne diseases among injection drug users: A national survey on the regulation of syringes and needles." *Journal of the American Medical Association* 277:53–62.

Gottfried, Robert S. 1983. *The Black Death.* New York: Free Press.

Graves, Edmund J., and Maria F. Owings. 1998. "1996 summary: National Hospital Discharge Survey." *Advance Data* 301.

Gray, Bradford. 1991. *The Profit Motive and Patient Care.* Cambridge, MA: Harvard University Press.

Greenberg, Brigitte. 1999. "Study questions use of growth hormones." *Arizona Republic* March 14:A27.

Greenberg, Michael R. 1987. "Health and risk in urban-industrial society." Pp. 3–24 in *Public Health and the Environment: The United States Experience,* edited by Michael R. Greenberg. New York: Guilford.

Greenfield, S., W. Rogers, M. Mangotich, M.F. Carney, A. R. Tarlov. 1995. "Outcomes of patients with hypertension and non-insulin dependent diabetes mellitus treated by different systems and specialties." *Journal of the American Medical Association* 274:1436–1444.

Greenhouse, Steven. 1999. "AMA's delegates decide to create union of doctors." *New York Times* June 24:A1+.

Grob, Gerald N. 1997. "Deinstitutionalization: The illusion of policy." *Journal of Policy History* 9:48–73.

Gronfein, W. 1985. "Incentives and intentions in mental health policy: A comparison of the Medicaid and community mental health programs." *Journal of Health and Social Behavior* 26:192–206.

Gussow, Zachary, and George S. Tracy. 1968. "Status, ideology, and adaptation to stigmatized illness: A study of leprosy." *Human Organization* 27:316–325.

Gwyther, Marni E., and Melinda Jenkins. 1998. "Migrant farmworker children: Health status, barriers to care, and nursing innovations in health care delivery." *Journal of Pediatric Health Care* 12:60–66.

Haan, M., G. Kaplan, and T. Camacho. 1987. "Poverty and health: Prospective evidence from the Alameda County study." *American Journal of Epidemiology* 125:989–998.

Haas, Jack, and William Shaffir. 1987. *Becoming Doctors: The Adoption of a Cloak of Competence.* Greenwich, CT: JAI.

Habermas, Jürgen. 1981. "New Social Movements." *Telos* 49:33–37.

Hadley, Jack, Earl P. Steinberg, and Judith Feder. 1991. "Comparison of uninsured and privately insured hospital patients." *Journal of the American Medical Association* 265:374–379.

Hafferty, Frederic W. 1991. *Into the Valley: Death and the Socialization of Medical Students.* New Haven, CT: Yale University Press.

_____. 1998. "Beyond curriculum reform: Confronting medicine's hidden curriculum." *Academic Medicine* 73:403–407.

Hafferty, Frederic W,. and Ronald Franks. 1994. "The hidden curriculum, ethics teaching, and the structure of medical education." *Academic Medicine* 69:861–867.

Hahn, Harlan. 1985. "Toward a politics of disability definitions, disciplines, and policies." *Social Science Journal* 22(October):87–105.

Hall, Oswald. 1949. "Types of medical careers." *American Journal of Sociology* 55:243–253.

Halpern, S.A. 1990. "Medicalization as a professional process: Postwar trends in pediatrics." *Journal of Health and Social Behavior* 31:28–42.

Hammond, Ross. 1998. *Addicted to Profit: Big Tobacco's Expanding Global Reach.* Washington, DC: Essential Action.

Harlan, Sharon L., and Pamela M. Robert. 1998. "The social construction of disability in organizations: Why employers resist reasonable accommodation." *Work and Occupations* 25:397–435.

Harris Poll. 1994. "Americans with disabilities make gains in education but are no more likely to be working in 1994 than they were in 1986." 45(July):4–7.

_____. 1996. "Majority continues to support women's right to choose abortion." 56(September 9):4.

_____. 1997a. "Physician assisted suicide: Two to one majority disagree with Supreme Court decision." 36(August 4):4.

_____. 1997b. "Confidence in institutions falls to lowest level recorded in last 30 years." 7(February):2.

Haug, Marie. 1988. "A re-examination of the hypothesis of physician deprofessionalization." *Milbank Quarterly* 66(supplement 2):48–56.

Hayward, Mark D., and Melonie Heron. 1999. "Racial inequality in active life among adult Americans." *Demography* 36:77–91.

Helman, Cecil G. 1986. "'Feed a cold, starve a fever': Folk models of infection in an English suburban community, and their relation to medical treatment." Pp. 211–232 in *Concepts of Health, Illness and Disease: A Comparative Perspective,* edited by Caroline Currer and Meg Stacey. Leamington Spa, UK: Berg.

Helzer, John L., Lee N. Robin, Mitchell Taibleson, Robert A. Woodruff, Theodore Reich, and Eric D. Wish. 1977. "Reliability of psychiatric diagnosis: A methodological review." *Archives of General Psychiatry* 34:129–133.

Hendlin, Herbert, Chris Rutenfrans, and Zbigniew Zylicz. 1997. "Physician-assisted suicide and euthanasia in the Netherlands: Lessons from the Dutch." *Journal of the American Medical Association* 277:1720–1722.

Henshaw, Stanley K. 1995. "Factors hindering access to abortion services." *Family Planning Perspectives* 27(2):54–59.

_____. 1998. "Abortion incidence and services in the U.S., 1995–1996. *Family Planning Perspectives* 30(6):263–270, 287.

Henshaw, Stanley K., Susheela Singh, and Taylor Haas. 1999. "The incidence of abortion worldwide." *International Family Planning Perspectives* 25(supplement):S30–38.

Hepler Charles D., and Linda M. Strand. 1990. "Opportunities and responsibilities in pharmaceutical care." *American Journal of Hospital Pharmacy* 47:533–543.

Herbert, Bob. 1999. "Children in crisis." *New York Times* June 10:A31.

Herzlinger, Regina. 1997. *Market Driven Health Care.* Reading, MA: Addison-Wesley.

Higgins, Paul C. 1992. *Making Disability: Exploring the Social Transformation of Human Variation.* Springfield, IL: Charles C. Thomas.

Hilbert, Richard A. 1984. "The acultural dimensions of chronic pain: Flawed reality construction and the problem of meaning." *Social Problems* 31:365–378.

Hilts, Philip J. 1993. "Rise of TB linked to a U.S. failure." *New York Times* October 8:A14.

_____. 1996. *Smokescreen: The Truth Behind the Tobacco Industry Cover-up.* Reading, MA: Addison-Wesley.

_____. 1999. "In tests on people, who watches the watchers?" *New York Times* May 25:D1+.

Himmelstein, David U., and Steffie Woolhandler. 1994. *The National Health Program Book: A Source Guide for Advocates.* Monroe, ME: Common Courage.

Himmelstein, David U., Steffie Woolhandler, Ida Hellander, and Sidney M. Wolfe. 1999. "Quality of care in investor-owned versus not-for-profit HMOs." *Journal of the American Medical Association* 282:159–163.

Hintz, Raymond L., Kenneth M. Attie, Joyce Baptista, and Alex Roche. 1999. "Effect of growth hormone treatment on adult height of children with idiopathic short stature." *New England Journal of Medicine* 340:502–507.

Hochschild, Arlie. 1983. *The Managed Heart: The Commercialization of Human Feelings.* Berkeley: University of California Press.

Horn, Joshua S. 1969. *"Away with all Pests . . .": An English Surgeon in People's China.* London: Paul Hamlyn.

Horwitz, Allan V. 1982. *Social Control of Mental Illness.* New York: Academic.

Horwitz, Allan V., and Jeffrey S. Mullis. 1998. "Individualism and its discontents: The response to the seriously mentally ill in late twentieth century America." *Sociological Focus* 31:119–133.

Houghton, John Theodore, L. Meria Filho, B.A. Callander, and N. Harris (eds.). 1996. *Climate Change 1995: The Science of Climate Change.* New York: Cambridge University Press.

Hsiao, William, Dovwe Yntema, Peter Braun, Daniel Dunn, and Christine Spencer. 1988. "Measurement and analysis of intraservice work." *Journal of the American Medical Association* 260:2361–2370.

Hubbard, Ruth, and Mary Sue Henefin. 1985. "Genetic screening of prospective parents and of workers." *International Journal of Health Services* 15:231–251.

Hughes, David, 1988. "What nurse knows best: Some aspects of nurse/doctor interactions in a casualty department." *Sociology of Health and Illness* 10:1–22.

Hunt, Charles W. 1989. "Migrant labor and sexually transmitted disease: AIDS in Africa." *Journal of Health and Social Behavior* 30:353–373.

_____. 1996. "Social vs. biological: Theories on the transmission of AIDS in Africa." *Social Science and Medicine* 42:1283–1296.

Hurwitz, Eric L., Ian D. Coulter, Alan H. Adams, Barbara J. Genovese, and Paul G. Shekelle. 1998. "Use of chiropractic services from 1985 through 1991 in the United States and Canada." *American Journal of Public Health* 88:771–776.

Iglehart, John K. 1999. "The American health care system: Expenditures." *New England Journal of Medicine* 340:70–76.

Inlander, Charles B., and Eugene I. Pavalon. 1990. *Your Medical Rights: How to Become an Empowered Consumer.* Boston: Little, Brown.

James, W. Philip T., Michael Nelson, Ann Ralph, and Suzi Leather. 1997. "Socioeconomic determinants of health. The contribution of nutrition to inequalities in health." *British Medical Journal* 314(7093):1545–1549.

Janis, Irving L., and Leon Mann. 1977. *Decision Making: A Psychological Analysis of Conflict, Choice, and Commitment.* New York: Free Press.

Janoff, Barry. 1999. "Centers of attention." *Progressive Grocer* 78(4):75–80.

Jeffery, Roger, Patricia Jeffery, and Andrew Lyon. 1984. "Female infanticide and amniocentesis." *Social Science and Medicine* 19:1207–1212.

Johnson, William G., and James Lambrinos. 1987. "The effect of prejudice on the wages of disabled workers." *Policy Studies Journal* 15:571–590.

Jones, James. 1993. *Bad Blood: The Tuskegee Syphilis Experiment,* revised edition. New York: Free Press.

Jonsen, Albert R., Andrew L. Jameton, and Abbyann Lynch. 1978. "History of medical ethics: North America in the twentieth century." Pp. 992–1004 in *Encyclopedia of Bioethics,* edited by Warren T. Reich. New York: Free Press.

Journal of the American Medical Association. 1990. "Christian Scientists claim healing efficacy equal if not superior to that of medicine." 264:1379–1381.

Kahn, Joseph P. 1989. "Radner's humor good medicine, even though it couldn't save her." *Arizona Daily Star* (May 30):1C–2C.

Kanter, Rosabeth Moss. 1972. *Commitment and Community: Communes and Utopias in Sociological Perspective.* Cambridge, MA: Harvard University Press.

Katz, Jay. 1984. *The Silent World of Doctor and Patient.* New York: Free Press.

Kaufman, Martin. 1971. *Homeopathy in America: The Rise and Fall of a Medical Heresy.* Baltimore: Johns Hopkins University Press.

Kaye, H. Stephen, Mitchell P. LaPlante, Dawn Carlson, and Barbara L. Wenger. 1996. "Trends in disability rates in the United States, 1970–1994." *Disability Statistics Abstract* (17). Washington, DC: U.S. Department of Education, National Institute on Disability and Rehabilitation Research.

Kaysen, Susan. 1993. *Girl, Interrupted.* New York: Random House.

Kellerman, Arthur L., and Bela B. Hackman. 1990. "Patient 'dumping' post-COBRA." *American Journal of Public Health* 80:864–867.

Kellerman, Arthur L., Frederick P. Rivara, Norman B. Rushforth, Joyce G. Banton, Donald T. Reay, Jerry T. Francisco, Ana B. Locci, Janice Prodzinski, Bela B. Hackman, and Grant Somes. 1993. "Gun ownership as a risk factor for homicide in the home." *New England Journal of Medicine* 329(15):1084–1091.

Kellogg, J.H. 1880. *Plain Facts for Young and Old.* Burlington, IA: Segner and Condit.

Kelly, Susan E., Patricia A. Marshall, Lee M. Sanders, Thomas A. Raffin, and Barbara A. Koenig. 1997. "Understanding the practice of ethics consultation: Results of an ethnographic multi-site study." *Journal of Clinical Ethics* 8:136–149.

Kemper, Peter. 1992. "Use of formal and informal home care by the disabled elderly." *Health Services Research* 27:421–451.

Kennedy, Randy. 1998. "Study finds that hospitals overwork young doctors." *New York Times* May 19:A22.

Kessler, Ronald C., James S. House, and J. Blake Turner. 1987. "Unemployment and health in a community sample." *Journal of Health and Social Behavior* 28:51–59.

Kessler, Ronald C., Katherine A. McGonagle, Shanyang Zhao, Christopher B. Nelson, Michael Hughes, Suzann Eshleman, Hans-Ulrich Wittchen, and Kenneth S. Kendler. 1994. "Lifetime and 12-month prevalence of DSM-III-R psychiatric disorders in the United States. Results from the National Comorbidity Survey." *Archives of General Psychiatry* 51:8–19.

Kessler, Ronald C., and Harold W. Neighbors. 1986. "A new perspective on the relationships among race, social class, and psychological distress." *Journal of Health and Social Behavior* 27:107–115.

Kessler, Ronald C., Shanyang Zhao, Steven J. Katz, Anthony C. Kouzis, Richard G. Frank, Mark Edlund, and Philip Leaf. 1999. "Past-year use of outpatient services for psychiatric problems in the National Comorbidity Survey." *American Journal of Psychiatry* 156:115–123.

Kidder, David. 1988a. "The impact of hospices on the health-care costs of terminal cancer patients." Pp. 48–68 in *The Hospice Experiment,* edited by Vincent Mor, David S. Greer, and Robert Kastenbaum. Baltimore: Johns Hopkins.

_____. 1988b. "Hospice services and cost savings in the last weeks of life." Pp. 69–87 in *The Hospice Experiment,* edited by Vincent Mor, David S. Greer, and Robert Kastenbaum. Baltimore: Johns Hopkins University Press.

Kiesler, Charles A., and Amy E. Sibulkin. 1987. *Mental Hospitalization: Myths and Facts About a National Crisis.* Newbury Park, CA: Sage.

Kiple, Kenneth F. 1993. *Cambridge World History of Human Disease.* New York: Cambridge University Press.

Kirk, Stuart A. 1992. *The Selling of DSM: The Rhetoric of Science in Psychiatry.* New York: Aline de Gruyter.

Kirp, David L. 1989. *Learning by Heart: AIDS and Schoolchildren in America's Communities.* New Brunswick, NJ: Rutgers University Press.

Klass, Perri. 1987. *A Not Entirely Benign Procedure: Four Years as a Medical Student.* New York: Putnam's.

Kleinman, Lawrence C., Howard Freeman, Judy Perlman, and Lillian Gelberg. 1996. "Homing in on the homeless: Assessing the physical health of homeless adults in Los Angeles County using an original method to obtain physical examination data in a survey." *Health Services Research* 31:533–549.

Kligman, David. 1999. "Institutions: Mental or penal?" *Arizona Republic* May 9:A22.

Klima, Edward S., and Ursula Bellugi. 1979. *The Signs of Language.* Cambridge, MA: Harvard University Press.

Knafl, Kathleen, and Gary Burkett. 1975. "Professional socialization in a medical specialty: Acquiring medical judgment." *Social Science and Medicine* 9:397–404.

Kolata, Gina. 1985. "Heart panel's conclusions questioned." *Science* 227:40–41.

———. 1994. "Pharmacists paid to suggest drugs." *New York Times* July 29:A1.

Kolata, Gina, and Kurt Eichenwald. 1999. "For the uninsured, drug trials are health care." *New York Times* June 22:A1+.

Kolder, Veronika E.B., Janet Gallagher, and Michael T. Parsons. 1987. "Court-ordered obstetrical interventions." *New England Journal of Medicine* 316: 1192–1196.

Kolko, Gabriel. 1999. "Ravaging the poor: The International Monetary Fund indicted by its own data." *International Journal of Health Services.* 29:51–57.

Koopman, Cheryl, Sherman Eisenthal, and John D. Stoeckle. 1984. "Ethnicity in the reported pain, emotional distress, and requests of medical outpatients." *Social Science and Medicine* 18(6):487–490.

Koren, Gideon, and Naomi Klein. 1991. "Bias against negative studies in newspaper reports of medical research." *Journal of the American Medical Association* 266:1824–1826.

Koren, Gideon, Heather Shear, Karen Graham, and Tom Einarson. 1989. "Bias against the null hypothesis: The reproductive hazards of cocaine." *Lancet* 2(8677):1440–1443.

Kozol, Jonathan. 1988. *Rachel and Her Children: Homeless Families in America.* New York: Crown.

Kramer, Peter. 1993. *Listening to Prozac.* New York: Viking.

Krauss, Herbert H., Christopher Godfrey, Jean Kirk, and David M. Eisenberg. 1998. "Alternative health care: Its use by individuals with physical disabilities." *Archives of Physical Medicine and Rehabilitation* 79:1440–1447.

Kristof, Nicholas D. 1997. "For Third World, water is still a deadly drink." *New York Times* January 9:A1+.

Kübler-Ross, Elizabeth. 1969. *On Death and Dying.* New York: Macmillan.

Kunitz, Stephen J. 1996. "The history and politics of U.S. health care policy for American Indians and Alaskan natives." *American Journal of Public Health* 86:1464–1473.

Kuo, JoAnn, and Kathryn Porter. 1998. "Health status of Asian Americans: 1992–1994." *Advance Data from Vital and Health Statistics* No. 298.

Kurz, Demi. 1987. "Emergency department responses to battered women: Resistance to medicalization." *Social Problems* 34:69–81.

Kutner, Nancy G. 1987. "Social worlds and identity in end-stage renal disease (ESRD)." *Research in the Sociology of Health Care* 6:33–71.

Laing, Ronald D. 1967. *The Politics of Experience.* New York: Ballantine.

Lancet. 1990. "Marketing of breast milk substitutes." 335(8698):11511–11552.

Lane, Harlan. 1992. *The Mask of Benevolence: Disabling the Deaf Community.* New York: Knopf.

LaPlante, Mitchell P., Jae Kennedy, H. Stephen Kaye, and Barbara L. Wenger. 1996. "Disability and employment." *Disability Statistics Abstract* (11). Washington, DC: U.S. Department of Education, National Institute on Disability and Rehabilitation Research.

Lappé, Frances Moore, Joseph Collins, and Peter Rosset. 1998. *World Hunger: Twelve Myths,* second edition. New York: Grove.

Lassey, Marie L., William R. Lassey, and Martin J. Jinks. 1997. *Health Care Systems Around the World: Characteristics, Issues, Reforms.* Upper Saddle River, NJ: Prentice Hall.

LaVeist, Thomas A. 1993. "Segregation, poverty, and empowerment: Health consequences for African Americans." *Milbank Quarterly* 71:41–64.

Lawrence, Ruth A. 1995. "Commentary: Breastfeeding is more than just good nutrition." Pp. 395–404 in *Breastfeeding: Biocultural Perspectives,* edited by Patricia Stuart-Macadam and Katherine A. Dettwyler. New York: Aldine De Gruyter.

_____. 1997. *A Review of the Medical Benefits and Contraindications to Breastfeeding in the United States (Maternal and Child Health Technical Information Bulletin).* Arlington, VA: National Center for Education in Maternal and Child Health.

Leape, Lucian L. 1992. "Unnecessary surgery." *Annual Review of Public Health* 13:363–383.

_____. 1994. "Error in medicine." *Journal of the American Medical Association* 272:1851–1857.

Leavitt, Judith Walzer. 1983. "Science enters the birthing room: Obstetrics in America since the eighteenth century." *Journal of American History* 70:281–304.

_____. 1986. *Brought to Bed: Childbearing in America, 1750–1950.* New York: Oxford University Press.

Leavitt, Judith Walzer, and Ronald L. Numbers. 1985. *Sickness and Health in America.* Madison: University of Wisconsin Press.

Lehrman, Ella-Joy. 1992. "Findings of the 1990 annual ACNM Membership Survey." *Journal of Nurse-Midwifery* 37:33–47.

Leichter, Howard M. 1997. "Rationing of health care in Oregon: Making the implicit explicit." Pp. 138–162 in *Health Policy Reform in America: Innovations from the States,* edited by Howard M. Leichter. Armonk, NY: M.E. Sharpe.

Lerner, Melvin J., and Dale T. Miller. 1978. "Just world research and the attribution process: Looking back and ahead." *Psychological Bulletin* 85:1030–1051.

Lewin, Tamar. 1993. "Nursing invades turf once ruled by doctors." *New York Times* Nov. 22:A1+.

Lewis, Caroline T. 1993. "Midwife-attended births." Pp. 247–250 in *Encyclopedia of Childbearing: Critical Perspectives,* edited by Barbara Katz Rothman. Phoenix, AZ: Oryx.

Libov, Charlotte. 1999. "Beat your risk factors." *Ladies' Home Journal* 116(4):86–90.

Lifton, Robert J. 1986. *The Nazi Doctors: Medical Killing and the Psychology of Genocide.* New York: Basic.

Light, Donald W. 1988. "Toward a new sociology of medical education." *Journal of Health and Social Behavior* 29:307–322.

_____. 1992. "The practice and ethics of risk-rated health insurance." *Journal of the American Medical Association* 267:2503–2508.

_____. 1994. "Excluding more, covering less: The health insurance industry in the United States." Pp. 310–320 in *Beyond Crisis: Confronting Health Care in the United States,* edited by Nancy F. McKenzie. New York: Penguin.

Light, Donald W., and Sol Levine. 1988. "Changing character of the medical profession: A theoretical overview." *Milbank Quarterly* 66(supplement 2):10–32.

Lindbloom, Erik. 1993. "America's aging population: Changing the face of health care." *Journal of the American Medical Association* 269:674–675.

Link, Bruce G. 1982. "Mental patient status, work, and income: An examination of the effects of a psychiatric label." *American Sociological Review* 47:202–215.

_____. 1987. "Understanding labeling effects in the area of mental disorders: An assessment of the effects of expectations of rejection." *American Sociological Review* 52:96–112.

Link, Bruce G., Francis T. Cullen, James Frank, and John F. Wozniak. 1987. "The social rejection of former mental patients: Understanding why labels matter." *American Journal of Sociology* 92:1461–1500.

Link, Bruce G., Francis T. Cullen, Elmer Struening, Patrick E. Shrout, and Bruce P. Dohrenwend. 1989. "A modified labeling theory approach to mental disorders: An empirical assessment." *American Sociological Review* 54:400–423.

Link, Bruce G., and Bruce P. Dohrenwend. 1989. "The epidemiology of mental disorders." Pp. 102–127 in *Handbook of Medical Sociology,* fourth edition, edited by Howard E. Freeman and Sol Levine. Englewood Cliffs, NJ: Prentice-Hall.

Link, Bruce G., Bruce P. Dohrenwend, and Andrew E. Skodol. 1986. "Socioeconomic status and schizophrenia: Noisome occupational characteristics as a risk factor." *American Sociological Review* 51:242–258.

Link, Bruce G., Frances P. Mesagno, Maxine E. Lubner, and Bruce P. Dohrenwend. 1990. "Problems in measuring role strains and social functioning in relation to psychological symptoms." *Journal of Health and Social Behavior* 31:354–369.

Link, Bruce G., Elmer L. Struening, Michael Rahav, Jo C. Phelan, and Larry Nuttbrock. 1997. "On stigma and its consequences: Evidence from a longitudinal study of men with dual diagnoses of mental illness and substance abuse." *Journal of Health and Social Behavior* 38:177–190.

Lips, Hilary M. 1993. *Sex & Gender: An Introduction.* Mountain View, CA: Mayfield.

Liska, Ken. 1997. *Drugs and the Human Body.* Upper Saddle River, NJ: Prentice Hall.

Litoff, Judy. 1978. *American Midwives: 1860 to the Present.* Westport, CT: Greenwood.

Little, Ruth E. 1998. "Public health in central and eastern Europe and the role of environmental pollution." *Annual Review of Public Health* 19:153–172.

Litwin, Mark S., David J. Pasta, Marcia L. Stoddard, James M. Henning, and Peter R. Carroll. 1998. "Epidemiological trends and financial outcomes in radical prostatectomy among Medicare beneficiaries, 1991 to 1993." *Journal of Urology* 160:445–448.

Liu, Xingzhu, and Junle Wang. 1991. "An introduction to China's health care system." *Journal of Public Health Policy* 12:104–116.

Lonsdale, Susan. 1990. *Women and Disability.* Basingstoke, UK: Macmillan.

Lorber, Judith. 1984. *Women Physicians: Careers, Status, and Power.* New York: Tavistock.

Loring, Marti, and Brian Powell. 1988. "Gender, race, and DSM-III: A study of the objectivity of psychiatric diagnostic behavior." *Journal of Health and Social Behavior* 29:1–22.

Ludmerer, Kenneth M. 1985. *Learning to Heal: The Development of American Medical Education.* New York: Basic.

Lu-Yao, Grace L., Dale McCloran, John Wasson, and John E. Wennberg. 1993. "An assessment of radical prostatectomy: Time trends, geographic variation, and outcomes." *Journal of the American Medical Association* 269:2633–2636.

Luker, Kristin. 1984. *Abortion and the Politics of Motherhood.* Berkeley: University of California Press.

Lurie, Peter, and Sidney M. Wolfe. 1997. "Unethical trials of intervention to reduce perinatal transmission of HIV in developing countries." *New England Journal of Medicine* 337(12):853–856.

Lynch, Michael. 1983. "Accommodation practices: Vernacular treatments of madness." *Social Problems* 31:152–164.

MacDorman, Marian F., and Gopal K. Singh. 1999. "Midwifery care, social and medical risk factors and birth outcomes in the USA." *Journal of Epidemiology and Community Health.* May:310–317.

Mairs, Nancy. 1986. *Plaintext.* Tucson: University of Arizona Press.

Manderscheid, Ronald W., and Mary Anne Sonnenschein (eds.). 1992. *Mental Health, United States, 1992.* Washington, DC: U.S. Government Printing Office.

Mann, Charles C. 1993. "The prostate-cancer dilemma." *Atlantic* November:102+.

Manuel, 1990. Barry M. 1990. "Professional liability—Ano-fault solution." *New England Journal of Medicine* 322:627.

Marmor, Theodore R., and Jerry L. Mashaw. 1994. "Canada's health insurance and ours: The real lessons, the big choices." Pp. 470–479 in *The Sociology of Health and Illness,* edited by Peter Conrad and Rachelle Kern. New York: St. Martin's.

Marmot, Michael G., Manolis Kogevinas, and Mary A. Elston. 1987. "Social/economic status and disease." *Annual Review of Public Health* 8:111–135.

Marmot, Michael G. and Martin J. Shipley. 1996. "Do socioeconomic differences in mortality persist after retirement?" *British Medical Journal* 313:1177–1180.

Martin, Emily. 1987. *The Woman in the Body.* Boston: Beacon.

Martin, Steven C., Robert M. Arnold, and Ruth M. Parker. 1988. "Gender and medical socialization." *Journal of Health and Social Behavior* 29:333–343.

McAdam, Douglas. 1982. *Political Process and the Development of Black Insurgency, 1930–1970.* Chicago: University of Chicago Press.

McCarthy, John D., and Mayer N. Zald. 1973. *The Trend of Social Movements in America: Professionalization and Resource Mobilization.* Morristown, NJ: General Learning.

McCord, Colin, and Harold P. Freeman. 1990. "Excess mortality in Harlem." *New England Journal of Medicine* 322:173–177.

McCrea, Francis B. 1983. "The politics of menopause: The 'discovery' of a deficiency disease." *Social Problems* 31:11–23.

McGinnis, J. Michael, and William H. Foege. 1993. "Actual causes of death in the United States." *Journal of the American Medical Association* 270:2207–2212.

McKeown, Thomas. 1979. *The Role of Medicine: Dream, Mirage, or Nemesis?* Princeton, NJ: Princeton University Press.

McKinlay, John B. 1994. "A case for refocussing upstream: The political economy of illness." Pp. 509–530 in *The Sociology of Health and Illness,* edited by Peter Conrad and Rachelle Kern. New York: St. Martin's.

McKinlay, John B., and Sonja J. McKinlay. 1977. "The questionable effect of medical measures on the decline of mortality in the United States in the twentieth century." *Milbank Memorial Fund Quarterly* 55:405–428.

McKinlay, John B., and John D. Stoeckle. 1989. "Corporatization and the social transformation of doctoring." *International Journal of Health Services* 18:191–205.

Mechanic, David. 1989. *Mental Health and Social Policy,* third edition. Englewood Cliffs, NJ: Prentice-Hall.

_____. 1995. "Sociological dimensions of illness behavior." *Social Problems* 41:1207–1216.

_____. 1997. "Managed mental health care." *Society* 35(1):44–52.

_____. 1998. "The changing face of mental health managed care." Pp. 7–14 in *Managed Behavioral Health Care: Current Realities and Future Potential,* edited by David Mechanic. San Francisco: Jossey-Bass.

_____. 1999. *Mental Health and Social Policy: The Emergence of Managed Care,* fourth edition. Boston: Allyn & Bacon.

Mechanic, David, and David A. Rochefort. 1990. "Deinstitutionalization: An appraisal of reform." *Annual Review of Sociology* 16:301–327.

_____. 1996. "Comparative medical systems." *Annual Review of Sociology* 22:239–270.

Melosh, Barbara. 1982. *"The Physician's Hand": Work Culture and Conflict in American Nursing.* Philadelphia: Temple University Press.

Melucci, Alberto. 1995. "The process of collective identity." Pp. 41–63 in *Social Movements and Culture,* edited by Hank Johnston and Bert Klandermans. Minneapolis: University of Minnesota Press.

Messer, Ellen. 1997. "Intra-household allocation of food and health care: Current findings and understandings—introduction." *Social Science and Medicine* 44:1675–1684.

Meyerowitz, Beth E., Janice G. Williams, and Jocelyne Gessner. 1987. "Perceptions of controllability and attitudes toward cancer and cancer patients." *Journal of Applied Social Psychology* 17:471–492.

Millenson, Michael L. 1997. *Demanding Medical Excellence: Doctors and Accountability in the Information Age.* Chicago: University of Chicago Press.

Miller, Robert H., and Harold T. Luft. 1997. "Does managed care lead to better or worse quality of care?" *Health Affairs* 16(5):7–25.

Millman, Marcia. 1976. *The Unkindest Cut: Life in the Backrooms of Medicine.* New York: Morrow.

Mills, C. Wright. 1959. *The Sociological Imagination.* New York: Grove.

Mirowsky, John, and Catherine E. Ross. 1980. "Minority status, ethnic culture, and distress: A comparison of blacks, whites, Mexicans, and Mexican Americans." *American Journal of Sociology* 86:479–495.

_____. 1989. "Psychiatric diagnosis as reified measurement." *Journal of Health and Social Behavior* 30:11–25.

Mishler, Elliot G. 1981. "Viewpoint: Critical perspectives on the biomedical model." Pp. 1–23 in *Social Contexts of Health, Illness, and Patient Care,* edited by Elliot G. Mishler. Cambridge: Cambridge University Press.

_____. 1990. "The struggle between the voice of medicine and the voice of the life-world." Pp. 295–307 in *The Sociology of Health and Illness: Critical Perspectives,* edited by Peter Conrad and Rachelle Kern. New York: St. Martin's.

Mizrahi, Terry. 1986. *Getting Rid of Patients: Contradictions in the Socialization of Physicians.* New Brunswick, NJ: Rutgers University Press.

Mollica, Robert L. 1997–1998. "Regulation of assisted living facilities: State policy trends." *Generations* 21:30–33.

Montgomery, Kathleen. 1992. "Professional dominance and the threat of corporatization." *Current Research on Occupations and Professions* 7:221–240.

_____. 1996. "Responses by professional organizations to multiple and ambiguous institutional environments: The case of AIDS." *Organizational Studies* 17:649–671.

Montgomery, Rhonda J.V. 1992. "Examining respite: Its promise and limits." Pp. 75–96 in *In-Home Care for Older People: Health and Supportive Services,* edited by Marcia G. Ory and Alfred P. Duncker. Newbury Park, CA: Sage.

Mor, Vincent. 1987. *Hospice Care Systems: Structure, Process, Costs, and Outcome.* New York: Springer.

Morbidity and Mortality Weekly Report. 1993. "Tuberculosis morbidity, United States, 1992." 42(36):696–697, 703–704.

_____. 1997. "Update: Blood lead levels—United States, 1991–1994. 46(7):141–46.

_____. 1998a. "Progress toward global measles control and regional elimination, 1990–1997." 47:1049–54.

_____. 1998b. "Tobacco use among U.S. racial/ethnic minority groups—African Americans, American Indians and Alaska Natives, Asian Americans and Pacific Islanders, Hispanics." 47(RR-18):1–16.

Morgan, Patricia A. 1988. "Power, politics, and public health: The political power of the alcohol beverage industry." *Journal of Public Health Policy* 9:177–197.

Moses, Evelyn B. 1997. *The Registered Nurse Population: March 1996: Findings from the National Sample Survey of Registered Nurses.* Rockville, MD: U.S. Department of Health and Human Services.

Mosher, James F. 1995. "The merchants, not the customers: Resisting the alcohol and tobacco industries' strategy to blame young people for illegal alcohol and tobacco sales." *Journal of Public Health Policy* 16:412–432.

Moss, N., and K. Carver. 1998. "The effect of WIC and Medicaid on infant mortality in the United States." *American Journal of Public Health* 88:1354–1361.

Murdock, George P. 1980. *Theories of Illness: A World Survey.* Pittsburgh, PA: University of Pittsburgh Press.

Murray, Christopher J.L., and Alan D. Lopez. 1996. *The Global Burden of Disease.* Cambridge, MA: Harvard University Press.

Murtaugh, Christopher M., Peter Kemper, Brenda C. Spillman, and Barbara Lepidus Carlson. 1997. "The amount, distribution, and timing of lifetime nursing home use." *Medical Care* 35:204–218.

Nader, Ralph. 1965. *Unsafe at Any Speed.* New York: Knightsbridge.

National Alliance for Caregiving, 1997. *Family Caregiving in the United States: Findings from a National Survey.* Bethesda, MD: National Alliance for Caregiving.

National Center for Health Statistics. 1998a. *Monthly Vital Statistics Report* 46(11S). Washington, DC: U.S. Public Health Service.

_____. 1998b. *Monthly Vital Statistics Report* 47(9). Washington, DC: U.S. Public Health Service.

_____. 1998c. "Current estimates from the National Health Interview Survey, U.S., 1995." *Vital and Health Statistics* Series 10, no. 199. Hyattsville, MD: U.S. Public Health Service.

_____. 1999. "Deaths: Final Report for 1997." *National Vital Statistics Report* 47(19).

National Hospice Organization. 1999. *Hospice Fact Sheet.* Arlington, VA: National Hospice Organization.

National Institute of Health Consensus Development Panel on Acupuncture. 1998. "Acupuncture." *Journal of the American Medical Association* 280:1518–1524.

National League for Nursing. Division of Research. 1997a. *Nursing Data Review.* New York: National League for Nursing.

_____. 1997b. *Nursing Datasource 1997.* Vol. 1. New York: National League for Nursing.

_____. 1997c. *Nursing Datasource 1997.* Vol. 2. New York: National League for Nursing.

National Low Income Housing Coalition. 1998. *Out of Reach: Rental Housing at What Cost?* Washington, DC: National Low Income Housing Coalition.

Natowicz, Marvin R., Jane K. Alper, and Joseph S. Alper. 1992. "Genetic discrimination and the law." *American Journal of Human Genetics* 50:465–475.

Navarro, Vincente. 1990. "Race or class versus race and class: Mortality differentials in the United States." *Lancet* 336(8725):1238–1240.

Naylor, C. David. 1995. "Grey zones of clinical practice: Some limits to evidence-based medicine." *Lancet* 345:840–842.

Neisser, Arden. 1983. *The Other Side of Silence: Sign Language and the Deaf Community in America.* New York: Knopf.

Nelkin, Dorothy, and Laurence Tancredi. 1989. *Dangerous Diagnostics: The Social Power of Biological Information.* New York: Basic.

Nersesian, William S. 1988. "Infant mortality in socially vulnerable populations." *Annual Review of Public Health* 9:361–377.

Nestle, Marion. 1993. "Food lobbies, the food pyramid, and U.S. nutrition policy." *International Journal of Health Services* 23:483–496.

Neubauer, Deane. 1997. "Hawaii: The health state revisited." Pp. 163–188 in *Health Policy Reform in America: Innovations from the States,* edited by Howard M. Leichter. Armonk, NY: M.E. Sharpe.

Neugebauer, Richard, Bruce P. Dohrenwend, and Barbara S. Dohrenwend, 1980. "Formulation of hypotheses about the true prevalence of functional psychiatric disorders among adults in the United States." Pp. 45–94 in *Mental Illness in the United States: Epidemiological Estimates,* edited by Bruce P. Dohrenwend, Barbara S. Dohrenwend, M. Schwartz Gould, Bruce G. Link, Richard Neugebauer, and Robin Wunsch-Hitzig. New York: Praeger.

New York Times. 1999a. "Registered nurses in short supply at hospitals nationwide." March 23:A14.

_____. 1999b. "Third of Hispanic Americans do without health coverage." April 9:A1.

Newman, Katherine S. 1999. *Filling from Grace: Downward Mobility in the Age of Affluence,* Berkeley: University of California Press.

Nichter, Mark, and Elizabeth Cartwright. 1991. "Saving the children for the tobacco industry." *Medical Anthropology Quarterly* 5:236–256.

Norris, Fran H., and Stanley A. Murrell. 1987. "Transitory impact of life-event stress on psychological symptoms in older adults." *Journal of Health and Social Behavior* 28:197–211.

North, Fiona M., S. Leonard Syme, Amanda Feeney, Martin Shipley, and Michael Marmot. 1996. "Psychosocial work environment and sickness absence among British civil servants: The Whitehall II Study." *American Journal of Public Health* 86:332–340.

Novack, Dennis H., Robin Plumer, Raymond L. Smith, Herbert Ochitill, Gary D. Morrow, and John M. Bennett. 1979. "Changes in physicians' attitudes toward telling the cancer patient." *Journal of the American Medical Association* 241:879–900.

Novello, Antonia C., Mark Rosenberg, Linda Saltzman, and John Shosky. 1992. "A medical response to domestic violence." *Journal of the American Medical Association* 267:3132.

Oddone, Eugene Z., Ronnie D. Horner, Richard Sloane, Lauren McIntyre, Aileen Ward, Jeff Whittle, Leigh J. Passman, Laura Kroupa, Robert Heaney, Susan Diem, and David Matchar. 1999. "Race, Presenting Signs and Symptoms, Use of Carotid Artery Imaging, and Appropriateness of Carotid Endarterectomy." *Stroke* 30:1350–1356.

Olshansky, S. Jay, Bruce Carnes, Richard G. Rogers, and Len Smith. 1997. "Infectious diseases—new and ancient threats to world health." *Population Bulletin* 52:2–47.

Omran, Abdel R. 1971. "The epidemiological transition." *Milbank Memorial Fund Quarterly* 49:509–538.

Organization for Economic Cooperation and Development. 1999. *Health Data 99.* Paris, France: Organization for Economic Cooperation and Development.

Orr, Robert D., and Eliot Moon. 1993. "Effectiveness of an ethics consultation service." *Journal of Family Practice* 36:49–53.

Osherson, Samuel, and Lorna AmaraSingham. 1981. "The machine metaphor in medicine." Pp. 218–249 in *Social Contexts of Health, Illness, and Patient Care,* edited by Elliot G. Mishler. Cambridge: Cambridge University Press.

Otten, Mac W., Steven M. Teutsch, David F. Williamson, and James F. Marks. 1990. "The effect of known risk factors on the excess mortality of black adults in the United States." *Journal of the American Medical Association* 263:845–850.

Owen, Anita L., and George M. Owen. 1997. "Twenty years of WIC: A review of some effects of the program." *Journal of the American Dietetic Association* 97:777–782.

Pan American Health Organization. 1998. *Health in the Americas*. Washington, DC: Pan American Health Organization.

Parens, Erik. 1998. *Enhancing Human Traits: Ethical and Social Implications*. Washington, DC: Georgetown University Press.

Parsons, Talcott. 1951. *The Social System*. New York: Free Press.

_____. 1964. *Essays in Sociological Theory*, revised edition. Glencoe, IL: Free Press.

Paul, Peter V. 1998. *Literacy and Deafness: The Development of Reading, Writing, and Literate Thought*. Boston: Allyn & Bacon.

Pawluch, Dorothy. 1983. "Transitions in pediatrics: A segmental analysis." *Social Problems* 30:449–465.

Peabody, John W. 1996. "Economic reform and health sector policy: Lessons from structural adjustment programs." *Social Science and Medicine* 43:823–835.

Pearlin, Leonard I. 1989. "The sociological study of stress." *Journal of Health and Social Behavior* 30:241–256.

Pearlin, Leonard I., and Carol S. Aneshensel. 1986. "Coping and social supports: Their functions and applications." Pp. 417–437 in *Applications of Social Science to Clinical Medicine and Health Policy*, edited by Linda H. Aiken and David Mechanic. New Brunswick, NJ: Rutgers University Press.

Perrone, Bobette, H. Henrietta Stockel, and Victoria Krueger. 1989. *Medicine Women, Curanderas, and Women Doctors*. Norman: University of Oklahoma Press.

Pescosolido, Bernice A. 1992. "Beyond rational choice: The social dynamics of how people seek help." *American Journal of Sociology* 97:1096–1138.

Pescosolido, Bernice A., Carol Brooks Gardner, and Keri M. Lubell. 1998. "How people get into mental health services: Stories of choice, coercion and 'muddling through' from 'first-timers.'" *Social Science and Medicine* 46:275–286.

Peterman, Thomas A., Rand L. Stoneburner, James R. Allen, Harold W. Jaffe, and James W. Curran. 1988. "Risk of Human Immunodeficiency Virus transmission from heterosexual adults with transfusion-associated infections." *Journal of the American Medical Association* 259:55–58.

Peyrot, Mark, James F. McMurry, Jr., and Richard Hedges. 1987. "Living with diabetes: The role of personal and professional knowledge in symptom and regimen management." *Research in the Sociology of Health Care* 6:107–146.

Phillips, Marilynn J. 1985. "'Try harder': The experience of disability and the dilemma of normalization." *Social Science and Medicine* 22:45–57.

_____. 1990. "Damaged goods: Oral narratives of the experience of disability in American culture." *Social Science and Medicine* 30:849–857.

Pill, Roisin, and Nigel C.H. Scott. 1982. "Concepts of illness causation and responsibility: Some preliminary data from a sample of working class mothers." *Social Science and Medicine* 16:43–52.

Plotnick, Robert D. 1992. "The effects of attitudes on teenage premarital pregnancy and its resolution." *American Sociological Review* 57:800–811.

Polednak, Anthony P. 1996. "Trends in U.S. urban black infant mortality, by degree of residential segregation." *American Journal of Public Health* 86:723–726.

Pollitt, Katha. 1990. "A new assault on feminism." *Nation* 250:409–418.

Population Reference Bureau. 1998. *World Population Data Sheet*. Washington, DC: Population Reference Bureau.

Presidential Commission on the Human Immunodeficiency Virus Epidemic. 1988. Report #88–436-P. Washington, DC: U.S. Government Printing Office.

Public Citizen's Health Research Group. 1997. *Hospital Emergency Rooms and Patient Dumping.* Washington, DC: Public Citizen.

Radloff, Lenore. 1975. "Sex differences in depression: The effects of occupation and marital status." *Sex Roles* 1:249–265.

Raisler, Jeanne, Cheryl Alexander, and Patricia O'Campo. 1999. "Breast-feeding and infant illness: A dose–response relationship?" *American Journal of Public Health* 89:25–30.

Ramsey, Paul. 1970. *The Patient as Person.* New Haven, CT: Yale University Press.

Ravenholt, R. T. 1993. "Tobaccosis." Pp. 176–185 in *Cambridge World History of Human Disease,* edited by Kenneth F. Kiple. New York: Cambridge University Press.

Rawls, John. 1971. *A Theory of Justice.* Cambridge, MA: Harvard University Press.

Reading, Richard. 1997. "Social disadvantage and infection in childhood." *Sociology of Health and Illness* 19:395–414.

Redlich, Fredrick C. 1978. "Medical ethics under National Socialism." Pp. 1015–1019 in *Encyclopedia of Bioethics,* edited by Warren T. Reich. New York: Free Press.

Reed, Alyson. 1997. "Trends in state laws and regulations affecting nurse-midwives: 1995–1997." *Journal of Nurse-Midwifery* 42:421–426.

Register, Cheri. 1987. *Living with Chronic Illness: Days of Patience and Passion.* New York: Free Press.

Reilly, Phillip. 1987. *To Do No Harm: A Journey Through Medical School.* Dover, MA: Auburn House.

Reinhard, Susan, and Allan V. Horwitz. 1996. "Caregiver burden: Differentiating the content and consequences of family caregiving." *Journal of Marriage and the Family* 57:741–750.

Remler, Dahlia K., Karen Donelan, Robert J. Blendon, George D. Lundberg, Lucian L. Leape, David R. Calkins, Katherine Binns, and Joseph P. Newhouse. 1997. "What do managed care plans do to affect care? Results from a survey of physicians." *Inquiry* 34:196–204.

Republic of China. 1998. *Republic of China Yearbook.* Taipei, Republic of China: Government Information Office.

Reverby, Susan. 1987. *Ordered to Care: The Dilemma of American Nursing.* New York: Cambridge University Press.

Rhoades, Everett R., John Hammond, Thomas K. Welty, Aaron O. Handler, and Robert W. Amler. 1987. "The Indian burden of illness and future health interventions." *Public Health Reports* 102:361–368.

Richards, Peter. 1977. *The Medieval Leper and His Northern Heirs.* Totowa, NJ: Rowman & Littlefield.

Riessman, Catherine K. 1983. "Women and medicalization: A new perspective." *Social Policy* 14:3–18.

Riley, Gerald F., Arnold L. Potosky, Carrie N. Klabunde, Joan L. Warren, and Rachel Ballard-Barbash. 1999. "Stage at diagnosis and treatment patterns among older women with breast cancer." *Journal of the American Medical Association* 281:720–726.

Risse, Guenter B. 1988. "Epidemics and history: Ecological perspectives and social responses." Pp. 33–66 in *AIDS: The Burdens of History,* edited by Elizabeth Fee and Daniel M. Fox. Berkeley: University of California Press.

Roback, Gene, Lillian Randolph, Bradley Seiman, and Thomas Pasko. 1999. *Physicians' Characteristics and Distribution in the United States.* Chicago: American Medical Association.

Robins, Lee N., and James L. Mills (ed.). 1993. "Effects of in utero exposure to street drugs." *American Journal of Public Health* 83:supplement.

Robinson, Chester A., Stephen J. Sepe, and Kimi F. Y. Lin. 1993. "President's child immunization initiative." *Public Health Reports* 108:419–426.

Roeder, Beatrice A. 1988. *Chicano Folk Medicine from Los Angeles, California. Vol. 34* in Folklore and Mythology Series. Los Angeles: University of California Press.

Rogers, Richard G., Robert A. Hummer, Charles B. Nam, and Kimberley Peters. 1996. "Demographic, socioeconomic, and behavioral factors affecting ethnic mortality by cause." *Social Forces* 74:1419–1438.

Rogler, Lloyd. 1991. "Acculturation and mental health status among Hispanics." *American Psychologist* 46:585–597.

Rooks, Judith. 1997. *Midwifery and Childbirth in America.* Philadelphia: Temple University Press.

Roos, Leslie L., Elliot S. Fisher, Ruth Brazauskas, Sandra M. Sharp, and Evelyn Shapiro. 1992. "Health and surgical outcomes in Canada and the United States." *Health Affairs* 11:56–72.

Rosenau, Pauline Vaillancourt. 1997. "Migration for medical care and pharmaceuticals: A research note on the NAFTA countries." *Social Science Quarterly* 78:578–590.

Rosenberg, Charles E. 1987. *The Care of Strangers: The Rise of America's Hospital System.* New York: Basic.

Rosenfield, Sarah. 1997. "Labeling mental illness: The effects of received services and perceived stigma on life satisfaction." *American Sociological Review* 62:660–672.

Rosenfield, Sarah. 1980. "Sex differences in depression: Do women always have higher rates?" *Journal of Health and Social Behavior* 21:33–42.

_____. 1989. "The effects of women's employment: Personal control and sex differences in mental health." *Journal of Health and Social Behavior* 30:77–91.

Rosenhan, David L. 1973. "On being sane in insane places." *Science* 179:250–258.

Rosenstock, Irwin M. 1966. "Why people use health services." *Milbank Memorial Fund Quarterly* 44:94–127.

Rotgers, Frederick, Daniel S. Keller, and Jon Morgenstern. 1996. *Treating Substance Abuse: Theory and Technique.* New York: Guilford.

Rothman, Barbara Katz. 1986. *The Tentative Pregnancy: Prenatal Diagnosis and the Future of Motherhood.* New York: Penguin.

_____. 1989. *Recreating Motherhood: Ideology and Technology in a Patriarchal Society.* New York: Norton.

Rothman, David J. 1971. *The Discovery of the Asylum.* Boston: Little, Brown.

_____. 1991. *Strangers at the Bedside: A History of How Law and Bioethics Transformed Medical Decision-Making.* New York: Basic.

_____. 1997. *Beginnings Count: The Technological Imperative in American Health Care.* New York: Oxford University Press.

Russell, Louise B. 1994. *Educated Guesses: Making Policy About Medical Screening Tests.* Berkeley: University of California Press.

Ryan, William. 1976. *Blaming the Victim,* revised edition. New York: Pantheon.

Sade, Robert M. 1971. "Medical care as a right: A refutation." *New England Journal of Medicine* 285:1288–1292.

Safran, Stephen P. 1998. "The first century of disability portrayal in film: An analysis of the literature." *Journal of Special Education* 31:467–479.

Sandhaus, Sonia. 1998. "Migrant health: A harvest of poverty." *American Journal of Nursing* 98:52–53.

Sandweiss, Stephen. 1998. "The social construction of environmental justice." Pp. 31–58 in *Environmental Injustices, Political Struggles: Race, Class, and the Environment,* edited by David E. Camacho. Durham, NC: Duke University Press.

Schappert, Susan M. 1993. "National ambulatory medical care survey: 1991 summary." *Advance Data from Vital and Health Statistics* Number 230.

Schechter, Susan. 1982. *Women and Male Violence.* Boston: South End.

Scheff, Thomas J. 1984. *Being Mentally Ill: A Sociological Theory,* revised edition. Chicago: Aldine.

Schneider, Joseph W., and Peter Conrad. 1983. *Having Epilepsy: The Experience and Control of Illness.* Philadelphia: Temple University Press.

Schneirov, Matthew, and Jonathan David Geczik. 1996. "A diagnosis for our times: Alternative health's submerged networks and the transformation of identities." *Sociological Quarterly* 37:627–644.

Schoendorf, Kenneth C., Carol J.R. Hogue, Joel C. Kleinman, and Diane Rowley. 1992. "Mortality among infants of black as compared with white college-educated parents." *New England Journal of Medicine* 326:1522–1526.

Schooler, Caroline, Ellen Feighery, and June A. Flora. 1996. "Seventh graders' self-reported exposure to cigarette marketing and its relationship to their smoking behavior." *American Journal of Public Health* 86:1216–1221.

Scotch, Richard K. 1989. "Politics and policy in the history of the disability rights movement." *Milbank Quarterly* 67:380–400.

Scott, Robert A. 1981. *The Making of Blind Men.* New Brunswick, NJ: Transaction.

Scott, Wilbur J. 1990. "PTSD in DSM-III: A case in the politics of diagnosis and disease." *Social Problems* 37:294–310.

Scull, Andrew. 1977. *Decarceration, Community Treatment and the Deviant: A Radical View.* Englewood Cliffs, NJ: Prentice-Hall.

_____. 1985. "Deinstitutionalization and public policy." *Social Science and Medicine* 20:545–552.

_____. 1989. *Social Order/Mental Disorder: Anglo-American Psychiatry in Historical Perspective.* Berkeley: University of California Press.

Scully, Diana. 1994. *Men Who Control Women's Health: The Miseducation of Obstetrician-Gynecologists.* New York: Teachers College Press.

Seligman, Martin E.P. 1975. *Helplessness: On Depression, Development and Death.* San Francisco: W.H. Freeman.

Sempos, Christopher T, James I. Cleeman, Margaret D. Carroll, Clifford L. Johnson, Paul S. Bachorik, David J. Gordon, Vicki L. Burt, Ronette R. Briefel, Clarice D. Brown, Kenneth Lippel, and Basil M. Rifkind. 1993. "Prevalence of high blood cholesterol among U.S. adults." *Journal of the American Medical Association* 269:3009–3014.

Sen, Amartya. 1999. *Development as Freedom*. New York: Knopf.

Shapiro, Joseph P. 1993. *No Pity: People with Disabilities Forging a New Civil Rights Movement*. New York: Random House.

Shearer, Gail. 1998. *Hidden from View: The Growing Burden of Health Care Costs*. Washington, DC: Consumers Union.

Shekelle, Paul G. 1998. "What role for chiropractic in health care?" *New England Journal of Medicine* 339:1074–1075.

Shenk, Joshua Wolf. 1999. "America's altered states: When does legal relief of pain become illegal pursuit of pleasure?" *Harper's Magazine* 298(1788):38–52.

Shilts, Randy. 1987. *And the Band Played On*. New York: St. Martin's.

Shortell, Stephen M., Teresa M. Waters, Kenneth W. B. Clarke, and Peter P. Budetti. 1998. "Physicians as double agents: Maintaining trust in an era of multiple accountabilities." *Journal of the American Medical Association* 280:1102–1108.

Siegel, Bernie. 1990. *Love, Medicine and Miracles,* revised edition. New York: Harper & Row.

Siegman, Aron W., and Theodore M. Dembroski (eds.). 1989. *In Search of Coronary Prone Behavior: Beyond Type A*. Hillsdale, NJ: Erlbaum.

Sigsbee, Bruce. 1997. "Medicare's resource-based relative value scale, a de facto national fee schedule: Its implications and uses for neurologists." *Neurology* 49:315–20.

Silverman, Milton, and Philip R. Lee. 1974. *Pills, Profits, and Politics*. Berkeley: University of California Press.

Simmons, Janie, Paul Farmer, and Brooke G. Schoepf. 1996. "A global perspective." Pp. 39–90 in *Women, Poverty, and AIDS: Sex, Drugs, and Structural Violence,* edited by Paul Farmer, Margaret Connors, and Janie Simmons. Monroe, ME: Common Courage.

Skocpol, Theda. 1996. *Boomerang: Clinton's Health Security Effort and the Turn Against Government in U.S. Politics*. New York: Norton.

Skolnick, Andrew. 1995. "Along U. S. southern border, pollution, poverty, ignorance, and greed threaten nation's health." *Journal of the American Medical Association* 273:1478–1482.

Smyke, Patricia. 1991. *Women & Health*. London: Zed.

Snyder, Thomas W. 1997. *Digest of Education Statistics, 1997*. Washington, DC: U.S. Government Printing Office.

Somervell, Philip D., Philip J. Leaf, Myrna M. Weissman, Dan G. Blazer, and Martha Livingston Bruce. 1989. "Prevalence of major depression in black and white adults in five U.S. communities." *American Journal of Epidemiology* 130:725–735.

Sontag, Susan. 1978. *Illness as Metaphor*. New York: Farrar, Strauss, and Giroux.

Specter, Michael. 1997. "Deep in the Russian Soul, a Lethal Darkness." *New York Times* June 8:A1+.

Spitzer, Robert L., and Joseph L. Fleiss. 1974. "A reanalysis of the reliability of psychiatric diagnosis." *British Journal of Psychiatry* 125:341–347.

Spitzer, Robert L., Janet B.W. Williams, and Andrew E. Skodol. 1980. "DSM-III: The major achievements and an overview." *American Journal of Psychiatry* 137:151–164.

Starr, Paul. 1982. *The Social Transformation of American Medicine*. New York: Basic.

_____. 1994. *The Logic of Health Care Reform: Why and How the President's Plan Will Work*. New York: Penguin.

Stein, Leonard. 1967. "The doctor-nurse game." *Archives of General Psychiatry* 16:699–703.

Stein, Leonard, David T. Watts, and Timothy Howell. 1990. "The doctor-nurse game revisited." *New England Journal of Medicine* 322:546–549.

Steinberg, Jacques. 1999. "Expanded school drug tests face a challenge." *New York Times* August 18:A14.

Steingart, Richard M. 1991. "Sex differences in the management of coronary artery disease." *New England Journal of Medicine* 325:226–230.

Stevens, Rosemary. 1989. *In Sickness and in Wealth: American Hospitals in the Twentieth Century*. New York: Basic.

Stewart, David C., and Thomas J. Sullivan. 1982. "Illness behavior and the sick role in chronic disease: The case of multiple sclerosis." *Social Science and Medicine* 16:1397–1404.

Stimmel, Barry. 1992. "The crisis in primary care and the role of medical schools: Defining the issues." *Journal of the American Medical Association* 268:2060–2065.

Stine, Gerald J. 1998. *Acquired Immune Deficiency Syndrome: Biological, Medical, Social, and Legal Issues*. Upper Saddle River, NJ: Prentice-Hall.

Stinson, Robert, and Peggy Stinson. 1979. "On the Death of a Baby." *Atlantic Monthly* 244(1):64–72.

Stokoe, William C. 1978. *Sign Language Structure*. Silver Spring, MD: Linstok.

Stolberg, Sheryl Gay. 1999a. "Trade Agency Finds Web Slippery With Snake Oil." *New York Times* June 25:A16+.

_____. 1999b. "The boom in medications brings rise in fatal risks." *New York Times* June 3:A1+.

Strahan, Genevieve W. 1997. "An overview of nursing homes and their current residents. Data from the 1995 National Nursing Home Survey." *Advance Data* 280.

Straus, Robert. 1957. "The nature and status of medical sociology." *American Sociological Review* 22:200–204.

Street, Richard. 1991. "Information-giving in medical consultations: The influence of patients' communicative styles and personal characteristics." *Social Science and Medicine* 32:541–548.

Stretesky, Paul, and Michael J. Hogah. 1998. "Environmental justice: An analysis of Superfund sites in Florida." *Social Problems* 45:268–287.

Sullivan, Deborah A., and Rose Weitz. 1988. *Labor Pains: Modern Midwives and Home Birth*. New Haven, CT: Yale University Press.

Sutton, John R. 1991. "The political economy of madness: The expansion of the asylum in progressive America." *American Sociological Review* 56:665–678.

Swedish Institute. 1997. *Social Insurance in Sweden*. Stockholm: The Swedish Institute. http://www.si.se.

_____. 1999. *The Care of the Elderly in Sweden*. Stockholm: The Swedish Institute. http://www.si.se.

Swiss, Shana, and Joan E. Giller. 1993. "Rape as a crime of war: A medical perspective." *Journal of the American Medical Association* 270:612–615.

Szasz, Thomas. 1970. *The Manufacture of Madness*. New York: Dell.

_____. 1974. *The Myth of Mental Illness,* revised edition. New York: Harper & Row.

Szasz, Thomas S., and Mark H. Hollander. 1956. "A contribution to the philosophy of medicine." *Archives of Internal Medicine* 97:585–592.

Taylor, Anna. 1998. "Violations of the international code of marketing of breast milk substitutes: Prevalence in four countries." *British Medical Journal* 316: 1117–1122.

Taylor, Humphrey. 1994. "The Canadian and U.S. health care systems compared." *Harris Poll* Number 25.

Taylor, William C., Theodore M. Pass, Donald S. Shepard, and Anthony L. Komaroff. 1987. "Cholesterol reduction and life expectancy." *Annals of Internal Medicine* 106:605–614

Tesh, Sylvia. 1988. *Hidden Arguments: Political Ideology and Disease Prevention Policy.* New Brunswick, NJ: Rutgers University Press.

Tessler, Richard, and Gail Gamache. 1994. "Continuity of care, residence, and family burden in Ohio." *Milbank Quarterly* 72:149–169.

Thoits, Peggy A. 1985. "Self-labeling processes in mental illness: The role of emotional deviance." *American Journal of Sociology* 91:221–249.

Thomas, Stephen B., and Sandra C. Quinn. 1991. "The Tuskegee Syphilis Study, 1932–1972: Implications for HIV education and AIDS risk education programs in the black community." *American Journal of Public Health* 81:1498–1504.

Tiefer, Leonore. 1994. "The medicalization of impotence: Normalizing phallocentrism." *Gender & Society* 8:363–377.

Tjaden, Patricia, and Nancy Thoennes. 1998 (November). "Prevalence, incidence, and consequences of violence against women: Findings from the national violence against women survey." *National Institute of Justice Research in Brief.*

Toole, Michael J., and Ronald J. Waldman. 1993. "Refugees and displaced persons: War, hunger, and public health." *Journal of the American Medical Association* 270:600–605.

Tosteson, Daniel C., S. James Adelstein, and Susan T. Carver (eds.). 1994. *New Pathways to Medical Education: Learning to Learn at Harvard Medical School.* Cambridge, MA: Harvard University Press.

Trupin, Laura, Douglas S. Sebesta, Edward Yelin, and Mitchell P. LaPlante. 1997. "Trends in labor force participation among persons with disabilities, 1983–1994." *Disability Statistics Report* (10). Washington, DC: U.S. Department of Education, National Institute on Disability and Rehabilitation Research.

Turshen, Meredith. 1989. *The Politics of Public Health.* New Brunswick, NJ: Rutgers University Press.

Tussing, A. Dale, and Martha A. Wojtowycz. 1997. "Malpractice, defensive medicine, and obstetric behavior. *Medical Care* 35:172–191.

Underwood, Felix J. 1926. "Development of midwifery in Mississippi." *Southern Medical Journal* 19:683–685.

U.S. Bureau of the Census. 1975. *Historical Statistics of the United States, Colonial Times to 1970.* Washington, DC: U.S. Government Printing Office.

_____. 1995. *Statistical Brief: Sixty-five Plus in the United States.* Washington, DC: U.S. Government Printing Office.

_____. 1998. *Statistical Abstract of the United States.* Washington, DC: U.S. Government Printing Office.

U. S. Bureau of Labor Statistics. 1998. *Occupational Outlook Handbook.* Washington, DC: Government Printing Office.

U.S. Conference of Mayors. 1998. *A Status Report on Hunger and Homelessness in America's Cities: 1998.* Washington, DC: U.S. Conference of Mayors.

U.S. Congress, Office of Technology Assessment. 1983. *The Impact of Randomized Clinical Trials on Health Policy and Medical Practice.* Background paper OTA-BP-H-22. Washington, DC: Government Printing Office.

U.S. Congressional Budget Office. 1993. *Trends in Health Spending: An Update.* Washington, DC: U.S. Government Printing Office.

_____. 1995. *The Effect of Managed Care and Managed Competition.* Washington, DC: U.S. Government Printing Office.

U.S. Department of Health and Human Services. 1990. *Indian Health Conditions.* Washington, DC: U.S. Government Printing Office.

_____. 1998. *Health, United States, 1998.* Washington, DC: U.S. Government Printing Office.

U.S. Environmental Protection Agency. 1992. *Environmental Equity: Reducing Risk for All Communities.* Washington, DC: U.S. Government Printing Office.

_____. 1996. *Environmental Health Threats to Children.* (EPA 175-F-96–001). http://www.epa.gov/ocepa111/NNEMS/oeecat/docs/1068.html.

U.S. General Accounting Office. 1991. *Canadian Health Insurance: Lessons for the United States.* Washington, DC: U.S. Government Printing Office.

_____. 1998. *Health Insurance Standards: New Federal Law Creates Challenges for Consumers, Insurers, Regulators.* Washington, DC: U.S. Government Printing Office.

Valenstein, Elliot S. 1986. *Great and Desperate Cures.* New York: Basic.

Van Olphen-Fehr, Juliana. 1998. *Diary of a Midwife: The Power of Positive Childbearing.* Westport, CT: Bergin & Garvey.

Vitek, C.R., and M. Wharton. 1998. "Diphtheria in the former Soviet Union: Reemergence of a pandemic disease." *Emerging Infectious Diseases* 4:539–550.

Vuckovic, Nancy, and Mark Nichter. 1997. "Changing patterns of pharmaceutical practice in the United States." *Social Science and Medicine* 44:1285–1302.

Waddell, Charles. 1982. "The process of neutralisation and the uncertainties of cystic fibrosis." *Sociology of Health and Illness* 4:210–220.

Waitzkin, Howard. 1981. "The social origins of illness: A neglected history." *International Journal of Health Services* 11:77–103.

_____. 1991. *The Politics of Medical Encounters.* New Haven, CT: Yale University Press.

_____. 1993. *The Second Sickness: Contradictions of Capitalist Health Care,* revised edition. New York: Free Press.

Waldron, Ingrid. 1994. "What do we know about causes of sex differences in mortality? A review of the literature." Pp. 42–54 in *The Sociology of Health and Illness: Critical Perspectives,* edited by Peter Conrad and Rachelle Kern. New York: St. Martin's.

Wall Street Journal. 1997. "Operating income leaps 20% on potent sales of Prozac." January 28:B4.

Wallace, Stephen P., Emily K. Abel, and Pamela Stefanowicz. 1996. "Longterm care and the elderly." Pp. 180–201 in *Changing the U.S. Health Care System: Key Issues in Health Services, Policy, and Management,* edited by Ronald M. Andersen, Thomas H. Rice, and Gerald F. Kominski. San Francisco: Jossey-Bass.

Wallerstein, Immanuel M. 1974. *The Modern World-System.* New York: Academic.

Wardwell, Walter I. 1979. "Limited and marginal practitioners." Pp. 230–250 in *Handbook of Medical Sociology,* edited by Howard E. Freeman, Sol Levine, and Leo G. Reeder. Englewood Cliffs, NJ: Prentice-Hall.

_____. 1988. "Chiropractors: Evolution to acceptance." Pp. 157–191 in *Other Healers: Unorthodox Medicine in America,* edited by Norman Gevitz. Baltimore: Johns Hopkins University Press.

Wayne, Leslie. 1999. "Colt's best defense." *New York Times* March 12:C1+.

Weil, Andrew, and Winifred Rosen. 1998. *From Chocolate to Morphine,* revised edition. New York: Houghton Mifflin.

Weinreb, Linda, Robert Goldberg, Ellen Bassuk, and Jennifer Perloff. 1998. "Determinants of health and service use patterns in homeless and low-income housed children." *Pediatrics* 102: 554–562.

Weissert, William G. 1991. "A new policy agenda for home care." *Health Affairs* 10: 67–77.

Weissman, Joel J., and Arnold M. Epstein. 1993. "The insurance gap: Does it make a difference?" *Annual Review of Public Health* 14:243–270.

_____. 1994. *Falling Through the Safety Net.* Baltimore: Johns Hopkins University Press.

Weitz, Rose. 1989. "Uncertainty in the lives of persons with AIDS." *Journal of Health and Social Behavior* 30:270–281.

_____. 1991. *Life with AIDS.* New Brunswick, NJ: Rutgers University Press.

_____. 1999. "Watching Brian Die: The Rhetoric and Reality of Informed Consent." *Health: An Interdisciplinary Journal for the Social Study of Health, Illness and Medicine* 3:209–227.

Werth, Barry. 1991. "How short is too short?: Marketing human growth hormone." *New York Times Magazine* June 16:14+.

Wertz, Dorothy C., and John C. Fletcher. 1998. "Ethical and social issues in prenatal sex selection: A survey of geneticists in 37 nations." *Social Science and Medicine* 46:255–273.

Wertz, Richard, and Dorothy Wertz. 1989. *Lying-In.* New Haven, CT: Yale University Press.

West, Candace. 1984. *Routine Complications: Troubles with Talk Between Doctors and Patients.* Bloomington: Indiana University Press.

White, Larry C. 1988. *Merchants of Death: The American Tobacco Industry.* New York: Morrow.

Whitehouse, Peter J., Eric Juengst, Maxwell Mehlman, and Thomas H. Murray. 1997. "Enhancing cognition in the intellectually intact." *Hastings Center Report* 27(3):14–22.

Whitt, Hugh P., and Richard L. Meile. 1985. "Alignment, magnification, and snowballing: Processes in the definition of 'symptoms of mental illness.'" *Social Forces* 63:682–697.

Wilkinson, Richard G. 1996. *Unhealthy Societies: The Afflictions of Inequality.* London: Routledge.

Williams, Christine. 1989. *Gender Differences at Work.* Berkeley: University of California.

_____. 1992. "The glass escalator: Hidden advantages for men in the 'female' professions." *Social Problems* 39:253–267.

Williams, David R. 1998. "African-American health: The role of the social environment." *Journal of Urban Health: Bulletin of the New York Academy of Medicine* 75(2):300–321.

Williams, David R., and Chiquita Collins. 1995. "U. S. Socioeconomic and racial differences in health: Patterns and explanations." *Annual Review of Sociology* 21:349–386.

Williams, David R., Yan Yu, James S. Jackson, and Norman B. Anderson. 1997. "Racial differences in physical and mental health: Socio-economic status, stress and discrimination." *Journal of Health Psychology* 2:335–351.

Williams, Peter C. 1984. "Success in spite of failure: Why IRBs falter in reviewing risks and benefits." *IRB: A Review of Human Subjects Research* 6(May/June):1–4.

Williams, Rhea P. 1988. "College freshmen aspiring to nursing careers: Trends from the 1960s to the 1980s." *Western Journal of Nursing Research* 10:94–97.

Wines, Michael. 1999. "Russians drown sorrows, and selves." *New York Times* June 28:A10.

Wingood, Gina M., and Ralph J. DiClemente. 1997. "The effects of an abusive primary partner on the condom use and sexual negotiation practices of African-American women." *American Journal of Public Health* 87:1016–1018.

Wintemute, Garen J. 1999. "Future of firearm violence prevention: Building on success." *Journal of the American Medical Association* 282:475–478.

Wirth, Louis. 1985. "The problem of minority groups." Pp. 309–315 in *Theories of Society: Foundations of Modern Sociological Theory,* edited by T. Parsons, E. Shils, K.D. Naegele, and J.R. Pitts. New York: Free Press.

Wise, Jacqui. 1998. "Companies still breaking milk marketing code." *British Medical Journal* 316:1111.

Wolpe, Paul Root. 1985. "Acupuncture and the American physician." *Social Problems* 32:409–424.

Woodwell, David A. 1997. "National ambulatory medical care survey: 1996 summary." *Advance Data from Vital and Health Statistics* 295:9.

Woolf, Stephen H. 1995, "Screening for prostate cancer with prostate specific antigen." *New England Journal of Medicine* 333:1401–1405.

World Bank. 1997. *Confronting AIDS: Public Priorities in a Global Epidemic.* New York: Oxford University Press.

_____. 1998. *Assessing Aid: What Works, What Doesn't, and Why.* New York: Oxford University Press.

World Health Organization. 1980. *International Classification of Impairments, Disabilities, and Handicaps.* Geneva, Switzerland: World Health Organization.

_____. 1985. "Appropriate technology for birth." *Lancet* 2(8452):436–437.

_____. 1993a. *World Health Statistics Annual, 1992.* Geneva, Switzerland: World Health Organization.

_____. 1993b. *Infant and Young Child Nutrition.* Geneva, Switzerland: World Health Organization.

_____. 1995. "The World Health Organization's infant-feeding recommendation." *Weekly Epidemiological Record* 70:119–120.

_____. 1997a. *Anti-Tuberculosis Drug Resistance in the World.* Geneva, Switzerland: World Health Organization.

_____. 1997b. *Fact Sheet No. 153: Female Genital Mutilation.* Geneva, Switzerland: World Health Organization.

_____. 1998a. *Report on the Tuberculosis Epidemic.* Geneva, Switzerland: World Health Organization.

_____. 1998b. *World Health Report, 1998.* Geneva, Switzerland: World Health Organization.

_____. 1998c. *Fact Sheet No. 154: Tobacco Epidemic: Health Dimensions.* Geneva, Switzerland: World Health Organization.

World Health Organization, Division of Child Health and Development. 1992. *Readings on Diarrhoea.* Geneva, Switzerland: World Health Organization.

World Health Organization, Division of Control of Tropical Diseases. 1998. *Malaria Prevention and Control.* Geneva, Switzerland: World Health Organization.

World Health Organization, Division of Reproductive Health. 1998. *Unsafe Abortion: Global and Regional Estimates of Incidence of and Mortality Due to Unsafe Abortion, with a Listing of Available Country Data,* third edition. Geneva, Switzerland: World Health Organization.

World Health Organization, Programme of Nutrition. 1998. *Progress Report.* Geneva, Switzerland: World Health Organization.

Wren, Christopher S. 1999. "Bid for alcohol in antidrug ads hits resistance." *New York Times* May 31:A1+.

Wysong, Jere A., and Thomas Abel. 1991. "Universal health insurance and high-risk groups in West Germany: Implications for U.S. health policy." *Milbank Quarterly* 68:527–560.

Yago, Glenn. 1984. *The Decline of Transit: Urban Transportation in Germany and U.S. cities, 1900–1970.* New York: Cambridge University Press.

Zimmerman, Mary K. 1993. "Caregiving in the welfare state: Mothers' informal health care work in Finland." *Research in the Sociology of Health Care* 10:193–211.

Ziporyn, Terra D. 1992. *Nameless Diseases.* New Brunswick, NJ: Rutgers University Press.

Zola, Irving K. 1972. "Medicine as an institution of social control." *Sociological Review* 20:487–504.

_____. 1985. "Depictions of disability—Metaphor, message and medium in the media: A research and political agenda." *Social Science and Medicine* 22:5–17.

_____. 1991. "Bringing our bodies and ourselves back in: Reflections on a past, present, and future 'medical sociology.'" *Journal of Health and Social Behavior* 32:1–16.

Zuger, Abigail. 1999. "Surgeons leaving the O.R. for the office." *New York Times* May 18:D1+.

Zussman, Robert. 1992. *Intensive Care: Medical Ethics and the Medical Profession.* Chicago: University of Chicago Press.

_____. 1997. "Sociological perspectives on medical ethics and decision-making." *Annual Review of Sociology* 23:171–189.

INDEX